The Quantum Dimension

The Paradigm Company

For information:
The Paradigm Company
3500 Mountain View Drive
Boise, Idaho 83704
(208) 322-4440

Email: paradigm@srnrl.com

Website: www.srnrl.com

First Edition - PDF format - March 2009

Second Edition - Print format - September 2015

Second Edition ISBN:
978-1517233099

The Quantum Dimension

Lawrence Dawson

Dedication

To the scientists Albert Einstein, Niels Bohr, Robert Millikan,
Max Planck, Edwin Hubble and J. S. Russell;
to the mathematicians Janne Rydberg,
Colin Maclaurin, James Maxwell and Oliver Heaviside
who saw darkly, as through a smoked glass,
but who prepared the way for a clearer vision.

"A new scientific truth does not triumph by convincing its opponents and making them see the light, but rather because its opponents eventually die, and a new generation grows up that is familiar with it." — Max Planck

Forward

During the summer and fall of 2008, twentieth century physics was shattered by the confirmation of a fourth quantum dimension. In that period, a new order of physical reality was experimentally confirmed. It could not have been anticipated nor fully acknowledged as possible by conventional three-dimensional physics. Four-dimensional quantum geometry experimentally derived Planck's Constant with a precision not seen since the famous Robert Millikan derivation of 1916. It did so by confirming a new order of radiation which had been theoretically predicted only three months earlier in one of the nation's most prestigious physics journal.

Starting in mid-2008, the negative light frequencies predicted by the author's quantum electron-string model as well as by a new soliton-based "1+1, ' kink' geometry" (*Journal of Physical Review, Feb. 2008*) were experimentally proven and measured. Black light, identified as negative radiation (N-radiation), was shown to drop the temperature in cotton fibers at a rate which derived Planck's Constant as a function of the number of hydrogen bonds in the molecule. This experiment confirmed four-dimensional quantum geometry and the model of the atom it had predicted. The recognition of quantum geometry as the actual structure of our space will force three major revisions upon twentieth century science.

First, Edwin Hubble's "expanding universe" is shown to be an "appearance" rather than factual. Quantum geometry proposes that any Euclidean line in free-space is forced into curvature by the intersecting quantum dimension. Redshift may be explained as the quantum-forced elliptical curvature of the line of opposition between our view from the earth and a foreign galaxy light source. Light is shown to travel this curved "trajectory" producing a variance with linear distance. The amount of curvature is related to distance by the quantum law of ellipses, generating a redshift which mathematically predicts distance. The formula for this curvature-to-distance is compared with Hubble's formula for recession velocities and shown to better fit Hubble's own originating data.

The mathematics for quantum curvature predicts no great age of the universe as the expanding universe does. The time factors differ. By quantum curvature, only about 248 million years of age are confirmed by astronomical redshift observations. This is true because maximum red shift for the visible universe is only "Z=1.571" which is consistent with the maximum of Z=1.4 from distant galaxies so far detected by the Sloan Digital Sky Survey. Further, when the quantum curvature model is fitted to Hubble's 1929 data table, the distance to the edge of the visible universe is shrunk by a factor of approximately 88 *times* over current estimates. Without an expanding universe, no mathematical contortions are required to fit great age with great distances.

The second "scientific fact" in need of revision is the presumed characteristic of the electromagnetic wave. The EM wave cannot conduct radiation across vast interstellar distances, but is shown to be local in character. The Heaviside formulation of Maxwell's field equation for the magnetic-permeability/ electric-permittivity of free space (currently the accepted SI standard for permeability and permittivity) establishes the electromagnetic wave. Quantum geometric analysis of the Heaviside formulation demonstrates that the electromagnetic waves can only exist in proximity to mass generating electrodynamic fields. However, the Heaviside formulation also demonstrates a direct interface between the electromagnetic wave and quantum space. Quantum space itself can conduct waveform energy across the vast distances of interstellar space as pulses in volume.

Vacuous space can conduct radiation because Einstein's concept of force attached to vacuum (his "cosmological constant" which reemerges with the elimination of the expanding universe) is essentially correct. The cosmological constant is shown to be the quantum time force which sustains the spacial quantum. By pushing upon the quantum time force in a regular frequency, waves composed of expanding and contracting vacuous volume are sent through the universe at the speed of light.

For space to be quantum, time must also be a quantum; composed of discrete time values separated by an extremely small variance. The time-quantum was also proved by the negative radiation study of cotton.

Thanks to Heaviside, the quantum time-force may be given in amp-Newtons (Newtons per SI amp). The electromagnetic wave is established by the flux in electric permittivity across the magnetic permeability of free space. Permeability is a constant (Heaviside; *Electro-Magnetic Theory*, p. 21) and the electric field must reach a permittivity threshold (capacitance *per* meter) in order for space to conduct the flux. In the Heaviside's formulation, the permeability constant *times* the permittivity threshold is equal to the inverse of the square of the speed of light. The inverse of the speed of light squared is also the formula for time-force as determined by quantum geometry. Heaviside's magnetic-permeability/ electric-permittivity exactly equals this time force and allows for a direct exchange between quantum space and the electromagnetic wave. The Heaviside formula gives the quantum time-force in amp-Newtons of time-variance squared:

$$\text{time-force of free space} = 1.11264\left(10^{-17}\right) \text{ amp-Newtons } of \ \Delta\text{second}^2/\text{meter}^2$$

Finally, the the Bohr/Schrodinger model for the shell/subshell electron schematic must be abandoned and the periodic table of elements requires revision. The consensus shell/subshell schematic was produced by an ill-defined application of Niels Bohr's electron orbit model and the Schrodinger wave-function. The number of electrons contained by subshells were determined by the number of "lobed orbitals" which the Schrodinger equation produces as modified by the quantum angular momentum number "*l:* "

However, the Schrodinger equation and its lobed orbitals is shown to mimic a nonlinear squared wave function which describes the change in wave form which occurs when an electron collapses out of a three-dimensional orbital operating in four-dimensional space into a two-dimensional orbit operating in molar three-dimensional volume. The Schrodinger eigenfunction converts the collapsed wave into a three dimensional orbital which retains the collapsed two-dimensional lobed form. His lobed orbitals are chimera and cannot explain the orbital adjustments which subshells need in order to incorporate more electrons. An analysis of the sodium "D-lines" under "Zeeman effect" demonstrates this to be the case.

The actual orbital is produced by the rotation of the electron bond at 90° to the proton bond when the electron reaches a boundary distance which must be maintained between a quantum and an Euclidean particle . This rotation is explained by the nature of particle charges as a projection onto a dimension not incorporated in a particle's volume; the electron having a quantum leg of volume while the proton is completely Euclidean. The rotation is accompanied by the multiplication of the electron's charge by a "Piezoelectrical nuclear capacitance." Rotation and multiplication produces a three dimensional orbital in four-dimensional space with the electron following a three-dimensional wave path and possessing a "1/2 spin" energy-constant as defined by the Bohr magneton and a wave-phase time constant. Subshell capacities are explained by electron-voltage changes across wave phases as caused by 1/2 spin. This is confirmed by D-line and Zeeman data.

Lawrence Dawson, 2009

Table of Contents

Table of Quantum Values

Description	symbol	Value	unit
Proton radius (fundamental quantum)	α	$0.50214(10^{-15})$	meters
Electron quantum radius (Alpha Space)	$\overline{\alpha}$	$0.50214(10^{-15})$	meters
Time variance separating α (time quantum)	ΔT	$1.6749506744(10^{-24})$	seconds
Time-Force (of free space)	F_t^2	$1.11264\left(10^{-17}\right)$ amp-Newtons	$\dfrac{\Delta sec.^2}{m^2}$
Root orbital radius (four dimensional)	Q_r	$2.8799076686(10^{-8})$	meters$^{4/3}$
root molar radius (Euclidean volume)	$Q_r^{mol.}$	$8.8275073063(10^{-11})$	meters$^{3/4}$
Orbital Tension Constant (kink tension against molar volume)	$\sqrt{2}\,k$	$\sqrt{2}\left(2.8044\ 10^{30}\right) = \dfrac{1^2}{\overline{\alpha}^2}$	units $\dfrac{\overline{\alpha}^2}{m^2}$
Planck Capacitance constant nuclear capacitance	C_P	$e^2/h = 38.74\ \mu F$	Farads
Rydberg root wavelength	λ_r	$91.14(10^{-9})$	meters
Orbital wave-phase time constant	t_ψ	$2.0679691463\ (10^{-14})$	seconds
1/2 spin energy constant (time-const)(Bohr magneton)	$t_\psi\,\mu_B$	$1.9178364785\ (10^{-37})$	Joules
1/2 spin electron-voltage constant: $(t_\psi\,\mu_B)/e$	$eV_{(1/2)s}$	$1.1970188584\ (10^{-18})$	electron volts
Intra-atomic 1/2 spin anomalous magnetic moment	γ	$\dfrac{30°}{\pi} = 9.5493°\ \overline{\theta}$ declination $\dfrac{\overline{\theta}}{180°} = 0.0530516477 = \gamma$	magnetic moment gain/loss

Four-Dimensional Electron Orbitals
and their energy distribution into shells/subshells

The universe has played a great practical joke and has made science the goat. It has palmed a "geometric card." A twisted mathematical hand has been dealt to science and has turned many of our bets into losers. The missing card is the quantum dimension and the unique system of mathematics it requires.

Our space is four dimensional, not three, but the extra, unrecognized dimension is utterly unlike anything our mathematicians have imagined extra dimensional space might be. It is a dimension which differs in measurement and intersection, which has unknown geometric characteristics relative to our known three dimensions, and this is the reason science has failed to recognize it. Science has been forced to address unrecognized characteristics of this extra dimension with nonlinear mathematics and bizarre geometric models which have failed to depict physical actuality.

This book is dedicated to revealing those unrecognized quantum mathematical principles and the reformation it will force upon our physical models. Quantum mathematics compose a system which seeks to replace contemporary "quantum mechanics." In contrast to quantum mechanics, mathematical descriptions of the quantum dimension have clear physical referents and operate by strict rules of causation. They are linear mathematics rather than nonlinear. The quantum dimension is everywhere observable and everywhere measurable, that is, if one knows where to look and how to measure.

The quantum dimension is composed by a quantum of much more precise definition and less vague usage than currently employed by physics. A quantum is defined as a non-divisible unit of space which can be further differentiated only by a mathematical principle called "the negation of subdivision."

This alternative quantum geometric system begins with the following proposition:
Any Euclidean line in non-solid space is "kinked" into curvature by the intersection of the quantum dimension. The intersecting quantum dimension composes a linear plane with the Euclidean line which is also "kinked" into curvature. This structure is a stand-alone unit of three-dimensional volume (a vacuum soliton) in four-dimensional space which is designated as the quantum squared. The Euclidean line and its "kinked" curvature is a planar component of this discrete unit of space and the quantum imposes a unit of measure upon the Euclidean planar surface equal to one half its own value. This unit of measure may be differentiated to "0" by negation of subdivision for the quantum squared at every possible point along the Euclidean line. A strictly Euclidean unit of measure must be differentiated by subdivision and cannot be differentiated to "0." Only the intersecting quantum dimension allows rational calculus differentiation of any Euclidean dimension. Differential calculus is a rational mathematical system only because of the existence of the quantum dimension.

The Failure of Science to Identify the Quantum Dimension
A four-dimensional geometry cannot operate anomalously, that is, as an occasional "abnormality" in three-dimensional space. Four-dimensional space must be mathematically demonstrated to operate systematically. The quantum dimension, as it has been encountered by twentieth century science, has always been treated as a three-dimensional anomaly.

In the late nineteenth and early twentieth century certain scientific measurements began to approximate whole number, non-divisible units subsequently called "quantums." Instead of the smooth continuum of distances— with unrestricted divisibility — which traditional, three-dimensional Euclidian geometry presumes, measurements began to acquire whole unit jumps. These quantum measurements were treated as three-dimensional abnormalities, rather than as identifying a fourth quantum dimension.

The important quantum measurements were three in number: 1) The Rydberg whole-number formulation of the spectrographic distribution of hydrogen radiation emissions; 2) Max Planck's whole-unit packets of "energy quanta" for monochromatic light frequencies— known as Planck's Constant; 3) Robert Millikan's measurement of x-ray ionized oil droplets, the complete mass of which changed in whole-units of electron "elementary charges."

However, none of these measurements were recognized as tapping into an extra, nontraditional geometric dimension— a dimension external to conventional three dimensional definition. Instead, a form of geometric schizophrenia was applied to three-dimensional scientific descriptions. Perhaps the most severe example of such geometric schizophrenia is the theory of "wave-particle duality" which emerge from the quantum discoveries.

Geometrically, a sine wave is a "flux" in density traveling through space. it is composed of high density, followed by low density or by a "relaxation trough." It is this "change in density" which is in motion. It is not the "medium" composing the density which is in motion[1]. This "density in motion" has a wavelength and a wave duration, both of which establish the wave's frequency.

In contrast, a "particle" is, geometrically, a solid. A solid can occupy different positions in space in obedience to the Newtonian laws of motion. A wave is density in motion measured by wavelength and frequency. A particle is a geometric solid which may be in motion as measured by velocity and momentary position. To say that they are the same thing is schizophrenic. It is the imposition of a mental image disassociated from possible objective observation. How can one apply a wave function to a particle without losing the concept of the wave function or of the particle or both? This question needs to be answered regardless of the fact that wave-particle duality has guided physics for nearly a hundred years.

"Wave Particle Duality" is Actually a Dimensional Shifting
The concept of a "wave-particle duality" was first proposed for the electron by Louis de Broglie in 1923[2]. De Broglie offered an equation for a "wave state" of the electron which was a function of the particle's momentum rather than the energy which identified the electron's associated light frequency. De Broglie's "wave-particle" formulation actually shows that a light-energy function which is dimensionally shifted from four dimensions to three gives a new wavelength. He calls the dimensionally shifted light-energy wavelength the particle's "wavelength"

He fails to see that both wavelengths are true radiation emissions by the electron. The light wavelength being output by the energy function results from a *four-dimensional orbital* and the wavelength of his momentum functions results when the electron is shifted from the four dimensional orbital to a three dimensional orbit. "Wave particle duality" is nothing more than a blind discovery of dimensional shifting from four dimensional space. De Broglie has used

---

[1] With the exception of the "transverse wave" in which medium particles are in motion at 90° to the wave.
[2] Wave Properties of Matter ; *dirac.phys.ncku.edu.tw/courses/general/QPhys/QM3.pdf*

a nonlinear differential equation to achieve this dimensional shifting. :
Bohr's formula for orbital electron volts and energy in an associated light wavelength "λ_L"

$$(\text{electron volts})(\text{elementary charge}) = eV(e) = \left(\text{frequency} = \frac{c}{\lambda_L} \right) h \ ; \ h = \text{Planck's Constant};$$

$$E = eV(e) = \frac{c}{\lambda_L} h = m(v^2/2) \ ; \qquad \lambda_L = \frac{c(h)}{m\left(v^2/2\right)}$$

De Broglie's "electron wavelength" is a function of the electron's momentum:

$$\text{momentum} = m(v) \ ; \qquad \lambda_e = \frac{h}{m(v)}$$

De Broglie has converted: $\quad "\lambda = \dfrac{c}{m\left(v^2/2\right)} h" \quad$ to $\quad "\lambda = \dfrac{1}{m(v)} h".$

The orbit of the electron will be shown to be a three dimensional orbital in four dimensional space which contains an acceleration/deceleration phase within the orbital. This acceleration/deceleration phase outputs light frequency.

The electron is orbiting in six transverse waves to the Euclidean plane of orbit. The quantum has inserted extra-dimensional space into the Euclidean three-dimensional graph. The kink is not the third Euclidean dimension, but a "1+1 dimension[3]" The kink inserts space into another dimension in order to force a linear Euclidean measurement into quantum curvature.

Two-Dimensional Orbit	Three-Dimensional Orbital
	' *kink*' *tension and Acceleration/Deceleration wave*

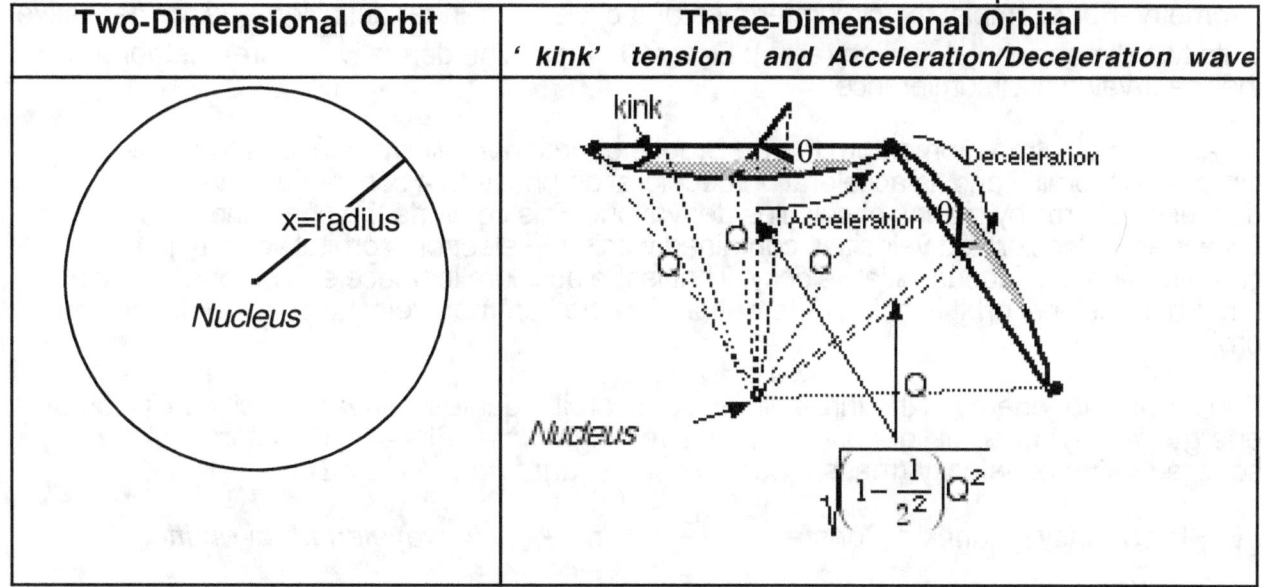

The energy equations for three-dimensional orbitals are the following:
Energy = orbital velocity energy + acceleration energy - deceleration energy

$$E = m\frac{v^2}{2} + (\text{Force})(\text{distance}) - (\text{Force})(\text{distance}) = f(h)$$

$$t = \frac{1}{2f} \ ; \quad d = \frac{\theta}{\pi/2}\left(\frac{Q\pi}{4}\right) = \frac{\theta Q}{2} \ ; \quad f = \text{light frequency} \ ; \quad h = \text{Planck's Constant}$$

---

[3] See the "1+1 dimensional kink" in *Solitons* by Sascha Vongehr, 1997;
physics.usc.edu/~vongehr/solitons_html/solitons.pdf

$$\text{Force} = F = \text{mass(acceleration)} = m\left(\frac{2d}{t^2}\right) = m\frac{2\theta}{\pi}\left(\frac{Q\pi}{2t^2}\right) = m\frac{\theta Q}{t^2}$$

$$F(d) = m\left(\frac{2d}{t^2}\right)d = m\left(\frac{2d^2}{t^2}\right) = m\frac{(\theta Q)^2}{2}4f^2 = 2mf^2\,\theta^2 Q^2 \; ; \; "\theta" \; \textit{measured in radians}$$

$$E = f(h) = m\frac{v^2}{2} + 2mf^2\,\theta^2 Q^2 - 2mf^2\,\theta^2 Q^2$$

$$\left\langle 2mf^2\,\theta^2 Q^2 - 2mf^2\,\theta^2 Q^2 \right\rangle \textit{acceleration / deceleration phase - energy outputting light}$$

Energy is gathered during the acceleration phase of the cycle by suppression of the kinks from the nucleus. This energy is then surrendered during the deceleration phase as light output at frequency.

There can be no common "moment of momentum" along the path of orbit as velocity changes for any point due to this acceleration/deceleration phase. A common momentum for points along the electron orbit can only be applied when the three-dimensional orbit is converted to a two dimensional orbit by calculus derivative. This is exactly what de Broglie has done to arrive at his equation for "electron wavelength."

De Broglie's electron "wave state" formula is a nonlinear equation which operates by eliminating a dimension by means of the calculus derivative. This is known principle in geometry. For example, the circumference of a circle— its linear definition— is the derivative of its two-dimensional definition (area): $D(\pi r^2) = 2\pi r$. The derivative of area establishes the linear value of circumference.

To arrive at electron momentum the de Broglie equation *implicitly* converts the three dimensional orbital and its acceleration/deceleration phase to a conventional two dimensional orbit by means of calculus derivation. This converts the changing accelerated/decelerated velocities contained within the electron's orbital velocity to a common and continuous initial velocity. That is the acceleration/deceleration phase of the three dimensional orbit is being differentiated to the common velocity of a two dimensional orbit.

The formula for energy in the three dimensional orbit equates a linear velocity definition of energy (*linear* speed of light *divided by* wave length *times* Planck's Constant) with a squared value of velocity (mass *times* velocity *squared divided by* 2) :

$$E = \text{(frequency)(Planck's Constant)} = \frac{c}{\lambda_L}h = m\frac{v^2}{2}; \; \lambda_L = \textit{wavelength of emitted light}$$

Velocity is the initial velocity of the electron and its energy conversion contains within it a wave function which contributes energy by acceleration and subtracts the same amount of energy by equivalent deceleration remaindering total energy as defined by the electron's initial velocity. Therefore, the orbit contains no common "moment of momentum." De Broglie has resolved this difficulty with the following nonlinear differential equation:

$$D(\frac{c}{\lambda_L}h) = D(m\frac{v^2}{2})$$

$$\frac{1}{\lambda_e}h = m(v) \; ; \; \lambda_e = \text{electron wavelength } \textit{(the nonlinear factor)}$$

De Broglie is arguing that to differentiate energy to momentum also differentiates "c" (in the frequency factor for Planck energy) to unity. Geometrically, momentum as the derivative of energy is a downward shift in one dimension. Orbital energy equals a *linear* function of the speed of light, wavelength and Planck's Constant. Orbital energy also equals a function of velocity *squared*. The linear function must also be shifted by derivative, just as the velocity-squared function must be shifted in dimension by derivative. The derivative of a single dimension is always unity: $D(x) = 1$. Therefore, the derivative of the speed of light is "1."

Setting the derivative of a linear function equal to the derivative of a square function threatens to eliminate the variable from the linear function. De Broglie solves the dilemma by proposing a nonlinear solution. The speed of light in the frequency formula is differentiated to unity, but the wavelength is shifted by differentiation to one which is a function of momentum. De Broglie simply asserts that this nonlinear wavelength (a function of electron momentum) is the "wavelength of the electron." Logically, it could just as well be the wavelength associated with a two dimensional orbit established in three dimensional space as arrived at by the calculus derivative of the quantum-defined three dimensional orbit established in four dimensional space.

To summarize:
Empirical science has established a linear velocity value for the energy in an electron orbit (speed of light *divided by* wavelength *times* Planck's Constant). This linear velocity value is set equal to the velocity squared energy value of the orbiting electron. De Broglie has proposed a linear wavelength as a function of the linear velocity of electron momentum by a nonlinear differential equation for the energy equality. He proposes this nonlinear equality as an electron's "wave function."

Proof of the De Broglie Formulation: Electron Scattering

The de Broglie formulation has found empirical support. In 1925, the Davisson-Germer experiment determined that a beam of electrons fired against a nickel plate defracted at angles similar to those of x-rays. X-rays are known to follow Braggs Law for defraction when striking some crystalline structures. The highest peak for the measured angles of defraction suggested that the electrons possessed a "momentum wavelength" of 0.21 nm[4]. The electron beam scattered off the nickel plate in a "wavelike manner" in that the angle of defraction was determined by the angle of incidence, length of crystalline bond and a proposed x-ray wavelength.

But does this prove that the electron possesses a "wave state" or does it merely prove that free electrons accelerated by voltage to a velocity possess a potential wavelength which will be determined by atomic structure? The maximum velocity which an electron can possess and still be retained by the three dimensional orbital within four dimensional space has been determined by quantum geometry. It is the electron velocity achieved by acceleration of 13.6037 eV. This converts to a "0.3325 nm" de Broglie wavelength. the "0.21 nm" wavelength of Davission-Germer is shorter and of higher electron velocity than achieved by 13.6037 eV acceleration.

Any free electron with a velocity giving a "momentum wavelength" longer than 0.3325 nm is likely to be reabsorbed back into the three-dimensional orbital structure. Any free electron with a velocity giving a "momentum wavelength" shorter than 0.3325 nm is likely to be scattered by Braggs Law for defraction of the momentum wavelength. As the Davisson-Germer experiment proved, scattered electrons have velocities which give

---

4 *http://hyperphysics.phy-astr.gsu.edu/Hbase/quantum/davger2.html*

13

momentum wavelengths shorter than 0.3325 nm.

Any orbital has an electron voltage which can be converted to the de Broglie wavelength. Non-orbital momentum (*the momentum possessed by a dimensionally shifted electron*) can be described by the elementary charge of the electron, the mass of the electron and the electron voltage being applied:

Since momentum doesn't exist within the orbital ;

dimensionally shifted momentum $= m(v) = \sqrt{2m(eV)(e)}$

De Broglie's electron wavelength $= \lambda = \dfrac{h}{m(v)}$; $\lambda = \dfrac{h}{\sqrt{2m(eV)(e)}}$

De Broglie's "electron wave-state" formula can be seen as a modification of an empirically determined and mathematically systematic formulation for the energy state of an orbiting electron. However, it was not recognized that this energy state occurred as a three dimensional wave within four dimensional space. The de Broglie dimensional shifting of energy to momentum provides an Euclidean two dimensional waveform with a new frequency.

Niels Bohr's electron voltage for his "base" electron orbit is 13.6037 eV. Below is the comparison of the dimensionally shifted de Broglie "momentum" wavelength vs. the three dimensional orbital's light energy wavelength.

$\lambda =$ wavelength; $h =$ Planck's Constant; $e =$ elementary charge; m=mass of electron	
De Broglie's momentum wavelength	**Light-energy wavelength**
$\lambda_e = \dfrac{h}{m(v)}$; $eV = 13.6037$ $\lambda_e = \dfrac{h}{\sqrt{2m(eV)(e)}} = 0.3325\left(10^{-9}\right)$ meters $(x\text{-}ray)$	$E = m\left(\dfrac{v^2}{2}\right) = \dfrac{c}{\lambda_L}h = eV(e)$ $\lambda_L = \dfrac{c(h)}{m\left(v^2/2\right)}$; $eV = 13.6037$ $\lambda_L = \dfrac{c(h)}{eV(e)} = 91.14\left(10^{-9}\right)$ meters (uv)

When Planck's Constant is modified by dimensionally shifted particle momentum it gives a completely different wavelength than does Planck's Constant when modified by three dimensionally determined particle energy. De Broglie assumed that the momentum modification of Planck was characteristic of the particle and not any light frequency output.

De Broglie was wrong about this. Without four-dimensional quantum geometry he could not recognize that velocity was not, in fact, an independent variable. There were only certain conditions under which his equation would be correct and he could not recognize what those conditions might be. He did not understand what changes in electron velocity by voltage acceleration would launch dimensional shifting and produce his new wavelength.

The electron orbital is three dimensional motion in four-dimensional space. While consensus quantum mechanics identifies the electron orbital as a three-dimensional "shell," it does not recognize that the three-dimensional orbital was operating in four dimensional space. As a consequence, consensus quantum mechanics knows nothing of the actual geometric structure of the orbital.

The actual orbital path of the electron follows a 1/ 2 sine wave along a plane set at 90° to the plane of orbit. It completes the sine wave on a second plane, also at 90° to the plane of orbit. However, the two planes containing the completed sine wave are set at 120° to one another, establishing the sine wave in three-dimensional space. There are six such planes and three sine waves in a complete orbit. The electron accelerates and decelerates along these sine waves—. establishing the frequency of light output— much as acceleration/deceleration across a vibrating string establishes a sound frequency.

The maximum light frequency which can be output by the electron orbitals is determined by the terminal velocity of acceleration. The highest energy orbital, and the limit for all orbitals, is that which can reach the speed of light at maximum-acceleration terminal velocity. This is Bohr's "base" orbital with 13.6037 eV. Since terminal velocity equals electron initial velocity plus acceleration velocity, if the electron's initial velocity is greater than that required by the "base" orbital, the electron. can no longer inhabit four dimensional space. Its wave will "collapse" in three dimensional space and be forced to enter two dimensional space.

De Broglie has actually discovered what happens when an electron's initial velocity is accelerated by greater than 13.6037 electron volts. The wave being conducted in three dimensional space will collapse and be forced onto a two dimensional plane. The wave no longer has a three dimensional path. It must be expressed in two-dimensions.

This requires that the wave loop back upon itself. A three-dimensional wave in four-dimensional space collapsing to a two dimensional wave in three-dimensional space will become a nonlinear squared wave function[5]. Whereas the wave had been described on six intersecting planes in four dimensional space, it now is restricted to one plane in three dimensional space . The wave can no longer operate in three dimensions, but must loop back upon itself within a two-dimensional plane restricted to the orbital diameter of the three-dimensional wave. The nonlinear squared wave function describes this transformation. The sine wave along the "time axis" (x axis) becomes a "lobe like" orbit which gathers energy (amplitude) which it drops to return back upon itself, releasing the energy as a "burst."

Nonlinear Squared Wave-Function Descriptive of the De Broglie Transformation

$$y^2 = \sin x^2$$

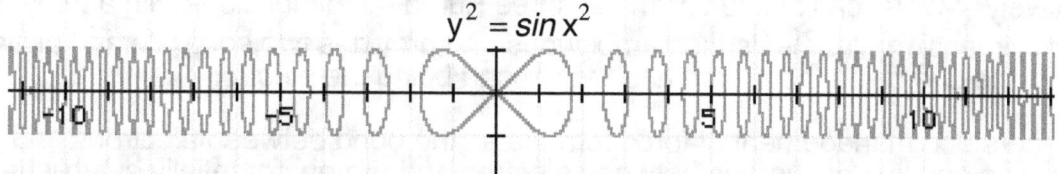

The sine wave becomes "orbital," returning back upon itself, and outputting "particle-like" linear radiation-bursts at accelerated frequency reminiscent of synchrotronic x-ray

The wave has become closed, returning back upon itself, and releasing bursts of radiation at accelerating frequency. These "bursts" of frequencies are accelerated to x-ray and have a different order of energy when compared to the continuous waves output by intact three dimensional orbitals. Potential energy is no longer equal to "frequency *times* Planck's Constant." The different energies between gamma rays and x-rays, often of the same order of frequency, demonstrates that radiation energies have lost Planck consistency.

The de Broglie formulation does not describe a "wave function" for the particle. It describes a changed condition governing the output of radiation by the particle as imposed by the

[5] This is in keeping with de Broglie's formulation which is shown to be a dimensional shifting by use of a nonlinear differential equation.

quantum dimension. De Broglie's "momentum wavelength" is descriptive of electron behavior when the electron's velocity has been accelerated (by electron voltage) past its capacity to be absorbed into a three dimensional orbital.

Four-Dimensional Atomic Structure

A three dimensional orbital operating in four-dimensional space must begin with new definitions for the electron and proton and their respective charges. These new definitions are not conceptual artifices but derive testable hypotheses. The structure so identified mathematically describe observed electron behaviors which the "consensus[6]" quantum mechanical model has failed to do for the last 80 years.

Suppose the electron is a quantum particle in the sense that one leg defining its volume is the quantum dimension and that its negative charge is a potential attachment to the third Euclidean dimension "missing" from its volume. Suppose further that the proton is an Euclidean particle in the sense that its volume is defined by three Euclidean dimensions and that its positive charge is a potential attachment to the quantum dimension which it is seeking to "add" to its volume. The charges of the particles would be geometric projections towards the fourth dimension relative to the particle's geometric volume definition.

Such configurations of particle volume and particle charge are the inevitable outcome of particles defined as "the square of the quantum squared." The unit of area designated the "quantum squared" cannot be composed by two quantum axii since there is only one quantum dimension. That is, the "quantum squared" must be composed by one quantum axis and a second Euclidean axis. The "square of the quantum squared" provides three Euclidean dimensional axii and one quantum axis.

"Mass" is composed by fundamental particles[7] of four-dimensional definition. All such four-dimensional particles must be composed as the "square of the quantum squared" or "$(Q^2)^2$." The "$(Q^2)^2$" provides two separate and distinct definitions of volume, with the fourth dimension not incorporated in volume being remaindered as charge. Since there is only one quantum dimension, "$(Q^2)^2$" volume can be defined by the one quantum dimension and two Euclidean dimensions with a remaindered Euclidean dimension. Alternatively "$(Q^2)^2$" can define volume as three Euclidean dimensions with a remaindered quantum dimension. "$(Q^2)^2$" is thus ambiguous and provides two separate and distinct solutions, the one being the proton and the second being the electron.

A misunderstood management of force formulates the bond between electrons and protons. The bonding of charges between electron and proton potentially establishes the quantum squared. The proton's quantum charge attaches to the quantum electron and the electron's Euclidean charge attaches to the Euclidean proton. The attached particles are attracted towards one another, not to merge, but to establish the quantum squared unit of space.
The quantum must be sustained by force and the quantum squared unit of space requires a "squaring of force."

The quantum requires force to establish. A quantum is defined as two points separated by some distance. Since there are no points within the quantum space of separation, there is no geometric structure to the distance. The only thing keeping the points separated must be

---

[6] "Consensus science" is defined as science composed of unproved hypotheses which become socially accepted by the body of science without reference to empirical testing.
[7] Consensus "particle physics" is addressed by quantum geometry.

force. The force of separation between the two quantum points is a mathematical function of the distance.

Force, in this sense, is defined as a "repellant" which pushes not mass but geometric points through space. Any quantum point must contain force which will be designated "F_Q." Any quantum point can establish a second quantum point at distance "Q" by the contained force "F_Q." However, if a second quantum point did exist at distance "Q," it would also contain the force "F_Q." If the two quantum points were "Q" apart, the force separating them would become:

$$\text{"}F_Q \times F_Q = F_Q{}^2.\text{"}$$

That is, the force separating two quantum points must be "force squared." "Force squared" is a mathematical function of "Q squared":

$$F_Q^2 = f(Q^2)$$

Because it requires force squared, the quantum unit of measure cannot be a "stand alone" unit of space. It must establish a unit of area.

Any distance "Q" is now separate by "F_Q^2," and this "force squared" can potentially establish a unit of area by identifying a third point. somewhere on the periphery of a circle made by rotating an equilateral triangle around the original quantum line:

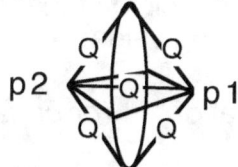

The resultant set of triangles are all equal sided units of area.

The problem is that the new peripheral point must not be a quantum point. If the new peripheral point were also a quantum point, it too would contain "F_Q" and would establish another "F_Q2" bond with point "1" and point "2." The equilateral triangle would be reiterated twice more then again and again, the "nth" reiteration producing 2 to the "n" power of equilateral triangles. Each one of these "2^n" triangles would have either point "1" or point "2" as a corner.

Space simply cannot be reiterated in this manner since it is restricted in size by point "1" and point "2" but the number of reiterations is not restricted. The process would construct two geodesic spheres which would intersect at the above circle. On the surfaces of sphere "1" and sphere "2" points could not overlap or be nearer than "Q" to one another. By geodesic construction, there are a finite number of equilateral triangles of sides equal to the radius which can be fitted into the surface of a sphere.

The only way that quantum area can be constructed with an infinite number of quantum reiterations is if the equilateral triangle proposed above had one Euclidean side composed of a continuum of points, the line providing a force value of "F_Q" to each point. This continuum of points is opposed to a single quantum point also with a force value of "F_Q." That is, the quantum must be single dimension intersecting a second Euclidean dimension, each quantum thus established having a force value of "$F_Q{}^2$."

The Preliminary Quantum-Squared Plane
Plane is summation of quantums

Bisect Euclidean line "X=Q"

$$\sqrt{\left(1-\frac{1}{2^2}\right)Q^2}$$

point of origin

Now I propose that each point along the continuum of the Euclidean line forms a quantum with the quantum point of origin and that these in turn form a continuum of quantums which compose a plane or the unit of quantum area designated "the preliminary quantum squared." Further, I propose that each one of the quantums composing that plane has a force value of "$F_Q^2 = f(Q^2)$."

However, the infinite set of quantums formed by the point of origin and the continuum of points along the Euclidean line "x=Q" do not have a distance of separation equal to "Q." These quantums, formed with less than "Q" distances of separation. are designated "differentiated quantums" and are given the symbol "$d(Q)$." By the principle of the negation of subdivision for the quantum squared, they have the following value:

$$d(Q) = \sqrt{\left(1-\frac{1}{x_n^2}\right)Q^2} \quad ; \quad d(Q^2) = \left(1-\frac{1}{x_n^2}\right)Q^2 \quad ; \quad x_n = n + \text{a partial of "1" } (or\ of\ n+1-n)$$

The differentiated quantums do not have a distance of "Q," yet all quantums along the Euclidean line have a force value of "$F_Q{}^2$." Distance is no longer a function of Force squared. The only solution to this is to "kink" the Euclidean line into curvature such that the distance between any point on the periphery of curvature and the point of origin does equal "Q." Further, by the Pythagorean Theorem: "$F_Q{}^2 = f(Q^2) = f(\text{kink}^2) + f(d(Q^2))$":

The "Kinked" Quantum Squared   **The Preliminary Quantum Squared**

$$\sqrt{\frac{1}{2^2}Q^2} = \frac{Q}{2}$$

$kx=\pi Q/2$

Kx="kinked" x

$$\sqrt{\left(1-\frac{1}{x_n^2}\right)Q^2}$$

$$\sqrt{\left(1-\frac{1}{2^2}\right)Q^2}$$

Point of Origin

$$\sqrt{\left(1-\frac{1}{2^2}\right)Q^2}$$

$$\sqrt{\left(1/2^2 - 1/x_n^2\right)Q^2} \text{ identifies all possible points along "x=Q/2."}$$

Each differentiated quantum "kinks" the Euclidean line at 90° for a value determined as:
    "kink^2=Q^2 − d(Q^2)" *by the Pythagorean Theorem.*
This "kinks" the Euclidean line into a half-circle curvature with a radius equal to "Q/2^8." Further, it establishes the quantum distance as "Q" for all points on the circumference with each

---

[8] For proof of equality see *Dawson's Theorem* in Appendix.

quantum so established having a force value of "$F_Q{}^2 = f(Q^2) = f(kink^2) + f(d(Q^2))$."

This quantum squared formulation defines the wave structure of the electron orbital. Further, the quantum squared formulation should replace the Schrodinger eigenfunction which is currently accepted, by consensus, as defining orbital wave-structure.

I will offer no critique of Schrodinger mathematics. Rather, I will demonstrate that the consensus electron shell/subshell schematic, as built upon Schrodinger, provides the wrong number of subshell electron capacities and misidentifies orbital energy states.

I will show this with an analysis of the sodium "D-lines"— the curious spectrographic light doublet split from the 589.2937 nm wavelength— which are output by energized sodium. The Schrodinger-defined shell/subshell schematic for the eleven sodium electrons offers an impossible "conceptual" explanation of the D-lines. In contrast, the four-dimensional quantum model uses quantum-identified shell/subshell electron voltages and quantum-identified "electron 1/2 spin" characteristics, as measured by the Bohr magneton, to precisely derive D-line frequencies by measured orbital and spin energy states. The inability of the consensus "quantum mechanical" orbital model to correctly identify orbital/spin energy states is decisively demonstrated.

I have said that the quantum dimension constructs the electron orbital by management of the charge-bond between the electron and the nucleus and that particle charges are potential projections onto the fourth-dimension relative to a particle's volume.

Further, in the analysis of the de Broglie "wave-particle duality," I have said that four-dimensional quantum geometry identifies an acceleration/deceleration phase for orbiting electrons. Acceleration/deceleration of orbital velocity occurs in a transverse wave which is set at 90° to the plane of the electron's orbit. This wave is in motion transversely (at a 90° angle) to the direction of electron's travel.

The Three-Dimensional Orbital
'kink' tension and Acceleration/Deceleration wave

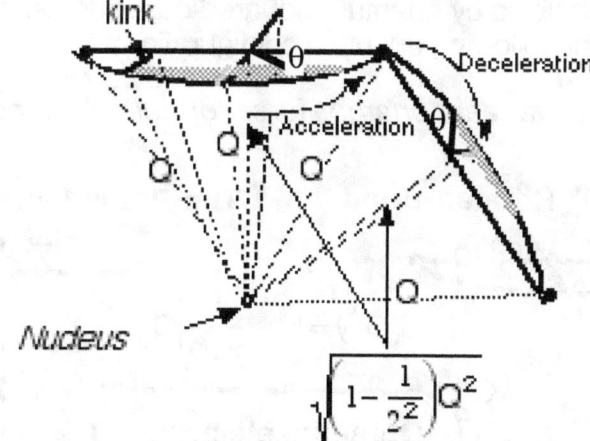

The transverse waveform is established by the "dimensional kinking" of an Euclidean line, as required by the quantum geometric construction which has been explained above.

However, tension establishes acceleration/deceleration. The electron accelerates from and decelerates toward its initial velocity across this wave path. The soliton mathematical process of "dimensional kinking[9]" identifies the tension. The "kinking" of an Euclidean dimensional line into curvature by the intersecting quantum dimension represents pressure or tension upon the Euclidean line.

That "kink tension" is established in the following manner: Suppose there is a mechanical-electrical force (Piezoelectrical force) which surrounds the nucleus and opposes the electron when the electron/proton charges have attracted the electron close enough to begin merging with the proton. The electron's bond is forcibly detached from the proton and turned perpendicular to the bond of the proton by that force. In this manner, the electron bond is forced into a position outside the nucleus. The electron's bond of attraction would now reside completely outside the nucleus and be vectored at right angles to the intersecting proton bond.

Such a detachment and rotation of bonds is not only possible but essential if both particles are four dimensional. The bonded particles are rushing towards one another in an attempt to merge and complete a four-dimensional definition. But that merger is not possible because volume can never be defined by four dimensions.

The merger is not possible because the electron cannot geometrically enter the space occupied by the proton. The electron is only a two-dimensional Euclidean "plate" rotating its quantum dimensional component in empty space. By definition, the quantum dimensional axis requires that the space within which it is being rotated be a geometric vacuum which contains no points.

This is not the case with the Euclidean proton, however. Any line contained within its volume is composed of a continuum of points. The electron can approach no closer to the proton than the length of its quantum radius. At that point, the covalent bond which has been enforcing closure between the two particles is forcibly separated and the detached bonds rotated to a 90° angle with one another.

The particle charges are now supplying a counter force to the force of attraction.

This counter force is symbolized by quantum-squared construction as "F_Q^2." This geometrically constructs the following quantum configuration:

$$F_Q^2 = f(Q^2) = \frac{e^- e^+}{Q^2} \quad \text{from standard charge - force formula for bonded distance}$$

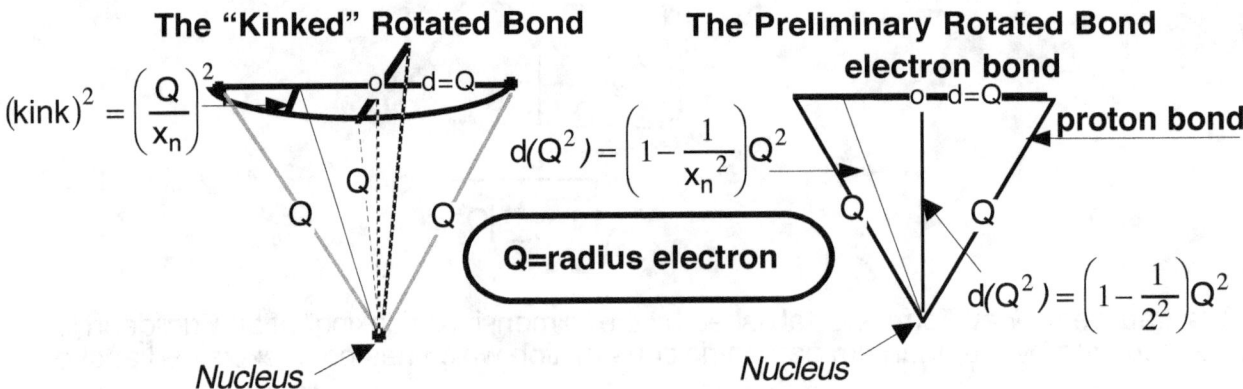

The "Kinked" Rotated Bond

$$(kink)^2 = \left(\frac{Q}{x_n}\right)^2$$

$$d(Q^2) = \left(1 - \frac{1}{x_n{}^2}\right)Q^2$$

Q=radius electron

The Preliminary Rotated Bond
electron bond
proton bond

$$d(Q^2) = \left(1 - \frac{1}{2^2}\right)Q^2$$

Nucleus   *Nucleus*

The quantum dimension must sustain a distance of separation between the rotated bond

[9] See the "1+1 dimensional kink" in *Solitons* by Sascha Vongehr, 1997; physics.usc.edu

and any point on the surface of the proton equal to the quantum radius of the electron. Since the electron is now "falling" along a tangent line to the proton, this is not possible. This "radius of separation" is only true for the end points along the rotated bond. It is not true of any point in between which establishes a "differentiated quantum distance" with the point of origin on the surface of the proton. A solution might be to force the rotated bond into an orbital curvature, but this would only expand the electron bond relative to the proton bond. The only solution which will retain the force equalities between electron and proton bonds is to "kink" the rotated bond into curvature at a 90° angle to the bond line by conventional quantum-squared construction.

The "kink^2" plus the differentiated quantum squared "$d(Q^2)$" equals "Q^2" by the Pythagorean theorem.

The electron following this path of "kinked" curvature will always undergo a force value which equals the electron charge *times* the proton charge *divided by* the electron's quantum radius squared (Q2). This is the force of attraction value, by standard formula, at the point at which the bond rotates and reverses the force.

An electron separated from the proton at a distance of its quantum radius is not an "orbital." What could possibly multiply the counter force rotating the electron bond by a sufficient amount to place it at a real orbital distance? The only means by which the orbital distance can be increased while retaining the reversed bond force is to multiply the charge of the electron by an amount equal to the multiple of the radial distance:

$$F_Q^2 = \frac{e^- e^+}{Q^2} = \frac{(ne^-)e^+}{nQ^2}$$

If the new orbital is to become "nQ^2," then the electron elementary charge must be "ne^-." The elementary charge of the bonded electron must be multiplied in order to expand the orbital radial.

Several things have occurred electrodynamically which make such a multiplication of the electron's elementary charge possible. The proton's own charge has moved from the particle's center to its surface. From every point on the proton's surface an electro-potential difference with the electron must be maintained. In short, detachment and rotation of the electron's bond has introduced an electron voltage field which envelopes the proton.

The attached proton's motion is constrained to oscillation by a four-dimensional potential-difference field and by its own bond attachment to the physical electron. The four-dimensional field may be thought of as field "invading" three dimensional space, one dimension of which is being pushed apart or separated by "kinks." The two-dimensional plane of orbit has been "kinked" into curvature, adding a new third dimension unaccounted for by Euclidean geometry. Additional space has somehow been inserted between dimensions by the force of "kinking." The motion of the attached proton occurs within this four-dimensional matrix and applies a force of suppression against the kink.

Nuclear force applied to suppressing the kinks can also be applied to expanding the radius and thus increasing the potential difference (electron voltage) between the nucleus and the rotated electron bond. As the above equation shows, expansion of the quantum-squared radius requires an equivalent multiplication of the electron's charge in order to keep the quantum force intact. To multiply the electron charge keeps the force along the expanded rotated bond equal to the original. However, the potential difference between the proton and the electron charges is increased by the multiplication of the electron's charge. Potential difference is voltage. To expand the radial by multiplying the charge increases electron voltage.

The kinetic energy of the nucleus is Piezoelectrical in that it increases the radius of the electron orbital and this increases the electron-voltage potential difference between electron and nucleus. However, expanding the radius and increasing electron voltage requires the multiplication of the elementary charge of the electron. Nuclear Piezoelectrical force does, in fact, multiply the charge of the electron.

Nuclear Capacitance and Electron Charge Multiplication

I will demonstrate, using Niels Bohr's electron-voltage formulation, that the frequency of light associated with an electron orbital equals the multiple of the elementary charge of the electron. Frequency is shown to be the multiple of the electron's elementary charge required by the orbital distance. Further, frequency also multiplies the electron's quantum radius (squared) to produce the orbital radial (squared) associated with the light frequency. These two facts combine to maintain the quantum force formula established at the point of electron bond rotation:

f = frequency ; r = orbital radial length ; α = electron's quantum radius

$$F_Q^2 = \frac{e^- e^+}{\alpha^2} = \frac{(fe^-)e^+}{f\alpha^2} = \frac{(fe^-)e^+}{r^2} \quad ; \quad f = \frac{r^2}{\alpha^2} \quad \textit{By quantum geometry: Proof later}$$

"Frequency-as-charge-multiple" is established by the electron voltage of the orbital and an electrical capacitance generated by the electron's quantum radius. That radius is the quantum space of separation which must be sustained between the surface of the proton and the electron field. This space of separation establishes a capacitance force which is continuously available to the nucleus and which converts Piezoelectrical nuclear kinetic energy to a multiple of the electron's charge. This is demonstrated by the following set of equations:

Bohr's electron voltage equation:
Energy=(voltage)(charge); Energy is Planck energy for light associated with orbital
h = Planck's Constant ; f = light frequency ;

eV = electron volts ; e = elementary charge

$E = f(h)$;   also

$E = eV(e)$

$eV(e) = f(h)$

$eV = \dfrac{f(h)}{e}$

The formula also implies an electrical capacitance transformation[10] :

E=(charge)^2/ Capacitance

eC = electron Capacitance

$$\frac{f(h)}{e^2} = \frac{1}{eC} \quad ; \quad eC = \frac{e^2}{f(h)} \quad ;$$

Frequency is shown to be a function of electron voltage and an electron capacitance constant *divided by* the elementary charge. The function identifies frequency as a multiple

---
[10] See Chapter 5; *Nuclear Capacitance, an Overlooked Quantum-Field Definition*

of the elementary charge. The eC constant is designated "Planck Capacitance" as it is the electron capacitance with a frequency of "1" and therefore describes eC for the distance at which the elementary charge is not multiplied. This distance is, of course, the electron's quantum radius. It is minimum distance which an electron can approach the proton. At the distance of the electron's quantum radius the electron bond must be rotated at 90° to the bond of the proton.

$$\text{Planck Capacitance} = C_P = \frac{e^2}{1\,(h)} \quad ; f = \frac{1}{eC}\left(\frac{e^2}{1\,(h)}\right); \quad \frac{1}{eC} = \frac{eV}{e}$$ [11]

$$f = \frac{eV\left(C_p\right)}{e}$$

Voltage times capacitance *equals* charge. Therefore "$eV\left(C_p\right)$" is a charge value. When divided by the elementary charge, it produces a multiple of the elementary charge. Therefore, electron voltage of the field *times* the capacitance produced by the space of separation necessarily sustained between the field and the proton produces a charge value which is a multiple of the elementary charge.

Light frequency equals the number of times the field's electron voltage, as operated upon by Planck Capacitance, multiplies the elementary charge of the electron. Alternatively, the electron's elementary charge multiplied by frequency produces the charge resulting from electron voltage as operated upon by Planck Capacitance.

Planck Capacitance is the capacitance generated by the necessary quantum distance between the proton surface and the electron field. This necessary distance is equal to the quantum radius of the electron. Planck Capacitance is the capacitance of the fundamental quantum squared; the minimum distance which the quantum squared can take. All quantum squared distance values must be multiples of the fundamental quantum-squared.[12] Planck Capacitance is determined to have a value of 38.74 μF (38.74 micro Farads).

To summarize: An attached electron rushing towards a proton by the force of respective charges can close no further than the quantum radius of the electron. At this point, the bond of the electron is forcibly rotated to 90° of the proton bond. The standard formula for bonded-charge force applies to this rejection and rotation. The standard formula is:

"force=(electron charge)(proton charge)/ (distance)2."

This force at the moment of bond rotation defines the quantum-squared force of separation. It occurs at the distance of the electron's quantum radius. The rotation of the bond has established a potential difference between the charge of the electron and the charge of the proton. This potential difference is defined as electron voltage.

By multiplying the electron's charge and the "(distance)2" by the same amount, a new orbital distance is established with the same force value but with a greater potential difference between the electron and proton charges. Greater orbital distance requires an equivalent increase in electron voltage as caused by a multiplication of the electron's charge.

Niels Bohr's formula for electron voltage was used to show that associated light frequency is

---

[11] Ibid.

[12] designated α^2 in quantum geometry. $\alpha = 0.50214(10^{-15})$ meters

the amount by which the electron's charge must be multiplied to reach the orbital distance. Frequency is also the inverse of the time constant defining acceleration/deceleration across the orbital wave. In this, the orbital wave acts like a tensioned string which also has a acceleration/deceleration time constant which defines the pitch of the string.

Acceleration/deceleration energy is independent of the orbital's potential energy. It is the amount of energy acquired during acceleration and given up as light by deceleration. The amount of acceleration/deceleration energy is determined by the portion of the peak-to-peak wavelength over which acceleration/deceleration is occurring. The portion of the total wave path dedicated to acceleration/deceleration is determined by the kinetic energy available to the nucleus to compress kinks. As the wave path approaches the crossover point between wave cycles the kinks become less forceful and less nuclear energy is required to accelerate the electron at that point.

I realize that I am adding complexities at an alarming rate, but there is no other way to introduce a broad overview of four-dimensional orbital process. I beg the reader to indulge me in this.

Orbital potential energy is the potential amount of energy available to the the orbital excluding acceleration/ deceleration energy . Potential energy equals Planck's Constant *times* frequency. In a study of negative radiation (kinks accelerate toward light wave pressure, not away from it[13]) the author and the Snake River N-Radiation Lab determined that Planck's Constant is the rate at which energy can be transferred between the nucleus and the electrons[14]. That is, nuclear kinetic energy can multiply the elementary charge of an orbital electron at an energy-exchange rate equal to Planck's Constant.

The elementary charge multiple *times* the exchange rate of Planck's Constant *equals* energy. If nuclear kinetic energy is greater than the frequency multiple of the electron's charge then the excess energy can be applied (at the Planck's Constant exchange rate) to suppressing kinks and initiating an acceleration/ deceleration phase *at frequency*. That is, the acceleration/deceleration time-factor is still a function of the charge multiple. As we have seen, the acceleration/deceleration phase will not effect the orbital's potential energy[15] .

Time is a factor in the formulation identifying radiological frequency as equal to a multiple of the electron's elementary charge. Radiological frequency is measured in Hertz which is the number of waves which pass a point in space in one second of time. If a different unit of time were used, the frequency would change. However, the electron's charge is measured in Coulombs which also has a time factor. A Coulomb is the amount of electric current at one amp which passes a point in one second of time. If the time unit for frequency were reduced, the time unit defining the Coulomb would be similarly reduced. The new reduced time frequency would be the multiple of the reduced time Coulomb. The physical value of frequency in seconds and frequency in the reduced time unit would be the same and could be exchanged by the formula: frequency/ sec=(frequency/ time-unit)(sec/ time-unit)

The above has been a broad stroke overview of the electrodynamic and kinetic forces which establish the energy exchange between nucleus and electron and which output

---

[13] See *Negative radiation pressure exerted on kinks;* Forgacs , Lukacs , and Romanczukiewicz; Phys.Rev.D77:125012,2008

[14] See *The Drop in Temperature by N-Irradiation of Cotton and the Derivation of Planck's Constant* in Appendix.

[15] For the examination of light absorption by the acceleration/ deceleration of kinks and the relationship to orbital potential energy see: *The Absorption of Light Energy by Kink Impedance* in Appendix.

light.[16]

While it is theoretically possible that the electron's charge can be multiplied by the mathematical value "x" (defined as a single value among all possible values), this is not true of the quantum squared. It is a known fact in physics that electron orbital distances, now defined as "(*frequency*)($Q2$) ," are quantum distances. Therefore, the basic quantum squared can only be multiplied by a discrete number of quantum-determined frequencies. The variance between the continuum of values which can multiply the electron's charge and the discrete number of frequencies which can multiply the basic quantum is the reason that the acceleration/deceleration phase is established within the orbital wave:

$$F_Q^2 = \frac{e^- e^+}{\alpha^2} = \frac{\left(\left(x = f(f) = \psi + f\right) \bullet e^- \right) e^+}{f\alpha^2} \quad ; \quad f = \text{frequency} \; ; \; f\alpha^2 = \text{orbital distance}$$

$$\psi = 2mf^2\,\theta^2\left(f\alpha^2\right) - \left[2mf^2\,\theta^2\left(f\alpha^2\right) \rightarrow \text{light energy}\right] \; ; E = 2mf^2\,\theta^2\left(f\alpha^2\right) \; by \; acceleration$$

Note: This is the single most important formulation for all of orbital mechanics. Since "ψ "can have a negative value, it explains how orbitals can be acquired with insufficient energy in the nucleus. They are acquired by " minus nuclear acceleration plus nuclear deceleration" across the kinks, remaindering electron velocity as the sole determiner of orbital energy

The fact that the orbitals of electrons are quantum is known. What is not known is why the orbitals must acquire those quantum distances, how orbital distances are structured by the quantum dimension and what the measures of those quantum distances are.

Orbital frequencies used to determine orbital electron voltages have been established by the Rydberg formula for hydrogen light-frequency emissions. The Rydberg formula is the following[17] :

$$f = \text{frequency} = \frac{c}{\lambda} \; ; \; f_r = \text{root frequency} = \frac{c}{\lambda_r} \; ; \; \lambda_r = 91.14 \; 10^{-9} \; \text{meters}$$

$$f = \frac{c}{\lambda} = \left(\frac{1}{n^2} - \frac{1}{n'^2}\right) \; f_r = \left(\frac{1}{n^2} - \frac{1}{n'^2}\right) \frac{c}{\lambda_r} \; ; \; 1 \le n \le 7 \; ; \; n+1 \le n' \le 8$$

$$eV = \frac{f(h)}{e} \; ; \; h = \text{Planck's Constant} \; ; \; e = \text{electron's elementary charge}$$

Rydberg's formula has been empirically shown to predict the following series of hydrogen light frequency emissions:

Lyman Series (ultraviolet). Formula: $\quad \frac{c}{\lambda} = \left(\frac{1}{(1)^2} - \frac{1}{(n')^2}\right)\frac{c}{\lambda_r} \; ; \; 2 \le n' \ge 9$

Balmer Series (visible light frequencies). Formula: $\quad \frac{c}{\lambda} = \left(\frac{1}{(2)^2} - \frac{1}{(n')^2}\right)\frac{c}{\lambda_r} \; ; \; 3 \le n' \ge 8$

Paschen Series (infrared spectrum). Formula: $\quad \frac{c}{\lambda} = \left(\frac{1}{(3)^2} - \frac{1}{(n')^2}\right)\frac{c}{\lambda_r} \; ; \; 4 \le n' \ge 8$

[16] For the full mathematical treatment of the subject see: *The Electrodynamic Formulations for Electron/Nuclear Energy Exchange* in Appendix

[15] For the examination of light absorption by the acceleration/ deceleration of kinks and the relationship to orbital potential energy see: *The Absorption of Light Energy by Kink Impedance* in Appendix.

Brackett Series (lower infrared spectrum). Formula: $\dfrac{c}{\lambda} = \left(\dfrac{1}{(4)^2} - \dfrac{1}{(n')^2} \right) \dfrac{c}{\lambda_r}$; $5 \leq n' \geq 8$

Pfund Series (far infrared spectrum). Formula: $\dfrac{c}{\lambda} = \left(\dfrac{1}{(5)^2} - \dfrac{1}{(n')^2} \right) \dfrac{c}{\lambda_r}$; $6 \leq n' \geq 8$

The Lyman Series of ultraviolet spectrographic lines showing the "root wavelength" from which Rydberg derived his formula for all hydrogen emissions

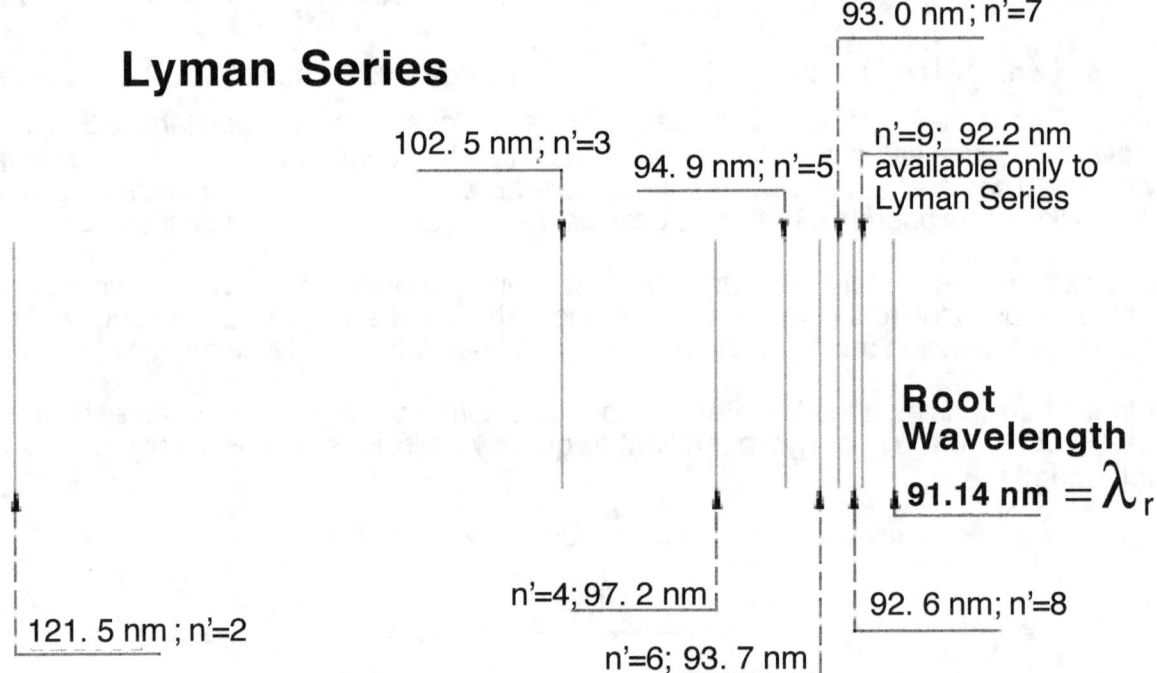

Lyman Series

93. 0 nm; n'=7

102. 5 nm; n'=3

94. 9 nm; n'=5

n'=9; 92.2 nm
available only to
Lyman Series

Root Wavelength

91.14 nm $= \lambda_r$

n'=4; 97. 2 nm

92. 6 nm; n'=8

121. 5 nm ; n'=2

n'=6; 93. 7 nm

The Rydberg Formula and Quantum Geometry

The Rydberg formula has never been recognized as the only way that quantum dimensional units of distance must be differentiated. A quantum is composed of only two points separated by "some" distance. Unlike an Euclidean distance composed of a continuum of points, the quantum distance is composed only of two points. To subdivide the quantum requires supplying additional points to the distance of separation which essentially multiplies the quantum.

The quantum distance can only be differentiated by the negation of subdivision, not by subdivision. That is, a subdivisional unit of distance can be subtracted from a quantum subdivision remaindering the original two points at a closer distance of separation. The negation subdivision must always be smaller than the subdivision. "(1/2-1/ 3)Q" is a proper quantum negation of subdivision which remainders a new quantum value of "(1/ 6)Q." With a linear quantum, new distance values equal to a subdivision can always be reached by negation of subdivision.

However. a quantum cannot exist alone and autonomously. It can only exist as a component of a unit of space identified as "the quantum squared (see page 18). The quantum must be multiplied by an Euclidean line to compose the quantum unit of area designated "the quantum squared." Since there is only one quantum dimension, squaring the quantum must depend upon an extraneous Euclidean dimension to do so.

It is the quantum squared, not the linear quantum, which must be differentiated by the negation of subdivision. Exact subdivisions of the quantum squared units of space can never be achieved by the negation of subdivision. This is a very important principle for the atomic model as it provides subdivisional shells with boundaries which can never be acquired by the negation of subdivision, but which contains subshells all of which are proper negations of the subdivisional shell.

The failure to recognize the hydrogen frequency distribution as the "negations of subdivisions" of a value contained within the distribution is partially the fault of Rydberg himself. He disguised the fact that his "constant" is really the inverse of the shortest wavelength which appears in the Lyman Series spectrograph above. The distribution for all wavelengths of hydrogen emissions are really a harmonic distribution, by quantum mathematical principles, from a "root" wavelength; the shortest wavelength and highest frequency in the series at 91.14 nm.

The root orbital is the quantum value for all the radiation-frequency orbitals because the root represents the maximum distance at which the three-dimensional orbital can function to emit and absorb light frequencies. At any greater distance, the maximum terminal velocity of the acceleration/deceleration wave-function would push electron velocities past the speed of light. All lower orbitals must be a negation of subdivision for the square of the root orbital's radius:

f_r = root frequency ; $f\left(\alpha^2\right)$ = orbital distance ; α = electron quantum radius

$$f\left(\alpha^2\right) = \left(\frac{1}{n^2} - \frac{1}{n'^2}\right) f_r\left(\alpha^2\right)$$

There are seven subdivisions "$1 \leq n \leq 7$" for frequency-based orbitals. This number of subdivisions will be shown to be established by a frequency limit.

The Rydberg distribution can only be understood within the context of a two-way energy exchange between the nucleus and electron orbitals. We have already discussed the energy exchange between nucleus and the orbital. Nuclear kinetic energy provided by vibration within an electron voltage field is applied to the multiplication of the electron's charge and an equivalent increase in orbital distance by multiplication of the square of the electron's quantum radius. The nucleus exchanges energy with the electron as a multiplication of charge.

In contrast, the energy exchanged between the electron and nucleus is a function of the kinks. The electron's Euclidean bond is kinked into curvature by the quantum bond of the proton. The kinks are a component of the "force squared" which is geometrically required to sustain the distance of separation for the quantum bond. If a counter force is expressed against the kinks— which are Euclidean in nature and subject to force— that force is also expressed against the quantum bond. By extension, it is also expressed against the attached proton. Force against the kinks denotes force against the nucleus.

The counter-force is radiation pressure[18] from a light source which is of the same frequency as the acceleration/deceleration phase of the orbital wave. Energy is supplied to the orbital by kink-impedance of radiation pressure. The kinks will accelerate away from the pressure,

[18] For a discussion of radiation pressure against kinks see *Negative radiation pressure exerted on kinks;* Forgacs , Lukacs , and Romanczukiewicz; Phys.Rev.D77:125012,2008

transferring energy to the nucleus at the rate of frequency *times* Planck's Constant. The fact that radiation energy is transferred to the nucleus at this rate is proven by Max Planck's black body studies[19].

The amount of energy which can be absorbed by the nucleus from impedance of light waves is restricted to "*f* (*h*)," as determined by Planck. However, a part of the total energy output by light-impedance must be dedicated to rebalancing kink tensions. Acceleration/deceleration energy must be split between management of kink tensions and transfer to the nucleus. The energy transferred to the nucleus is the orbital potential energy determined by the Planck energy value "*f* (*h*)." The amount of energy required to balance kink tensions can be mathematically determined. That is, light pressure must accelerate the kinks by a mathematically determined amount before energy can be transferred to the nucleus at the Planck energy rate[20].

The next to last orbital in the Rydberg series is the last which can actually absorb light energy and send it to the nucleus. The orbital is designated:

$$"f\left(\alpha^2\right) = \left(\frac{1}{6^2} - \frac{1}{7^2}\right) f_r\left(\alpha^2\right) \; ; \; \lambda = 12.367 \; \mu m"$$ *would require 89.10° of deflection.*[21]

This orbital requires 89.1° out of a possible 90° of acceleration/deceleration deflection in order to manage kink tension. That is, nearly the whole amount of available acceleration/deceleration kink tension must be dedicated to tension management in order to allow nuclear acquisition of Planck potential energy. The last orbital cannot exchange light frequency energy with the nucleus. The last and only nonfunctioning orbital is designated:

$$f\left(\alpha^2\right) = \left(\frac{1}{7^2} - \frac{1}{8^2}\right) f_r\left(\alpha^2\right) \; ; \; \lambda = 19.054 \; \mu m$$ *would require 137.29° of deflection.*[22]

This last orbital in the series cannot impede radiation because the required deflection angle of the acceleration/deceleration phase is greater than the 90° available to the orbital wave.

Rydberg is a distribution of quantum-assigned harmonically related radiation frequencies which have the energy capacity to sustain the orbitals. Radiation pressure can provide energy to the nucleus at the rate of "frequency *times* Planck's Constant." As we have seen, the nucleus must multiply the electron's charge by an amount equal to frequency. This multiple is supplied by the nucleus at the energy rate of exchange equal to of Planck's Constant. Radiation impedance by the kinks provides the nucleus with the same amount of energy by which the nucleus must multiply the electron charge to sustain the orbital. Alternatively stated, the Rydberg distribution is a set of orbitals supplied by radiation with the amount of energy which is required to sustain the orbitals.

The Rydberg distribution must be understood as the set of frequencies allowed by the principles of quantum geometry. Those frequencies must be distributed by the negation of subdivision of the proton's quantum bond (squared) with the quantum defined by the maximum frequency which can be impeded and the number of subdivisions (later called "shells") determined by the minimum frequency which can be impeded. Minimum frequency is determined as the maximum negational number *minus* "1" set equal to a subdivision which is negated one time and above which all frequencies can be impeded.

[19] Max Planck, *"On the Law of Distribution of Energy in the Normal Spectrum"*. Annalen der Physik, vol. 4, p. 553 ff.
[20] See *The Absorption of Light Energy by Kink Impedance* in the Appendix.
[22] For a discussion of radiation pressure against kinks see *Negative radiation pressure exerted on kinks*; Forgacs , Lukacs , and Romanczukiewicz; Phys.Rev.D77:125012,2008

The lowest frequency in the distribution (n=7; n'=8) cannot, itself impede frequency. Its orbital energy state must be sustained by harmonic impedance of all higher frequencies in the distribution[23] .

Without recognizing four-dimensional process and structure, consensus science has patched together a shell/subshell model of electron orbitals which vaguely approximates the actual structure. A set of consensus-determined shell's reproduce the Rydberg formulations for the Lyman, Balmer, Paschen, Brackett and Pfund series. These shells are positioned correctly with respect to the nucleus and are distributed with the correct number of subshells representing Rydberg negations . The Lyman-like "7th" shell is furthest from the nucleus and contains "7" subshells ("s'" through "i") just as Lyman contains "7" negations of the Lyman subdivision (n=1; n'= 2 through 8) . The Balmer-like "6th" shell is second furthest from the nucleus and contains "6" subshells ("s" through "h") as Balmer contains "6" negations of the Balmer subdivision (n=2; n'= 3 through 8). The Paschen-like "5th" shell contains "5" subshells("s" through "g"); Paschen negations (n=3; n'=4 through 8). The Brackett-like "4th" shell contains "4" subshells ("s." "p," "d." "f "); Brackett negations (n=4; n'=5, 6, 7, 8). The Pfund-like "3rd" shell contains "3" subshells ("s," "p," "d"); Pfund negations (n=5; n'= 6, 7, 8). Last frequency-impedance series; "2nd" shell contains "2" subshells ("s," "p"); series negations (n=6; n'=7, 8). Terminal series; "1st" shell contains "1" subshell ("s"); series negation (n=7; n'=8).

The consensus shell/ subshell schematic has faithfully reproduced the structure of the Rydberg "negation of subdivision" distribution from the root orbital. However, this is not true of the energy distribution[24]. The consensus is wrong. The "s" subshell is thought by consensus to be the lowest energy subshell for any shell. However, the reverse is true. The "s" subshell is actually the greatest energy subshell in any shell. The consensus mistake caused physics to lose its way and fail to identify actual shell/subshell energy states.

The "s" subshell is the one available to all shells. The Rydberg equivalent of the "s" subshell is the "n'=8" negation number which can negate all "n" subdivisions. The "n'=8" is the highest and most energetic orbital and is available to any shell. It is the most energetic because it subtracts the smallest value from the "n" subdivision. It is always greater than the "p" subshell with a negation number of "n'=7":

$$("s" \text{ subshell}) = \left(\frac{1}{n^2} - \frac{1}{8^2}\right) f_r(\alpha^2) > ("p" \text{ subshell}) = \left(\frac{1}{n^2} - \frac{1}{7^2}\right) f_r(\alpha^2)$$

The consensus shell/subshell has reversed this order of energy, identifying the "s" subshell as less energetic than the "p" subshell, if both are contained in the same shell.

The Rydberg empirically determined order of shell/subshell energies were reversed by consensus to make the orbital schematic compatible with the Schrodinger/Bohr explanation of radiation output by electrons.

Rydberg's formula for the distribution of the light frequencies output by hydrogen was used by Niels Bohr explain his system of electron orbits. He explained light output as electron voltage gain by alleged "electron falls" between those orbits.

Bohr combined Max Planck's "packets of energy" for light frequencies with Rydberg to

[23] See *Quantum Harmonics by Negation of Subdivision vs. Euclidean Harmonics by Subdivision* in Appendix.

---

[22] Ibid.

establish the electron voltage standard:

(Rydberg frequency)(Planck's Constant)/ (elementary charge)=(electron voltage)

To arrive at his formula, Bohr simply applied a standard electrodynamic equality:

(voltage)(charge in coulombs)=Energy

He proposed that the force of charge attraction for an orbiting electron was the equivalent of force of gravitation for an orbiting body and that electron voltage of the "base" orbit (root orbit) was the equivalent of maximum potential energy for the orbiting electron:

$$F = \frac{e^- e^+}{r^2} \ (bond\ force) \quad \text{is the equivalent of} \quad F = \frac{m_1 m_2}{r^2} \ (gravitational\ force)$$

Rydberg light-frequency output was explained as electron voltage gains by changes in orbits. He proposed that a "fall" to a lower energy orbit resulted in gain in electron voltage. To sell this idea that a fall to a lower energy orbit output electron voltage gain, he used the conceit that potential energy is the absence of kinetic energy and thus a minus value and lower energy.

$$-\frac{f_r(h)/e}{n^2} = -\frac{13.60\ eV}{n^2} \quad (the\ equivalent\ of\ potential\ energy\ for\ "nth"\ orbital)$$

Therefore, the "falls" between orbits would explain Rydberg frequencies as electron voltage gains:

Rydberg formula converted to electron volts:

$$\left(\frac{h}{e}\right)f = \left(\frac{h}{e}\right)\frac{c}{\lambda} = \left(\frac{1}{(n=1)^2} - \frac{1}{(n'=2)^2}\right)\frac{c}{\lambda_r}\left(\frac{h}{e}\right) = \left(\frac{1}{1^2} - \frac{1}{2^2}\right)13.6037\ eV$$

Bohr example of "fall" from "n = 2" to "n = 1" outputting eV gain:

$$-\frac{13.60\ eV}{(n=2)^2} - \left(-\frac{13.60\ eV}{(n=1)^2}\right) = \left(\frac{1}{1^2} - \frac{1}{2^2}\right)13.6037\ eV$$

Bohr has misstated his equality and, as a consequence, has confused orbital energy states. His argument that quantum orbits possess a "negative potential energy" is really a statement that they are force-partials of the whole bonding force:

Bohr's Actual Equality

partial of total bond force remaindered at orbit $= \left(\frac{1}{1^2} - \frac{1}{n^2}\right)\left(e^- e^+ = e^2\right)$

ΔF by orbit $= \left(\frac{1}{1^2} - \frac{1}{n^2}\right)e^2$; orbit force $= \frac{e^2}{n^2}$; $eV(eC) = e$

converted to electron volts; $\Delta F = \left(\frac{1}{1^2} - \frac{1}{n^2}\right)(eV\ eC)e$; $\frac{\Delta F}{(eC)e} = \left(\frac{1}{1^2} - \frac{1}{n^2}\right)eV$

$$\left(\frac{1}{1^2} - \frac{1}{n^2}\right)\frac{e^2}{(1.1778\ 10^{-20}\ Farads)(e)} = \left(\frac{1}{1^2} - \frac{1}{n^2}\right)(13.6037\ eV)$$

The only way an orbit's bonding force could have a negative value, as Bohr proposes, is if the orbital force it is somehow negating a total bonding force. That is, the force at the orbit is determined by negation of division rather than by division. Subtracting orbital force acquired by division from some whole force remainders a partial of the whole established at the orbit. Bohr's negative electron voltage formulation is really a negation of subdivision and can be reduced to a quantum negative integral.

The derivative of a unit of space used as a divisor, as with the Bohr formulation, cannot be positively integrated because the integral requires division by "0." The mathematician Colin Maclaurin tried to overcame the impossibility of integration of divisor derivatives by using estimates of the integral with his derivative series.[25]

However, the derivative of the "quantum-squared negation of subdivision" can be integrated negatively.[26] Further, this negative integral separates the quantum dimension from Euclidean space. Bohr's equations for orbits defined by negative force implicitly form the "quantum-squared negation of subdivision": $(1/1^2-1/n^2)(e^2)$. Bohr's formula. therefore, can also be shown to separate the quantum dimension from Euclidean space.

Negative Integration[27] of the Derivative of the "Quantum-Squared Negation of Subdivision"

$$y = \partial(1 - 1/x_n^2)$$

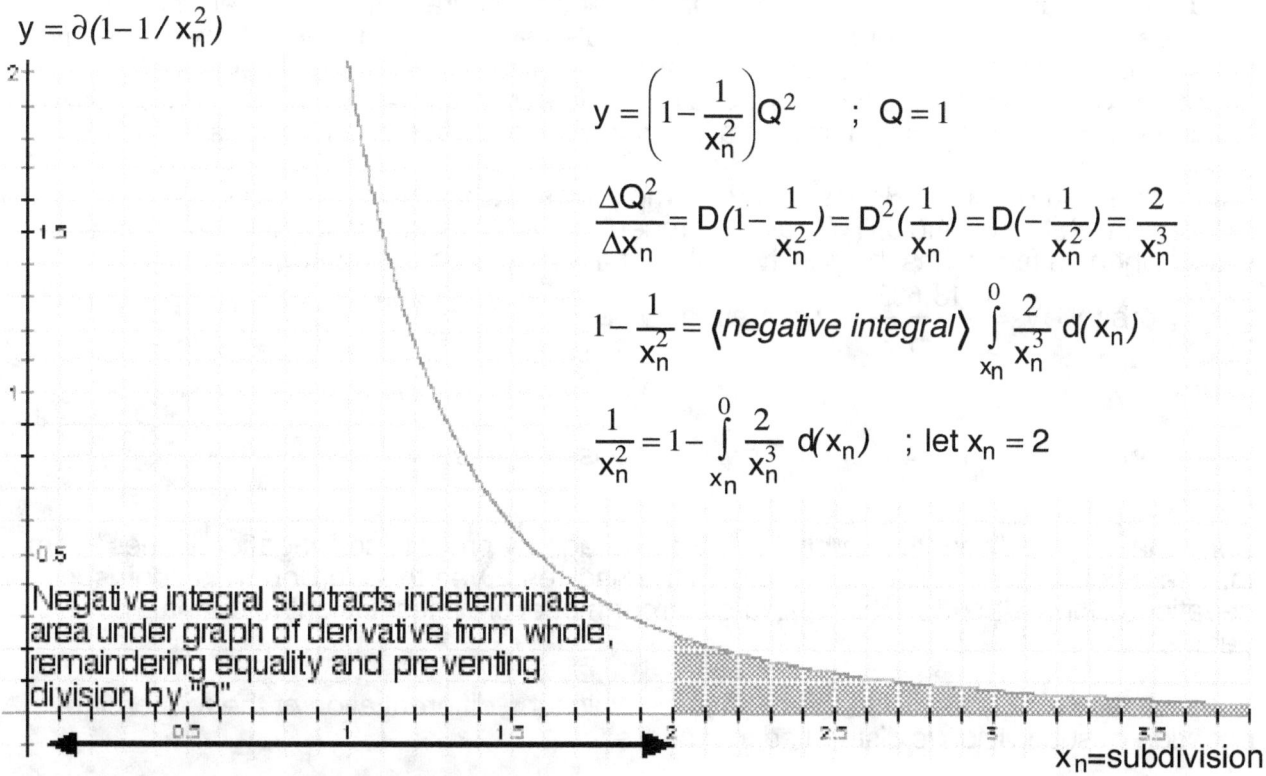

$$y = \left(1 - \frac{1}{x_n^2}\right)Q^2 \quad ; \quad Q = 1$$

$$\frac{\Delta Q^2}{\Delta x_n} = D\left(1 - \frac{1}{x_n^2}\right) = D^2\left(\frac{1}{x_n}\right) = D\left(-\frac{1}{x_n^2}\right) = \frac{2}{x_n^3}$$

$$1 - \frac{1}{x_n^2} = \langle negative\ integral\rangle \int_{x_n}^{0} \frac{2}{x_n^3}\ d(x_n)$$

$$\frac{1}{x_n^2} = 1 - \int_{x_n}^{0} \frac{2}{x_n^3}\ d(x_n) \quad ; \quad let\ x_n = 2$$

Negative integral subtracts indeterminate area under graph of derivative from whole, remaindering equality and preventing division by "0"

x_n=subdivision

As Applied to Bohr's Formula:

[25] Dawson, Lawrence *An Exact Quantum Formula for the Periphery of an Ellipse and the Detection of Systemic Error in the Maclaurin Derivative Series;* Master's Thesis; The Virtual University.

[26] See *Dawson's Theorem* in Appendix.

[27] The shaded area under the graph (if "x" were extended infinitely) represents the portion of the whole area under the graph which is equal to "$1/x^2$." That is, Euclidean units of space composed of a "continuum" or infinite number of points are converted to discrete quantum values by the negative integration of the derivative of the negation of subdivision. Proof that the negative integral converts Euclidean units, which have no restrictions on divisibility, into non divisible quantum units is the following:

let $x = 1$; $1 - \frac{1}{1^2} = \int_{1}^{0} \frac{2}{x^3}\ d(x)$; $\int_{1}^{0} \frac{2}{x^3}\ d(x) = 0$. Negative integration is the subtraction of indeterminate area between "1" and "0." "Area=0" indicates that there are no points along the x axis between "1" and "0." Therefore, the unit "1" is and must be a quantum separating point "0" and point "1."

$$Q^2 = \frac{F^2}{(eC)e} = \frac{e^2}{\left(1.1778 \ 10^{-20} \ \text{Farads}\right)\left(1.60217733 \ 10^{-19} \ \text{coulombs}\right)} = 13.6037 \text{ eV}$$

$$\frac{\Delta Q^2}{\Delta x_n} = D\left(1 - \frac{1}{x_n^2}\right)Q^2) = D(-\frac{Q^2}{x_n^2}) \qquad ; \qquad "x_n = n"$$

Change in "Q2" over change in "n" (Q2=13.6037 eV in the Bohr formula) is equal to both the derivative of the "negation of the quantum squared" as well as the derivative of the quantum-squared "negation value."

Change in the quantum squared value by using an "n2" divisor produces an ambiguous derivative which makes it difficult to determine what "n" actually is. "N" can be the negation value for the negation of subdivision or "n" might be the minus value of a Euclidean subdivision. Stated otherwise, "n" might be a quantum differentiation or a negative form of Euclidean differentiation. Bohr might have been alerted by the fact that force was negative, but he could not have known that the derivative of the negation of subdivision for the quantum squared is the second derivative for the linear subdivision, that such a derivative is integrable by negation which demonstrates an increase in force as "n" increases.

Without such knowledge, Bohr's formula simply presents the negation value as representative of the whole. This falsely presents orbital electron voltage as the "negation value" when, in fact, it has the positive value of the "negation of subdivision":

$$\text{orbital eV} \neq -\frac{13.6037}{n^2} \quad or \ negative \ value$$

$$\langle but \rangle$$

$$\text{orbital eV} = \left(1 - \frac{1}{n^2}\right)(13.6037 \text{ eV}) \ and \ positive \ value$$

Bohr can only get "negative" orbital electron voltage by presenting "negation values" from the "negation of subdivision" as the orbits themselves. Even though Bohr has confused negation values with orbits, his "falling electron" model provides insight to quantum geometry.

Bohr has unwittingly identified the method by which the differentiation of the root orbit by negation of subdivision is distributed into shells:

$$\text{Bohr's } "-\frac{Q^2}{n^2}" \text{ is actually } "\left(1 - \frac{1}{n^2}\right)Q^2"$$

$$\text{Bohr's fall from } "n' \text{ to } n; n' > n" \text{ is } "\left(-\frac{Q^2}{n'^2}\right) - \left(-\frac{Q^2}{n^2}\right) = \left(\frac{1}{n^2} - \frac{1}{n'^2}\right)Q^2"$$

$$\text{It actually is } "\left(1 - \frac{1}{n'^2}\right)Q^2 - \left(1 - \frac{1}{n^2}\right)Q^2 = \left(\frac{1}{n^2} - \frac{1}{n'^2}\right)Q^2"$$

$$\text{both} "\left(1 - \frac{1}{n'^2}\right)Q^2" \text{ and } "\left(1 - \frac{1}{n^2}\right)Q^2" \text{ are Lyman Series orbitals.}$$

Electron voltage gain between Lyman Series orbitals explains the distribution into the shells. Bohr's falls, properly understood, are actually the shell distribution pattern for the shell/subshell electron schematic. Further, the Bohr falls, as properly understood, provide exact electron voltages for the shell/subshell schematic. Both the energy states of the shell/subshell schematic and the origin of the distribution pattern have been lost to consensus science by Bohr's confusion of "negation values" with orbits.

Bohr, of course, did not say that falls between Lyman Series orbitals distributed the electron into lower order shells/subshells. Bohr said that falls between Lyman Series "negation values," treated as orbits, produced an electron voltage gain which was output as a light photon with a frequency determined by Rydberg. Electron voltage variance from the fall was assigned to light, not to orbitals. Knowledge of both the mechanics of light emissions and of orbital energy states suffered as a consequence. The consensus shell/subshell schematic was composed by the blind.

The problem of a one-time gain in electron voltage outputting continuous light was solved by further confusion. Again "wave-particle duality" was dragged out to explain how a one-time electron voltage gain outputting a single "photon" could be converted to a stream of such light particles. A wavelike orbit was needed to provide a continuous output of "photons," a wavelike orbit which has been supplied by the Schrödinger eigenfunction.[28]

As with the de Broglie "wave-particle" equation, the Schrödinger wave-function is a nonlinear differential equation which implicitly dimension shifts to describe four-dimensional space in three dimensional terms. Planck's Reduced Constant *times* the imaginary unit *times* the derivative of a wave function equals the "Hamiltonian operator" times the wave function.

The Basic Schrödinger Wave Function

$$(i)(\hbar)\frac{\partial}{\partial t}\psi(r,t) = \hat{H}\psi(r,t)$$

Although there are complexities to this equation which we need not address here, the nonlinear equation basically shows that the derivative of a Planck-defined wave function by its time value (the inverse of its frequency) produces a calculable operator for the wave, the Hamiltonian. The imaginary unit "*i* " is Einstein's time value for his "space-time continuum.[29] "

Since Einsteinian space is four-dimensional (albeit with an incorrect fourth dimension) the Schrödinger derivative is dimensionally shifting the wave-function from four dimensional space to three. As such, it is implicitly and unknowingly descriptive of a dimensional shifting of the wave which is actually possessed by the three dimensional orbital contained in four dimensional space. The wave is being shifted to some alternate waveform in three dimensional space. We have already seen what such a dimensional shifting entails in our analysis of de Broglie's "wave-particle duality."

The dimensional shifting of the actual wave contained by the three dimensional orbital to a two dimensional plane accommodated by three dimensional space can only be accomplished by the nonlinear squared wave function. The wave is actually shifted out of its orbital form where it is outputting light by a straight forward acceleration/deceleration energy exchange similar to a tensioned string, to some kind of fanciful and unrealistic "lobed orbital" which appears to be outputting particle-like, closed-wave, self-contained bursts of "light"

The Nonlinear Squared Wave-Function also Applies to the Schrödinger

[28] Schrödinger, Irwin (December 1926). "*An Undulatory Theory of the Mechanics of Atoms and Molecules*" . Phys. Rev. 28 (6): 1049–1070. doi:10.1103/PhysRev.28.1049

[29] See *Relativity,* Einstein, Albert ; Three Rivers Press, ISBN 0-517-88441-0 ;

Dimensionally-Shifted Wave

$$y^2 = \sin x^2$$

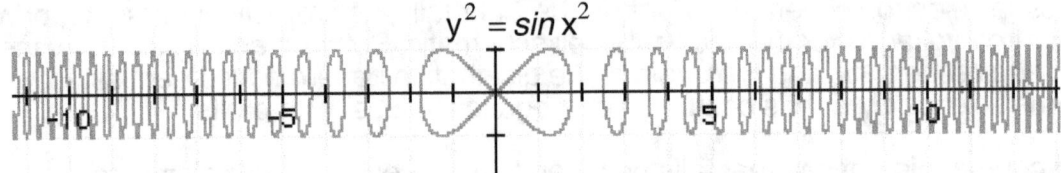

Notice the "lobed orbit" appears to be outputting "closed-wave bursts" similar to particles. These are "photons." They are only chimera, dimensional distortions of actual 3-D waves. The de Broglie nonlinear differential equation describes actual physical process. It describes what happens when an electron's velocity exceeds the capacity of the root orbital to contain that velocity. The nonlinear squared wave function describes the collapse from four dimensional space into three dimensional space and gives a somewhat accurate picture of the process.

No so for Schrödinger's nonlinear differential wave-function. The Schrödinger dimensionally shifted wave has been manipulated to create "lobed" models of three dimensional orbitals, models which have been used to construct an inaccurate atomic structure.

The Linear Hamiltonian Solution to Schrödinger's Nonlinear Wave-Function Reflects the "Lobed Orbits" of the Squared Wave-Function Shown Above	A Schrodinger-Proposed Electron "Lobed Orbital" Based upon the Hamiltonian Solution
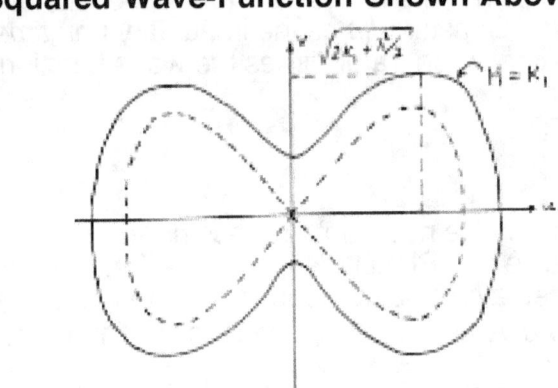	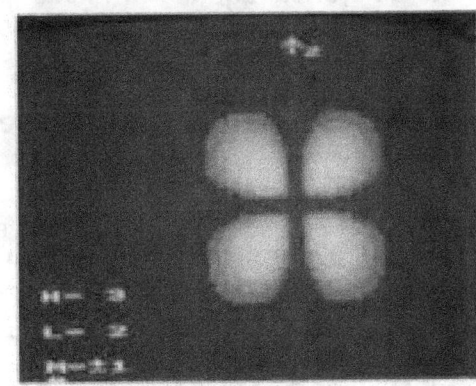
Level curve H=K₁ of Hamiltonian	*H=3 (three dimensional), ℓ =2*

[30] [31]

Even if Schrödinger is described by the consensus as being as fundamental to quantum mechanics as Newton is to motion mechanics, the situation is simply otherwise.

> *In physics, specifically quantum mechanics, the Schrödinger equation is an equation that describes how the quantum state of a physical system changes in time. It is as central to quantum mechanics as Newton's laws are to classical mechanics.*[32]

The difference between Schrödinger and Newton is that the Schrödinger equations are interpretive, while Newtonian mechanics are predictive. Schrödinger's wave-functions are accepted as providing energy solutions without empirical confirmation. The wave-functions are applied to modeling atomic structure and, if an empirically encountered radiological event

[30] *the Geometry of Nonlinear Schrödinger Standing Waves*; Newton, Paul and Watanabe, Shinya ; IMA Preprint Series; Aug. 1991; Dept. of Mathematics, University of Illinois, Urbana.

[31] *Quantum Numbers to Periodic Tables: The Electronic Structure of Atoms;* The Chemogenesis Web Book ; available on internet.

[32] *The Schrodinger Equation;* http://en.wikipedia.org/wiki/Schr%C3%B6dinger_equation

must be explained, the Schrödinger based model is made to fit by the simple expedient of using the frequency empirically provided as the time factor for the Schrödinger's wave-function which has been assigned by the model. The Schrödinger-based model is then said to have "explained" the radiological event.

The Schrödinger dimensionally-shifted orbital can never predict radiological events since "time" for the wave is not a mathematical function of orbital distance as it is with the quantum-dimensional orbital:

Let "$\overline{\alpha}$" = the quantum radius of the electron (a constant); let r=the radius of the orbital

$$F_Q^2 = \frac{e^- e^+}{\alpha^2} = \frac{\left(\langle x = f(f) = \psi + f \rangle \bullet e^- \right) e^+}{f\alpha^2} \quad ; \quad f = \text{frequency} \; ; \; f\alpha^2 = \text{orbital distance}$$

$$f\left(\alpha^2\right) = r^2 \; ; \quad f = \frac{r^2}{\alpha^2} \; ; \quad f = \frac{1}{t}$$

$$\frac{r^2}{\alpha^2} = \frac{1}{t}$$

Both "r" and "t" are dependent variables and the wave "ψ" time-factor is set by "r."

The Schrödinger equation is the following:

$$(i)(\hbar)\frac{\partial}{\partial t}\psi(r, t) = \hat{H}\psi(r, t)$$

The time-factor "t" for the wave "ψ" t is not dependent upon "r."
It is unimportant whether "r" in the Schrödinger equation is an orbital radius or wave amplitude. For the actual three-dimensional orbital, wave amplitude always equals radius divided by 2 (see page 18) . The important point to note is that the Schrödinger dimensionally shifted wave cannot retain light-frequency information. This is demonstrated by the graph of the squared wave function against its wave of origin. The consistent frequency of the original wave, as measured along the "x" time-axis, has become inconstant for the dimensionally-shifted squared wave function.

$$y = \sin x \qquad \text{graphed against} \qquad y^2 = \sin x^2$$

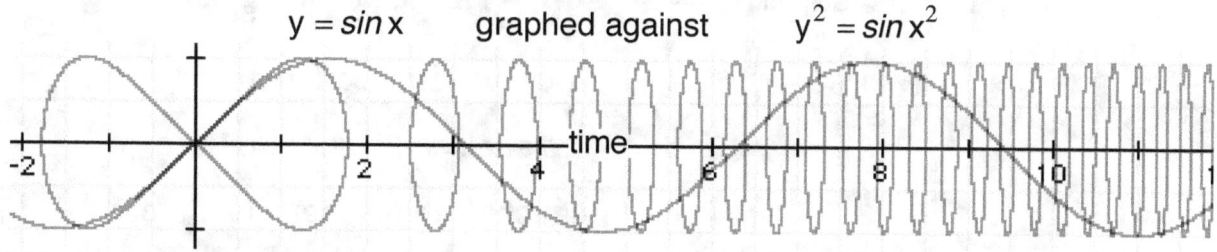

1/2 frequency is retained only for the "lobed orbit." It is no longer retained for the emissions. Frequency accelerates along the time axis for the dimensionally-shifted squared wave function. A dimensionally shifted wave cannot retain light-frequency information.
The Schrodinger nonlinear differential equation dimensionally shifts the linear wave function contained by the three dimensional orbital, itself lodged in four dimensional space, to a squared wave function in three dimensional space. This fact is proven by Schrödinger's own mathematics which equates the nonlinear equation with a calculable Hamiltonian operator. The Hamiltonian solution to Schrödinger's nonlinear wave-function faithfully reflects the "lobed orbit" of the squared wave-function. By increasing the Hamiltonian operator from "1" to "3" Schrodinger produces a three dimensional lobed orbital in three dimensional space. However, the dimensionally shifted lobed orbital cannot retain light-frequency information. Frequency must be assigned from outside the equation. Schrödinger's lobed

orbitals cannot predict frequency but can only interpret light-frequencies which have been empirically determined outside the equation.

Sodium D-Lines: Establishing Four-Dimensional Structure over Bohr/Schrodinger
Two atomic models have been presented. The three dimensional orbital in four-dimensional space with its acceleration/deceleration internal wave outputting light has been contrasted with Bohr's falling electrons between orbits, orbits which have been modified by the Schrodinger wave-function to output streams of "photons."

The two models are proposed for the shell/subshell schematic of the periodic table which can accommodate either model. Specifically, the four-dimensional structure proposes a different order of subshell energy and provides an exact distribution of electron voltage for the shell/ subshell schematic. In contrast, Bohr/Schrodinger relies upon the Schrodinger function to describe "quantum energy states" as the need arises. Most shell/subshell energies are treated as relational rather than exact.

Further, electron shell/subshell fill-in for the elements of the periodic table differ between the models. The number of electrons which can be accommodated by the various subshells are different. The number of electrons accommodated is dependent upon the quantum angular momentum number "ℓ" which defines the number of "lobed orbitals" for the Schrodinger dimensionally-shifted wave-function and the number of "sub-waves" for the three-dimensional transverse wave in the quantum geometric model. The number of electrons accommodated is determined by a square function of "ℓ" for the Shrodinger "squared wave-function" and a linear function for the quantum geometric transverse wave:

ℓ = subshell quantum "angular momentum number"

By Schrodinger orbitals: **By quantum three-dimensional orbitals:**
subshell electrons $= 2 + 2(2\ell)$ subshell electrons $= 2 + 2(\ell)$

Subshell label	ℓ	Max electrons		Shells containing it	Historical name
s	0	Schrod. consens. 2	Quant. 3-D orb	Every shell	sharp
p	1	6	4	2nd shell and higher	principal
d	2	10	6	3rd shell and higher	diffuse
f	3	14	8	4th shell and higher	fundamental
g	4	18	10	5th shell and higher	
h	5	22	12	6th shell and higher	
i	6	26	14	7th shell and higher	

The sodium "D-lines" provide an exact empirical test for both models. The "D-lines" are a naturally occurring light "doublet" emitted by energized sodium (588.995 nm and 589.5924 nm). The "D-lines" are apparently split from 589.2937 nm wavelength and are

36

known to be closely associated with the Zeeman effect[33].

Naturally-Occurring Sodium Spectrographic D-Lines[34]	Zeeman Splitting of Sodium D-Lines by Application of a Magnetic Field
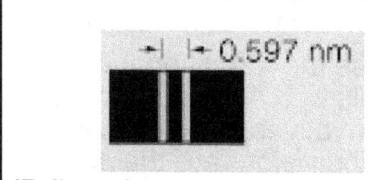 *(D-line change in eV)* $\Delta eV_{dbl} = \pm0.0010659039$	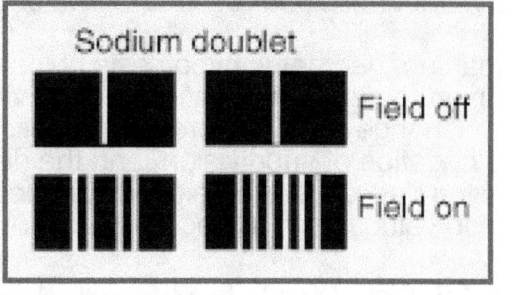

Two things need to be explained about the sodium D-lines. First, from where in the orbital shell/subshell schematic are they emitted and why the "589.2937" nm split. Second, why is further "Zeeman effect" splitting so unique? Why does the higher frequency D-line split into three doublets and the lower frequency D-line split into two doublets?

We will begin with the Bohr/Schrodinger model. Time and distance are independent variables for the Schrodinger wave function. Therefore, the Schrodinger wave function cannot assign frequency to Planck's Constant by orbital distance (frequency is the inverse of wave time). The Schrodinger nonlinear differential equation assigns "interpretive energy" to an orbital which has been coupled, by fiat, with a light emission.

This is what has occurred with the Bohr/Schrodinger "Interpretation" of D-lines emissions. Orbital distances have been assigned by a presumed shell/subshell schematic for sodium. Sodium's eleven electrons are said to occupy:

$1s^2\ 2s^2p^6\ 3s^1$ shell/subshells.

D-line light emissions are explained as "falls" between subshells within this presumed schematic. Specifically, by "falls with 1/2 and 3/2 spins" between the "p" subshell in the 3rd shell and the "s" subshell in the 3rd shell[35]. Energy is assigned to the "quantum state" by the Schrodinger equation using D-lines frequencies. There is no dependent relationship between the chosen shell/subshells and D-lines frequencies. Consequently, there is no way to disprove the Schrodinger interpretation.

Schrodinger converts rational four-dimensional space into irrational three-dimensional space. This is confirmed by Heisenberg uncertainty which holds that electron momentum (velocity) and electron position cannot be simultaneously held within the Schrodinger wave function.

However, the Schrodinger equality does predict electron capacity for the subshells by using the quantum angular number "ℓ." Subshell capacity is the one variable which the Schrodinger nonlinear equation has actually predicted and is, therefore, the one variable with which Schrodinger can be empirically tested.

Schrödinger's predicted capacity can be tested if the D-lines are explained by the attempt of an electron to climb to a subshell which has already reached its electron capacity. The D-line wavelength shows where the electron might be "stuck" between subshells and the

---

[33] The "Zeeman effect" is the splitting of spectrographic line into two or more lines by a magnetic field.

[34] Source: daviddarling.info/encyclopedia/D/D_lines.html

[35] hyperphysics.phy-astr.gsu.edu/Hbase/quantum/sodzee.html#c2

37

electron voltage for the "stuck orbital" can be determined. If the variance between D-line electron voltage and destined subshell electron voltage can identify the number of electrons in the subshell, then the Schrodinger predicted capacity can be tested. The quantum geometric model provides exactly such a test which is confirmed both mathematically and by the number of Zeeman splits of the original D-lines.

The quantum geometric hypothesis holds that Bohr's "falls" confuse the "negation values" of the "quantum-squared negation of subdivision" establishing orbitals with "negative force orbits". Change in orbital force is a function of two ambiguous derivatives. The derivative of "the negation of subdivision" and the derivative of "negative subdivision" are the same. That is the derivative is ambiguous and led Bohr to confuse quantum whole-number "negation values" with a "negative force orbit."

When corrected, Bohr's "electron falls" become a different thing. They no longer produce gains in electron voltage which output Rydberg light frequencies, but are Lyman Series differentials which distribute electron voltage into lower order shells/subshells.
Niels Bohr didn't realize that his "negative electron voltage" orbits were misidentified and that this would prove more significant than his attempt to explain orbital distances as restricted to whole number multiples. Nothing similar to Bohr's "negative force orbits" existed in classical mechanics.

The quantum dimension easily identifies why geometric structure is a function of whole numbers, yet the quantum dimension also operates within the principles of physical reality, including classical force and energy. Bohr's "negative force orbits" were a serious mistake which prevented science from recognizing actually electron energy states for well over a hundred years. Quantum geometry proposes the correction. Bohr's electron "falls" have nothing to do with light emissions and everything to do with orbital energy states.

Bohr "Falls" Within the Lyman Series Explain the Shell/Subshell eV Distribution

$$\frac{\Delta eV}{\Delta x_n} = D\left(1 - \frac{1}{x_n^2}\right) 13.6037\ eV\,) = D\left(-\frac{13.6037\ eV}{x_n^2}\right)\ (Bohr\ negative\ force) = \frac{2}{x_n^3} 13.6037\ eV$$

$$\frac{\Delta eV}{\Delta n^2} = \left(\frac{1}{1^2} - \frac{1}{n'^2}\right) 13.6037\ eV - \left(\frac{1}{1^2} - \frac{1}{n^2}\right) 13.6037\ eV = \left(\frac{1}{n^2} - \frac{1}{n'^2}\right) 13.6037\ eV\ ; \qquad n' > n$$

Lyman Shell 7	Balmer Shell 6	Paschen Shell 5	Brackett Shell 4	Pfund Shell 3	Impedance Shell 2	Non-imped. Shell 1
Subshell	Subshells	Subshells	Subshells	Subshells	Subshells	Subshells
s	s	s	s	s	s	s
p	p	p	p	p	p	t = 1/ f > t_ψ *True*
d	d	d	d	d		***Base-State Subshells***
f	f	f	f	*Variance with Lyman "f"*	*Variance with Lyman "d " -*	*Variance with Lyman "p"*
g	g	g	*Variance with Lyman "g"*			
h	h	*Variance with Lyman "h"*				
i	*Variance with Lyman "i"*					

38

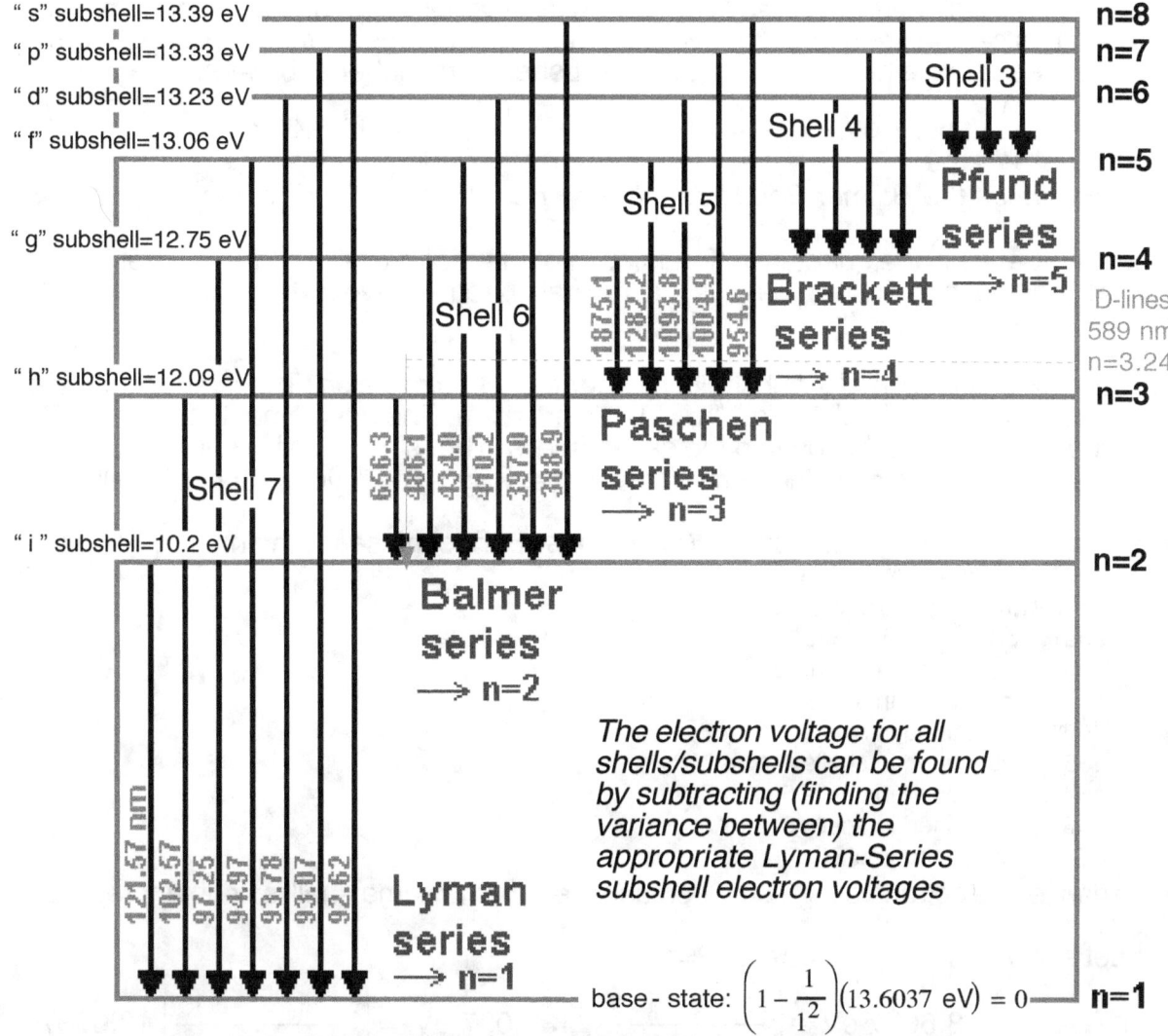

The Lyman Series Distribution into Shells/Subshells by "Falls" Between Lyman Series Subshells

Bohr misidentified "quantum negation numbers establishing subshells." He considered them to be "negative force orbits." This misidentification resulted in incapacity. He was rendered incapable of recognizing what whole-number differentiated electron orbitals actually are. Neither could he determine what the electron voltage associated with those orbitals must be. Bohr's mistake rendered him incapable of recognizing four-dimensional electron/proton structure. As a consequence consensus science, built upon the Bohr model, suffers the same incapacity.

Electron voltage is the potential difference between the charges of the electron and proton. It can be increased only by the multiplication of the electron's charge. The electron-charge multiple equals the light frequency output by the orbital. Electron-voltage is equal to frequency *times* "4.13567 10^{-15} eV" which is the potential difference at the alpha-space distance [$\overline{\alpha}$=0.50214 (10^{-15}) meters] which must separate electron and proton and at which the electron's bond must be rotated relative to the proton's bond.

Electron voltage is distributed into shells/subshells by "falls" between subshells in the

Lyman Series. Such Lyman Series falls account for the full range of the Rydberg frequency distribution as can be seen in the above illustration. For example, the visible spectrum Balmer Series is explained by falls to the "i" subshell from higher subshells:

$$\left(1-\frac{1}{n^2}\right) 13.6037 \text{ eV} - \left(1-\frac{1}{2^2}\right) 13.6037 \text{ eV} = \left(\frac{1}{2^2}-\frac{1}{n^2}\right) 13.6037 \text{ eV} \quad ;$$

Produces the Balmer Series "shell 6" $n > 2$; $n = s, p, d, f, g$ or h subshells

The Lyman Series is the quantum differentiation of the root orbital. The root orbital is assigned the quantum position because acceleration terminal-velocity across the orbital's transverse wave reaches the speed of light.

The root orbital is assigned 7 subshells by negation of subdivision. These subshells are given the labels "s," "p," "d," "f," "g," "h," "i" by convention. The assignment of subshells is determined by their distribution into the seven shells. The "s" subshell is contained in every shell. The "i" subshell is contained only in the highest and last :"Lyman Series" shell[36] :

Each subshell has a negation number (the number Bohr confused with an "orbit" in his "negative force" misinterpretation).

Quantum negation number as assigned to subshells

s=8   contained in all shells
p=7   contained in shells 2 thru 7
d=6   contained in shells 3 thru 7
f=5   contained in shells 4 thru 7
g=4   contained in shells 5 thru 7
h=3   contained in shells 6 thru 7
i=2   contained in shell 7

The variance in electron voltage between subshells in the same shell is constant regardless of the shell:

Let subshell 1=n' ; let subshell 2=n'-1

$$\left(\frac{1}{n^2}-\frac{1}{n'^2}\right) 13.6037 \text{ eV} - \left(\frac{1}{n^2}-\frac{1}{(n'-1)^2}\right) 13.6037 \text{ eV} = \left(\frac{1}{(n'-1)^2}-\frac{1}{n'^2}\right) 13.6037 \text{ eV}$$

The variance in electron voltage between subshells "n' " and "n'-1" is always the same regardless of which "n" shell the two subshells may be a part.
The electron-voltages variance between subshells are the following:
Electron voltage gain between subshells is the same regardless of the shell.

Quantum negation number	Variance in electron volts between subshells					
s=8	s-p =0.065	Shells 2 through 7				
p=7	s-d = 0.165	p-d = 0.100	Shells 3 through 7			
d=6	s- f = 0.332	p- f = 0.266	d- f = 0.166	Shells 4 through 7		
f=5	s-g = 0.638	p-g = 0.572	d-g = 0.472	f-g = 0.306	Shells 5 through 7	
g=4	s-h = 1.299	p-h = 1.234	d-h = 1.133	f-h = 0.967	g-h =0.661	Shells 6 through 7
h=3	s- i = 3.188	p- i = 3.122	d- i = 3.022	f- i = 2.856	g - i = 2.55	h- i = 1.89 Shell 7
i=2						

The above table illustrates that the consensus Bohr/Schrodinger explanation of the sodium

---

[36] http://en.wikipedia.org/wiki/Electron_shell

D-lines cannot be correct. The 588.995 nm and 589.5924 nm D-lines have an electron voltage of 2.10501 eV and 2.10288 eV respectively with a 0.0021 eV variance.

The Bohr/Schrodinger consensus shell/subshell schematic for sodium's eleven electrons is the following:

$$1s^2 \ 2s^2 2p^6 \ 3s^1$$

The D-lines are said to be output by a "fall" from the "3p" to the "3s" subshells by the one electron in the "3" shell. However, by the above table it may be seen that there is only 0.065 eV variance between any two "s" and "p" subshells contained in the same shell. There is simply not enough electron voltage gain to explain light outputs of 2.10+ electron volts.

How can the Bohr/Schrodinger model support a 2.10 eV gain between the "3p" and "3s?" By using the Schrodinger equation supplied with the time values of D-line frequencies to describe the "quantum energy states" of alleged falls between the subshells. Since there is no relationship between shell/subshell orbital distances and electron voltage with Bohr/Schrodinger, energy states can be supplied by fiat. Interpretations can be given to events such as the D-lines without demonstrating "cause and effect" relationships.

We must now turn to the actual explanation of D-line emissions by sodium. The system of electron orbitals distributed by the Lyman Series is first and foremost a radiation energy exchange mechanism. Radiation is exchange to and from the nucleus by pressure upon extra-dimensional kinks which form the three dimensional orbital. The frequency of radiation is determined by the acceleration/deceleration time factor which is established by the orbital distribution and is a directly related to the multiplication of the electron's charge. Energy is exchanged with the nucleus at the rate of Planck's Constant as previously described.

For any element to output radiation, its nucleus must multiply the charge of the electron. When an element is heated, its nucleus acquires energy which it applies to electrons as a Piezoelectrical force multiplying electron charge. The nucleus must move the electron through the shell/subshell system to reach the shell which outputs the frequency band. The visible light frequencies are output by Balmer Series subshells. The Balmer Series is located in the sixth shell from the nucleus. To output visible light frequencies like the sodium D-lines the nucleus must move the electrons into the sixth shell.

Electrons fall into a "base state" if the nucleus is deficient in energy. Electrons are sustained in the base state by their own velocities and, possibly, a capacity to harmonically impede higher orbital frequencies. This base state is similar to but not the equivalent of the shell/subshell schematic for the periodic table of elements.

There are two reasons that the quantum geometric base state differs from the consensus periodic table shell/subshell schematic. First, the capacity of subshells to accommodate electrons are predicted differently, a difference which is to be tested by the D-lines. Second, the consensus schematic uses a system of artificial quantum numbers[37] to propose *relative* shell/subshell energy states while four-dimensional quantum geometry uses calculated numeric shell/subshell energy states.

The base state predicted for sodium by the quantum geometric model is the following:

$$1s^2 2p^4 2s^2 3d^3$$

---

[37] See "Madelung's rule" in *Quantum Numbers to Periodic Tables: The Electronic Structure of Atoms:* The Chemogenesis Web Book

The principle which governs base-state electron fill-in is the following: " electrons fill the lowest energy state to capacity and move upward." The lowest energy "1s" subshell can accommodate "2" electrons which are filled in first; then the four electron capacity (not Schrödinger's six) of the "2p" subshell is filled in; followed by the two electrons for the "2s;" the last three electrons going to the next highest shell/subshell being the "3d." These energy states are determined by the Lyman Series energy distribution for the shells and subshells. Each shell/subshell is given an exact electron voltage value by the Lyman Series distribution.

The base state does not fix orbitals as with the consensus schematic which only allows electron exchanges between subshells in the highest "valence" shell. The nucleus can reconfigure the shells/subshells when it acquires enough energy. In the case of sodium, this will occur when nuclear kinetic energy reaches the 0.166 eV required to move electrons to the "3d" subshell. "3d" is not filled to capacity so at 0.166 eV the nucleus can move additional electrons in. 0.166 eV is the electron voltage needed to move from the "base state" (eV=0 at lower shell boundary) of the 3rd shell to its first subshell, the "d" subshell.

Since the "3d" has a capacity of 6 electrons and only contains 3, three electrons from lower order shells/subshells can be moved upward leaving the following configuration:

$$2p^3 2s^2 3d^6$$

If the nucleus acquires 0.267 eV required by the next-higher "3p" subshell (from base state to 3p) it will reconfigure the shell/subshells by the following:

$$2s^1 3d^6 3p^4$$

In this manner an energized sodium nucleus could "walk" the electrons through higher-order shells and subshells. When nuclear kinetic energy reached the Balmer visible light (shell 6), it would encounter the first subshell ("h") which could accommodate all 11 electrons:

$$\text{electron distribution} = 6h^{11}$$

However, when nuclear kinetic energy supplies an additional 0.661 eV to move the electrons from subshell "h" (negation number 3) to the next higher subshell "g" (negation number 4) it is proposed that the following happens:

$$\text{let (irregular subshell; } n' \approx 3.24) = 6\left(^- g\right)$$

$$\text{electron distribution} = 6g^{10}, \; 6\left(^- g\right)^1$$

It is being proposed that sodium D-lines are explained as an attempt by nuclear kinetic energy to move sodium's 11 electrons from the "6h" subshell into the "6g" subshell. The subshell of destination ("6g") has the capacity for only 10 electrons. It is further proposed that the excess electron becomes stuck between the "h subshell; negation number 3" and "g subshell; negation number 4" in an irregular orbital with a negation number of approximately "3.24." An irregular subshell with a negation number of "3,24" would output light at approximately 589 nm wavelength which is the approximate wavelength of the D-lines. In addition it is concluded that electron voltage pressure from the "6g" subshell resists entry of the excess electron. This voltage pressure is expressed upon the kinked wave planes of the electron and this pressure forces that electron to output light as the D-lines doublet:

Rydberg Formula for Wavelength Applied to the Irregular Subshell "n'=3.24"

$$\frac{1}{\lambda} = \left(\frac{1}{2^2} - \frac{1}{3.24^2}\right) \frac{1}{91.14 \text{ nm}} \quad ; \quad \lambda = 588.99 \text{ nm}$$

Two lines of proof will be offered to support this hypothesis. First the variance in electron voltage between the "g" subshell and the irregular subshell, the electron voltage which is suppressing the rejected electron, can be compared with the "1/2 spin" energy of the electron moving along the orbital wave path as measured in Bohr magnetons. When this comparison is done, the calculations resolve to the correct number of electrons in the "g" subshell which are needed to explain D-lines emissions. Secondly, the "Zeeman splitting" of the D-lines identifies the number of electrons in the "g" subshell and this number is also that which is predicted.

Although the hypothesis thus presented can stand alone with its strong mathematical and empirical support, it still needs to be presented within the context of precedent physics. The model being tested varies from consensus atomic structure. The differences between the consensus atomic model and four-dimensional structure are explained by Bohr and Schrodinger error.

Bohr's misidentification of the quantum-required negational subdivisions as "negative force orbits" has been carried into the consensus model with the consequence that the relative energy states of subshells have been inverted and the consensus has suffered a systemic inability to identify actual subshell energy states. For example, the consensus shell/subshell schematic proposes that the "p" subshell is more energetic than the "s" subshell[38] when the reverse is true. The "p" subshell's negation number is "7" and the "s" is "8." As the subdivision proposed for Bohr's "negative force orbit," $1/7^2$ is greater than $1/8^2$ and the "p" subshell would be more energetic than the "s." As the negation of subdivision, however, the inverse holds sway. $(1-1/7^2)$ is less than $(1-1/8^2)$ and "p" is less energetic than "s." With the Bohr inversion, subshell energy states are inverted, relative and indeterminate.

Bohr's inverted subshell energy states have been used to establish a false reason that lower order subshells can accommodate a higher number of electrons. It is assumed that the alleged higher energy states of lower-order subshells allow those subshells access to a greater number of Schrodinger "lobed orbitals." Since the "lobed orbitals" slightly vary in orbital energy, they are thought to be able to evade the Pauli exclusion principle[39] for the subshell and thus allow more electrons to inhabit the subshell.

The facts are quite different. While the energy of lower-order subshells decreases, the variance in electron voltage between subshells increases. For example, the "s" subshell in the Lyman-Series shell has 98.44% of shell energy, the "p" 97.96% and the "d" 97.22%. However, the electron voltage variance between the "s" and "p" subshells is 0.065 eV and the electron voltage variance between the "p" and "d" subshells is 0.100 eV (see above table). As lower order subshell energy decreases, electron voltage variance between subshells increases.

Additional electron capacity for lower order subshells is produced by "doublets." The Zeeman effect upon the Rydberg distribution of hydrogen radiation emissions prove that shell/subshell orbitals can be forced by a magnetic field into a "doublet" which brackets the original orbital. That is, the electron will rise and fall between two orbitals which are slight more and slightly less energetic than the original. The orbit is "split," bracketing the original

---

[38] *Quantum Numbers to Periodic Tables: The Electronic Structure of Atoms:* The Chemogenesis Web Book; Mark R. Leach, mrl@meta-synthesis.com

[39] The Pauli exclusion principle holds that only two electrons orbiting in opposite directions can inhabit an orbital of set energy without interfering with one another. See *http://hyperphysics.phy-astr.gsu.edu/HBASE/quacon.html#quacon* for "Pauli exclusion."

and thus eliminating "exclusion principle interference. This possibility is presented by the effect of a weak magnetic field upon an original Rydberg spectrographic line which splits the original into two lines of wavelengths slightly greater and slightly less than the original.

Two Spectrographic Lines "Split" from "6h" subshell of 656.3 nm Wavelength

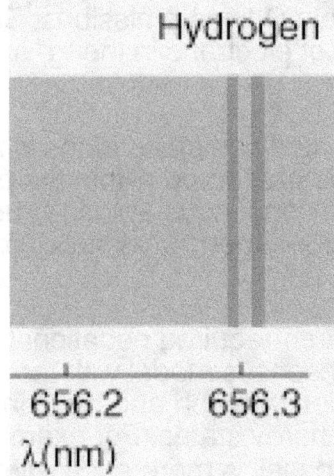

Hydrogen

656.2   656.3

$\lambda(nm)$

Under application of a magnetic field, the 656.3 nm spectrographic line (output by "6h" subshell; n=2, n'=3) is split into two lines at ±0.016 nm from the original. This represents an electron voltage difference of ±.000046 eV[40] from subshell eV.

An orbital doublet is built upon the premise that the quantum-squared orbital distance is directly related to electron voltage as a functions of light frequency. As frequency increases, so does orbital distance increase and both in direct relationship to an increase in electron voltage . Change in frequency, as represented by the doublet, implies a directly related change in electron voltage as well as a directly related change in orbital distance. The quantum formulae for these relationships are the following:

$$f = \frac{eV(C_p)}{e^-} \quad ; \quad f = \frac{r^2}{\alpha^2} \quad ; orbital\ radial^2 = r^2 \quad ; orbital\ frequency = f$$

$$f(e^-) = eV(C_p) \quad ; \quad f(\alpha^2) = r^2$$ [41]

$$\Delta f(e^-) = \Delta eV(C_p) \quad ; \quad \Delta f(\alpha^2) = \Delta r^2$$

The above doublet is generated by an electron voltage change of "±.000046 eV" from orbital electron voltage. This represents a frequency change of ± 1.113 (10¹⁰) Hz and an orbital radial change of ±5.295790976(10⁻¹¹) m or ± 0.49342201% of the original orbital distance of 1.0732782193(10⁻⁸) m.

The orbital radius is the quantum leg of the Q² orbital because it represents the proton's charge and the proton's charge is a projection into the quantum dimension. Since, the radius is a quantum distance which is sustained by force, it can respond instantaneously to changes in electron voltage[42].

[40] *Hydrogen Fine Structure* , hyperphysics.phy-astr.gsu.edu/Hbase/quantum/hydfin.html#c1
[41] See *Formulations for Light Frequency as the Multiple of the Electron Charge and Basic Quantum to Establish Orbitals* in Appendix.
[42] The projection of force is not restricted to the speed of light. Since quantum separation is sustained by force, quantum distances can adjust spontaneously and instantaneously to changes.

Zeeman effect doublets are clearly quantum in nature. They are explained by the capacity of the quantum to acquire new radial distances without regard to the velocity restriction for changes in Euclidean position. If the orbital radius were Euclidean, the electron could not obtain new orbital positions without the acceleration and velocity restrictions of the speed of light. However, as a quantum radius composed of a space of separation sustained by force, change in radial position is instantaneous with change in force. Space simply disappears or expands by eliminating or adding a whole number of the fundamental quantum $(\overline{\alpha}^2)$:

$$\frac{\Delta eV(C_p)}{e^-} = \Delta f = \frac{\Delta r^2}{(\overline{\alpha}^2)}$$

$$\Delta r^2 = \Delta f(\overline{\alpha}^2) = \frac{\Delta eV(C_p)}{e^-}$$

The change in the orbital radius "$r\,^2$" is the addition or subtraction of a whole number $(=\Delta f)$ of "$\overline{\alpha}^2$" quantum units which is produced by an change in "force" as defined by change in electron voltage (ΔeV). Since quantums are composed by force, the change in orbital radius is instantaneous by collapsing or inserting quantum units in the radial.

The orbital's transverse wave is composed by the kinking of the electron's Euclidean charge-bond in alternating directions. This kinking of the Euclidean-defined distance into curvature inserts new four-dimensional space into three dimensional space. It also composes the known "1/2 spin" of the electron, another phenomenon which has not been explained by consensus quantum mechanics.

The negative charge is being "kinked" producing 1/2 spin in alternating directions against any magnetic field. This in turn propagates electron voltage increase of pressure and electron voltage release of pressure against the kinks as expressed upon the alternating wave planes.

"6h" subshell producing 656.3 nm doublets in a weak magnetic field

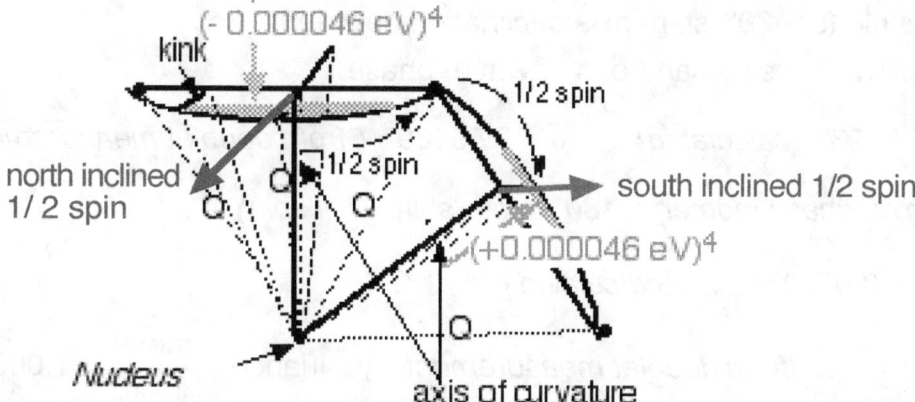

Electron voltage pressure against 1/2 spin energy in alternating wave planes produces an orbital doublet by instantaneous addition and subtraction of quantums. This doublet orbital brackets the original at orbital distances which are $\pm\Delta f\,\overline{\alpha}^2$ from the original orbital distance of "$r\,^2$." Such doublet orbitals allow subshells to acquire additional electrons without violating the Pauli exclusion principle.

The wave-path of any orbital composes the "1/2 spin" of the electron. Even though the wave contains an acceleration/deceleration phase which outputs light at frequency, it still has a consistent wave-phase time because acceleration/deceleration velocities cancel each other out, remaindering initial orbital velocity. That is, every orbital is characterized by two time values; the time value of the light frequency and a different and distinct wave-phase time. The wave-phase time is a constant for all orbitals at "$t_\psi = 2.067979114(10^{-14})$ seconds."[43]

The Wave-Phase Time Constant Predicts Zeeman Doublets

The "1.2 spin" has a magnetic moment measured in Bohr magnetons. Since the Bohr magneton has a unit value of "joules/t," 1/2 spin energy can be calculated by multiplying the magneton by wave-phase time. 1/2 spin energy is also a constant for all orbitals (see *Table of Quantum Values*). In turn, the energy constant can be converted into electron volts by dividing it by the elementary charge.

Wolfgang Pauli's "1/2 spin" of the electron is the one empirically-confirmed fact by consensus quantum mechanics. It was confirmed in the measurement by Kusch and Foley of the anomalous magnetic moment [44]. However, consensus physics measures the anomalous magnetic moment[45] of free particles the paths of which are bent by magnetic fields using accelerators. The anomalous magnetic moment is never measured within the atom. Quantum geometry reveals what the "1/2 spin" and its anomaly actually are when operating within the context of the atom. It does so by predictions of the changes in electron voltage (ΔeV) required to produce the Zeeman doublet.

Changes in doublet electron voltage can be predicted using the wave-phase time constant. Specifically, the doublet's change in electron voltage— taken to the fourth power (ΔeV^4)— is being expressed against the orbital wave-planes in four-dimensional space. This must be divided by the electron-voltage squared (eV^2) of the subshell. Subshell electron voltage must be squared because it is being expressed across the initial plane of the orbit. This value "$\Delta eV^4/eV^2$" is equal to the the 1/2 spin electron voltage (a constant) plus the intra-atomic anomalous magnetic moment (non g-factor anomaly):

The Intra-Atomic Anomalous Magnetic Moment

$\bar{\theta}$ = mean angle for $\Delta 30°$ spin-axis declination between

wave-plane transition and end of wave-phase.

$$\bar{\theta} = \frac{30°}{\pi} = 9.5493° \ \langle calculated \rangle \ ; \quad \bar{\theta} = 9.55169° \ \langle from\ doublet\ measurement \rangle$$

$\gamma = \langle$% gain magnetic moment; 180° axis-shift = full spin$\rangle$

$$\gamma = \frac{\bar{\theta}}{180°} = 0.0530516477 \langle calculated \rangle$$

$\gamma = 0.0530649632 \ \langle from\ doublet\ measurement \rangle \ ; \quad variance = \dfrac{meas.}{calc.} = 1.000251$

The declination of the spin axis produces an anomalous magnetic moment gain because

---

[43] See *Calculating the Half-Spin Time Constants* in the Appendix.

[44] Kusch, P.; Foley, H. M. (1948). "The Magnetic Moment of the Electron". Physical Review 74: 250–263.

[45] I realize that measurements of the anomalous magnetic moments of free particles by bending in a magnetic field are not restricted to electrons, but also include "muons" and other alleged electron-like heavy particles. However, consensus physics doesn't actually know what a "muon" is. It is a free electron, the charge of which has been multiplied by the nucleus during Beta decay.

the declination is towards a full spin. A "1/2 spin" is defined as a spin which does not reverse direction of rotation and therefore is expressing only 1/2 magnetic moment. If the axis of spin were rotated 100%, it would reverse the direction of rotation, completing a full spin and generating 100% magnetic moment. The mean angle of declination, as a percentage of 180° declination, represents anomaly as the percentage of gain.

Calculation for Required ΔeV to Produce Zeeman Doublet
(Incorporating the Calculated Intra-Atomic Anomalous Magnetic Moment)

$$1/2 \text{ spin eV} = eV_{(1/2)spn} = 1.1970188584 \times 10^{-18} \text{ eV}$$

$$\frac{\Delta eV_{dbl}^4}{eV_{sub}^2} = (1+\gamma)eV_{(1/2)spn} = (1.0530516477)(1.1970188584 \times 10^{-18}) = 1.2605226812 \times 10^{-18} \text{ eV}$$

$$\Delta eV_{dbl} = \sqrt[4]{eV_{sub}^2 \left(1.2605226812 \times 10^{-18} \text{ eV}\right)}$$

Note: ΔeV_{dbl} is the dimensionally shifted value "$eV^{3/4}$" or "$\sqrt[4]{eV^3}$". The "eV^2" value (subshell electron voltage) times the linear "eV" value (1/2 spin electron voltage) produces an "eV^3" value of which the fourth root is taken.

The formula does, in fact, work for the measured "656.3 nm" Zeeman doublets. The electron voltage for the "6h" subshell outputting 656.3 nm wavelength light is precisely "1.889139794 eV" (although this fact is not currently recognized by consensus physics). Using the above formula, this calculates to a "ΔeV" for the doublets of "±0.000046054 eV" which is correct.

A formula has been derived from the four-dimensional model which precisely predicts Zeeman hydrogen doublets as electron voltage changes from the light frequency. generating the doublet. Change in voltage is a function of an electron 1/2 spin anomalous magnetic moment which has been calculated by the model. The consensus Bohr/Schrodinger model offers no equivalent.

An attempt to prepare a full data set of Balmer Series Zeeman doublets has proven extremely frustrating. A review of the literature revealed that consensus science is loath to present observation and measurement, relying instead on interpretations of the data. Only the physics web site hosted by Georgia State University[46] included raw data with quantum mechanical interpretations. All of the rest of the literature discussing Zeeman doublets failed to disclose the actual doublet wavelengths under discussion. Quantum mechanical interpretations were substituted for actual measurement in all reports, interpretations which used the inappropriate "g-factor[47]" anomalous magnetic moment and presumed "Bohr falls" between misunderstood shells/subshells.

As noted earlier, the Schrodinger nonlinear equations allegedly identify "quantum energy states" but those equations can only interpret. They cannot predict. It is the current practice of consensus physics to let interpretation stand in the stead of observation. There is a difference between mathematics which identify process and therefore remain intimately connected to data and mathematics which only interpret and therefore substitute for data. It is the difference between quantum geometry and consensus quantum mechanics.

---

[46] http://hyperphysics.phy-astr.gsu.edu/hbase/HFrame.html

[47] I will let the reader familiarize themselves with incorrect physics. It is not my duty to teach primitive mythology.

If consensus physics is to retain a modicum of credibility, it must address the following question. If the shell/subshell schematic does not distribute electron-voltage and radiation-frequency as identified by the four-dimensional model, then why does a formula derived from those distributions and by that model accurately predict Zeeman effect doublets for hydrogen emissions? Is quantum mechanics to be defended against four-dimensional quantum geometry by data suppression? Is this "interpretive" consensus to now take precedent over evidence?

To summarize: the Euclidean bond of the electron is rotated at 90° and kinked into curvature by the quantum bond of the proton. This produces a wave-path for the electron along the kinked plane which establishes the electron's "1/2 spin." At every point along the wave-path, the electron has a magnetic moment as defined by the Bohr magneton. Further, the energy of spin is defined by the time it takes the electron to complete a 1/2 spin and is equal to the Bohr magneton multiplied by this time. The time of 1/2 spin is a wave-phase time constant. The time it takes to complete a 1/2 spin along the wave-path is shown mathematically to be the same for all orbitals.

The electron's spin is in balance with its axis at 90° to the kinked plane. However, the 1/2 spin is completed on two adjacent wave planes which cause it to move from a "north inclined" spin towards a "south inclined" spin (see illustration above). These adjacent wave planes establish the three dimensional wave as they are set at a 60° angle to one another. At the point of transition, the electron's axis of spin is declined 30° to each plane. The electron's axis of spin must adjust from this 30° declination back to the 90° state of balance during the "south inclined" phase of its 1/2 spin. The three dimensional character of the wave produces a magnetic moment anomaly because it provide a changing angle of declination to the axis of spin.

To calculate magnetic moment anomaly, the spin-axis angle of declination is treated as a percentage of a full 180° angle of declination. An angle of declination of 180° would reverse the direction of rotation and convert the 1/2 spin into a full spin. A full spin has double the magnetic moment of a 1/2 spin[48].

The angle of declination, as a percentage of 180° declination, produces a magnetic moment gain which is equal to the percentage. However, the angle of declination isn't constant. It changes from 30° to 0° over the course of the south-inclined phase of the spin. Only the mean angle of declination will suffice to define the anomalous magnetic moment gain. That mean isn't calculated as " 30°/2" but "30°/π.[49]"

This calculation produces a theoretical mean angle of declination which is very close to the experimentally calculated mean angle of declination using the 656.3 nm Zeeman doublets. There is only four thousandths of a variance between the two (see above).

The precision with which quantum geometry predicted the intra-atomic anomalous magnetic moment rivals the highly-touted precision with which quantum mechanics had predicted the "g-factor" anomalous magnetic moment for free electrons which are bent by magnetic fields in accelerators. Free electrons in accelerators are not confined to four-dimensional space as those same electrons would be when contained within orbitals. For this reason, using the "g-factor" anomaly cannot be accurate when applied to atomic structure. The intra-atomic

[48] A 1/2 spin has only one point of magnetic opposition while a full spin has two.
[47] I will let the reader familiarize themselves with incorrect physics. It is not my duty to teach primitive mythology.

anomalous magnetic moment derived by quantum geometry is relevant to shell/subshell structure while the quantum mechanical version is not.

I am not suggesting here that all of quantum mechanics be abandoned to the detriment of scientific interpretive power. I am suggesting that quantum mechanical interpretations which cannot be subjected to empirical testing as hypotheses, be abandoned as "artifacts" of a scientific culture of consensus. I further suggest that advocates of quantum mechanics refrain from substituting such interpretations for rigorous reports of actual data measurements. The validity of any conclusion is always dependent upon the rigor of the experimental methodology and the rigor of methodology can only be evaluated by a complete and thorough description of the origins of data.

The Schrodinger wave-function predicts the number of electron's which can be distributed into the subshells and , therefore, can produce a testable hypothesis. That is, it can be proven wrong empirically. If the number of electron in a subshell are found to be different than predicted by the Schrodinger orbital wave-function, then Schrodinger must be rejected. The Schrodinger wave-function for electron orbitals can no longer be accepted by fiat but must meet the test of a legitimate scientific hypothesis.

By the Schrodinger hypothesis, the number of electrons which can be accommodated by any subshell is determined by the number of lobed orbitals contained within the subshell's orbit and the number of lobed orbitals are, in turn, determined by the quantum angular momentum number "ℓ."

The Pauli exclusion principle restricting two electrons to any one orbit is said to be resolved by Schrödinger's lobed orbitals. These lobed orbitals are said to provide "noninterference" alternatives for the subshell orbit. Schrödinger's angular momentum number produces a set of lobed orbitals for each value. That is, the number of lobed orbitals produced by his nonlinear differential equation is equal to "2ℓ."

The Schrodinger Equation Produces "2ℓ" Lobed Orbitals

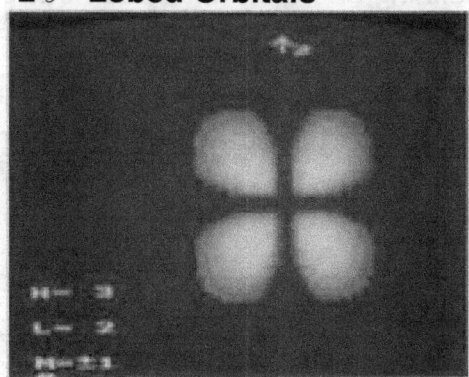

$H=3$ *(three dimensional), $\ell = 2$*[50]

However, I have demonstrated that Schrödinger's "lobed orbitals" are the geometric equivalent of a three dimensional electron wave-path in four-dimensional space which has been dimensionally shifted onto a plane in three dimensional space. The Schrodinger equality is the equivalent of the nonlinear squared wave-function "$y^2 = sin(x^2)$" which defines this shift. Schrödinger's eigenfunction produces a three dimensional orbital from the dimensionally-shifted two-dimensional wave , retaining the "lobed orbital form."

[50] *Quantum Numbers to Periodic Tables: The Electronic Structure of Atoms:* The Chemogenesis Web Book; Mark R. Leach,

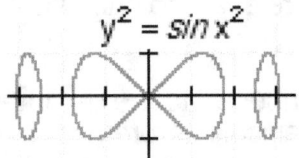

$y^2 = sin\ x^2$

Produces a lobed orbital form.

Schrodinger failed to recognize that he didn't have "two lobed orbital alternatives " but two halves of a single three-dimensional wave-phase for each value of ℓ.

Originating Wave Plotted Against Schrödinger's Dimensionally-Shifted Wave Function

Two Schrodinger lobed orbitals equal the wave-phase of a single doublet

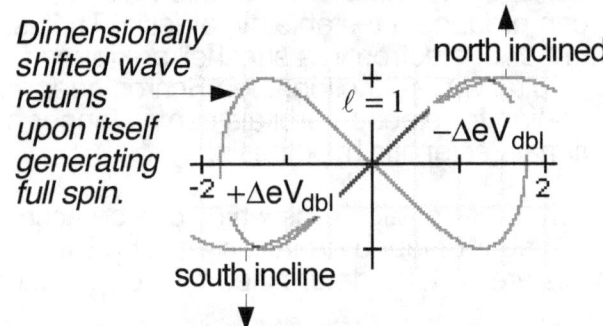

Schrödinger's dimensionally shifted lobed-orbital has converted the 1/2 spin of a three-dimensional wave-path in four dimensional space into a full spin in three dimensional space. The orbital doublet is completely eliminated by the dimensional shifting.

Schrödinger's dimensionally-shifted eigenfunction has eliminated the electron's 1/2 spin which is the very factor allowing for orbital doublets. Orbital doublets provide subshells with non -interference orbits and greater electron capacity. Schrodinger has replaced each doublet with two lobed orbitals. A single doublet defining one orbit has become two separate orbitals. Schrödinger's predicted subshell electron capacity is off by a factor of "2."

sub shell.	eV. variance with next higher	ℓ	Quantum Geometry		Schrodinger Wave Function	
			no. doublets ℓ = doublets	electron capacity $2 + (2)\ell$	no. lobed orbitals 2ℓ = lobed orbitals	electron capacity $2 + (2)2\ell$
s	0 eV	0	0	2	0	2
p	0.065 eV	1	1	4	2	6
d	0.100 eV	2	2	6	4	10
f	0.166 eV	3	3	8	6	14
g	0.306 eV	4	4	10	8	18
h	0.661 eV	5	5	12	10	22
i	1.89 eV	6	6	14	12	26

In the above table, the variance in electron voltage between subshells increase as the

subshells descend in order, allowing for a greater number of doublet orbits for lower-order subshells based upon the quantum angular momentum number "ℓ."

The Schrodinger eigenfunction has been applied to electron orbitals for 80 years by a consensus fiat. It has now been converted to a testable hypothesis. The hypothesis is proven wrong by the sodium D-lines. The Sodium D-lines under further division by a magnetic field, identify how many orbital doublets there are in the "6g" subshell to which the excluded electron outputting the D-lines is trying to gain admission. There are exactly 5 possible orbital in the "g" subshell. There is the initial single "6g" orbital level and four sets of doublet for a total or five possible orbitals. Each orbital is able to accommodate 2 electrons. Therefore, the maximum capacity of the "g" subshell is 10, not the 18 predicted by Schrodinger. The number of electron's available to the "g" subshell is found by quantum geometry's "$2 + (2)\ell$," not by Schrödinger's "$2 + (2)2\ell$."

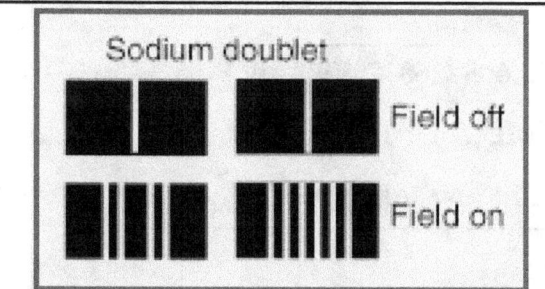

Sodium doublet

Field off

Field on

The low-end D-line further splits into one less doublet than the high-end D-line.

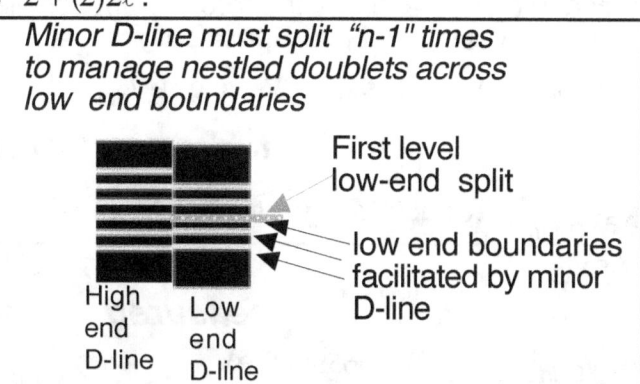

Minor D-line must split "n-1" times to manage nestled doublets across low end boundaries

First level low-end split

low end boundaries facilitated by minor D-line

High end D-line

Low end D-line

Under the additional energy of a magnetic field, the higher energy D-line doublet is split into three additional doublets . This represents the number of doublets which are opposing the rogue electron's attempt to enter the "6g" subshell. it is being opposed by the first level doublet (producing the D-lines) which has been split into a set of second level doublets, which are again split into third level doublets which are split into fourth level doublets. The "g" subshell is characterized by the initial orbit which has been split into four sets of doublets equaling five possible orbitals which can accommodating 10 electrons. Since sodium has 11 electrons, this represents an attempt of a highly energized nucleus to move the 11 electrons out of the "6h" subshell , which can accommodate 12 electrons, into the "6g" which can accommodate only 10.

Not only do the sodium D-lines inform as to the number of potential orbits contained in the "6g" subshell, but they also enlighten as to how doublet orbits must be managed by the atom. Notice in the above illustration that the "low-end" partner of the D-line doublet is further split under magnetic field influence into one less doublet than the "high-end" partner. The "high end" D-line splits into three further doublets, while the "low end" is further split into only two.

The sole function of low-end splits is to manage or facilitate the doublet orbital paths established by high-end splits. Doublet orbits are added by the further splitting of a doublet, adding pathways north and south of its own two doublet pathways. That is, doublets are nestled within one another. Both legs of the new doublet must be conducted by the legs of the originating doublet, the originating doublet which is now contained within the new doublet. The low-end of the originating doublet must always split by a different amount to facilitate pathways for the new doublet.

Notice in the above illustration that the low end D-line splits can be aligned with all low end legs of the major partner's splits. The further splitting by the minor partner has provided pathways for all the additional doublets.

The fact that additional new doublets require separate splitting by existent doublet partners affects the energy states governing doublet addition. This can be seen by an analysis of the variance in electron-voltage for the D-line doublet ($\Delta eV_{D\text{-}lns}$) and electron voltage for the "6g" subshell's primary doublet (ΔeV_{dbl}).

Using the four-dimensional formula to calculate "ΔeV_{dbl}" for the 6g primary doublet, it can be compared with the much greater energy in the D-line separation. The argument is made that the force of opposition restricting the D-line electron from entrance into the "6g" is a multiple of "ΔeV_{dbl}" and that the electron voltage change of the D-lines reflects this. The D-lines are forced into separation by the force of opposition from "6g" which is equal to a multiple of the primary "6g" doublet. Specifically:

$$\Delta eV_{D\text{-}lns} = 0.0010659039 \text{ eV} \quad ; 6g\Delta eV_{dbl} = \sqrt[4]{eV_{6g}^2\left(1.2605226812\ 10^{-18}\ eV\right)} = 5.3514\ 10^{-5} \text{ eV}$$

theoretically predicted

$$\frac{\Delta eV_{D\text{-}lns}}{\Delta eV_{dbl}} = \frac{\langle \text{doublet; high-end}\rangle\left(2\left(2^3\right)\right)\Delta eV_{dbl} + \langle \text{path; low-end}\rangle(1)(1)\left(2^2\right)\Delta eV_{dbl}}{\Delta eV_{dbl}} = 20$$

measured

$$\frac{\Delta eV_{D\text{-}lns}}{\Delta eV_{dbl}} = \frac{0.0010659039 \text{ eV}}{5.3513953922\ 10^{-5} \text{ eV}} = 19.9182422879 \quad ; \quad 19.9182422879 \neq 20$$

The change in electron voltage for the D-lines should be 20 times greater than the electron voltage change for the 6g doublet. This represents the number of splits and paths required by the 6g subshell to establish 4 additional noninterference orbits. However, the multiple is only measured at "19.92," not "20."

The variance is explained by the fact that the eV change determining the doublet splits is not the same as the eV change determining the pathways. The difference in eV is a function of the anomalous magnetic moment (quantum geometric version, not g-factor). The actual formula is:

$$\frac{\Delta eV_{D\text{-}lns}}{\Delta eV_{dbl}} = \frac{\left(2\left(2^3\right)\right)\Delta eV_{dbl} + (1)(1)\left(2^2\right)\Delta eV_{dbl}^{-0.57\gamma}}{\Delta eV_{dbl}} = 19.9182422879;$$

$$\gamma = \text{magnetic moment gain due to axis shifting} = 0.0530516477$$

$$-0.57\gamma = -0.57(0.0530516477) = -0.0302394392 \text{ gain } (a\ loss)$$

The negative phase of the orbital wave path has a negative gain-potential in magnetic moment. Gain in magnetic moment is due to shifting of the electron's 1/2 spin axis. It is positive and essential for the positive phase of the wave-path (consult illustrations).

However, the negative phase acquires a "−30° axis shift" at the point at which it transitions to the positive phase. The electron's axis along the negative-phase wave path has required the subtraction of this amount of declination during its course. The loss can be taken at the transition point or it can be spread along the wave-path.

Negative phase electron spin is against declining kink pressure. It is the relaxation phase and defines the low-end of the doublet split. The initial position of the negative phase within the primary doublet is set by the positive phase. That position is established by the factor in the above equation designated "$2(2^3)\,\Delta e\,V_{dbl}$" and is a component of the first "2."

However, the negative phase can further alter its doublet position by negative magnetic moment gain, that is, by adjusting the spread of the negative gain along its wave-path producing "$-0.57\,\gamma$." In this manner, the position within the doublet is altered to establish noninterference low-end doublet pathways. The variance between "20" and "19.92" in the above equalities represents hyperfine structure.

The Bohr/Schrodinger Electron Structure is Rejected by these Data

There is a clear mathematical relationship between the Sodium D-line doublet and the Zeeman doublet for the next higher frequency in the Balmer Series. That Balmer Series frequency has been assigned to the 6g subshell. The width of the D-line doublet, as measured by ΔeV, is a multiple of the 6g Zeeman-doublet ΔeV. This multiple is explained as the multiplier which would be needed to establish doublet orbits. The geometry required by such doublet orbits is revealed by the further splitting of the sodium D-lines under a Zeeman magnetic field.

The Construction of Doublet Orbitals by "ΔeV" (6g/D-Line Example)

Pathways are constructed by offsetting low end ΔeV from primary by enough to make high end and low end ΔeV equal for nested doublets. This is the meaning of "$\Delta eV_{dbl}^{-x\ \gamma}$"	
$\leftrightarrow$ ΔeV for primary doublet now offset 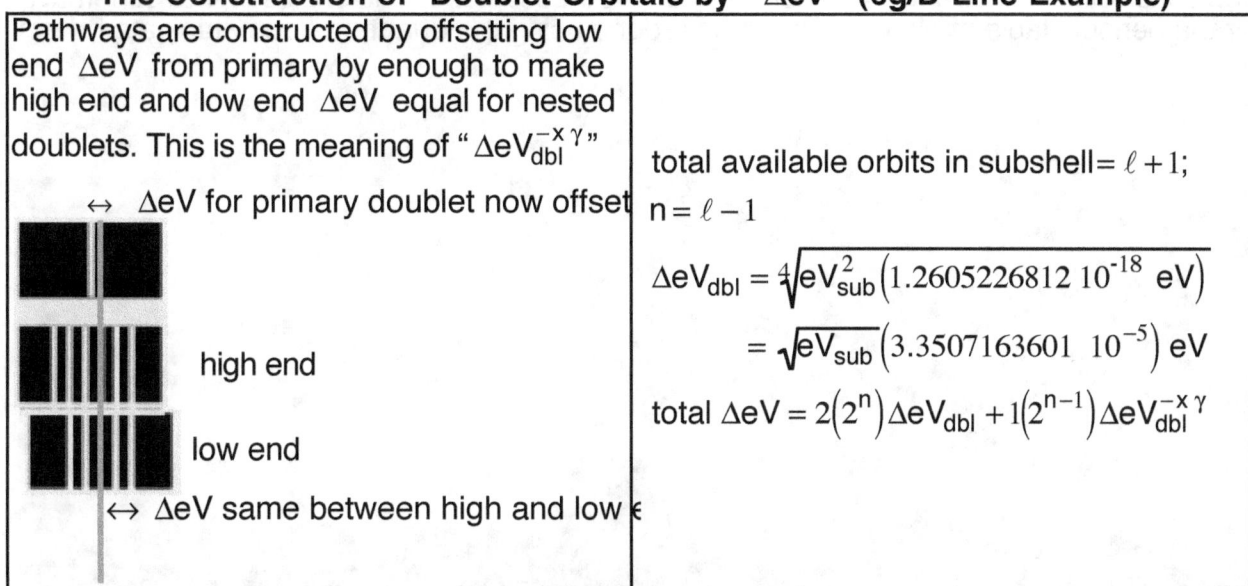 high end low end $\leftrightarrow$ ΔeV same between high and low	total available orbits in subshell $= \ell + 1$; $n = \ell - 1$ $$\Delta eV_{dbl} = \sqrt[4]{eV_{sub}^2\left(1.2605226812\ 10^{-18}\ eV\right)}$$ $$= \sqrt{eV_{sub}}\left(3.3507163601\ 10^{-5}\right)\ eV$$ $$\text{total } \Delta eV = 2\left(2^n\right)\Delta eV_{dbl} + 1\left(2^{n-1}\right)\Delta eV_{dbl}^{-x\ \gamma}$$

Zeeman doublets obviously produce a variation in an electron's orbit. They would seem to be good candidates for the orbital variation which subshells require in order to incorporate more than the two electron limitation established by the Pauli exclusion principle. However, this line of inquiry has been prevented by the complete domination of the Bohr/Schrodinger atomic model.

Bohr/Schrodinger provided a different explanation for orbital variations within subshells in order to avoid the Pauli limitations. Specifically, Schrödinger's wave-function produces varying numbers of lobed orbitals for subshells using his quantum angular moment number "ℓ." The Schrodinger explanation was accepted by consensus fiat and without experimental confirmation.

The Schrodinger lobed orbital predicts electron capacity for the subshells and this prediction has been used to construct the shell/subshell schematic for the periodic table of elements.

Unfortunately, the Schrodinger subshell capacities are proven wrong by doublet D-line analysis. The "g" subshell contains 10 electrons, as predicted by quantum geometry, not the 18 predicted by Schrodinger. The restriction of "g" subshell capacity to 10 is causing sodium to output the D-lines. The sodium nucleus attempts to place all 11 sodium electrons into the 6g but is prevented by subshell capacity, stranding the D-line electron. Analysis of D-line Zeeman doublets demonstrate that there are only four doublet orbits in the "g" subshell from which the D-line electron has been excluded. Four doublet orbits plus the base orbit makes five orbits. At two electrons per orbit, the 6g subshell has a ten electron capacity. Schrodinger has failed the empirical test of the hypothesis.

More could be said about the Schrodinger failure to correctly identify subshell capacities. For example, the attempted explanation of D-lines as "falls" between the "3p" and "3s" with different 1/ 2 spins is absurdly short of the required electron voltage. Yet this explanation is attempted because the Schrodinger model puts the eleventh sodium electron in the "valence" shell which is the third shell. There are no mechanics in Bohr/Schrodinger for energy exchanges between nucleus and electron and, therefore, no way to predict migrations by electrons into higher shells.

The conclusion that Bohr/Schrodinger is wrong is inescapable. The shell/subshell schematic for the periodic table of elements must be redone using the quantum geometric data herein presented.

Light-Harmonics, Electromagnetic Waves *and the Vacuum Soliton*

In 1888 the mathematician Janne Rydberg[51] demonstrated that the spectrographic lines of light emitted by hydrogen were arrayed by a whole number formula. Rydberg's formula could predict the spectrographic wavelengths in five bands of hydrogen radiation; from the Lyman-Series ultraviolet to the Pfund-Series far infrared. That Rydberg formula is the following:

RC = Rydberg's Constant = 109720 cm^{-1} ; $1 \leq n \leq 8$; $n' \geq n + 1$

$$\frac{1}{\lambda} = RC \left(\frac{1}{n^2} - \frac{1}{n'^2} \right) \ ;$$

" n" identifies the series (shells based upon subdivisions) ;

" n' " identifies wavelengths within the series (subshells; the negation of subdivision).
Radiation frequencies, which are output and absorbed by hydrogen, were shown to be functions of a whole number formula *times* a constant.

One of the first errors which twentieth century physics made was the failure to recognize that Rydberg's "Constant" was nothing more than the inverse of the shortest wavelength and highest frequency in the ultraviolet Lyman Series[52] . *All wavelengths output by hydrogen are harmonically related to that shortest wavelength by a formula which duplicates the negation of subdivision for the quantum squared:*

RC = Rydberg's Constant = 109720 cm^{-1} ;

λ_r = shortest wavelength in Lyman Series = 91.14 nanometers ; c = speed of light

$$RC = 109720 \ cm^{-1} = \frac{1}{91.14 \ nm} \ ; \ RC = \frac{1}{\lambda_r} \ ; \ frequency = \frac{velocity}{wavelength} = \frac{c}{\lambda}$$

$$\frac{c}{\lambda} = \frac{c}{\lambda_r} \left(\frac{1}{n^2} - \frac{1}{n'^2} \right) \ ; \ \ \ \frac{1}{\lambda} = \frac{1}{\lambda_r} \left(\frac{1}{n^2} - \frac{1}{n'^2} \right)$$

Rydberg's Constant is only the inverse of shortest wavelength (highest frequency) output by hydrogen. The conversion of Rydberg's Constant to root frequency is simple— any high-school math student could have accomplished it. Yet, twentieth century physics didn't recognize it.

Instead, Rydberg was explained by Niels Bohr's "falling electron" model. Bohr assigned "base" electron voltage to a hydrogen electron as a function of the elementary charge of the particle and the energy which Planck's Constant identified for Rydberg's Constant. He assigned this electron-voltage value as that of a "base orbit." He proposed a system of equidistant quantum orbits the electron voltages of which could be calculated as divisions by the orbital number (squared) of the "base" electron voltage. He demonstrated electrons falling between these quantum orbits would acquire enough electron voltage during the fall to output "bursts" of light at the Rydberg frequencies[53].

---

[51] 1854 – December 28, 1919). A Swedish physicist.
[52] See Appendix *"The Conversion of the Balmer and Rydberg Constants to the Root Wavelength."*
[53] Niels Bohr (1913). *"On the Constitution of Atoms and Molecules, Part II Systems Containing Only a Single Nucleus".* Philosophical Magazine 26: 476–502.

The Bohr "falling electron" model could mathematically explain all frequencies within the Rydberg distribution except one. The Bohr model could not explain the "root frequency" itself. It could not explain the "91.14 nm" wavelength upon which all other radiation frequencies in the Rydberg distribution are based. The "91.14 nm" wavelength occupies Bohr's "base" orbit (13.6 eV) to which an electron could potentially "fall" (from all other orbits) outputting Rydberg's Lyman-Series ultraviolet. There is no place for the electron to "fall" outputting the "91.14 nm" wavelength itself. This fact was disguised by the simple expedient of keeping Rydberg's Constant in place and ignoring the fact that it represented a root frequency.

Rydberg had actually discovered that all radiation emissions from hydrogen were a geometric frequency distribution from a single frequency, similar in many respects to the harmonic distribution discovered by Pythagorus.

Rydberg's harmonic distribution wasn't based upon differentiation by Pythagorean subdivision, but by a system of differentiation which is appropriate to the quantum; by "the negation of subdivision."

I am about to introduce the first principles of quantum geometry. The reader should be warned that there is no precedent for these principles in the current mathematical catechism nor can they be found in current mathematical texts. In mathematics these principles are known as "truisms;" as true by definition. They are demonstrated to be geometrically true because physical reality can be accurately measured by them. Any refusal to accept the geometric quantum and the accuracy of its measurement simply because it is unfamiliar and without canonical authority will render science to the status of ancient Egyptian art. A highly stylized "flat" pictorial representation of three-dimensional reality will be enforced by suppressing a more accurate illustration. The "quantum" is being supplied the following definition:

> Geometrically, a quantum is defined as two points separated by "some" distance. In this, the quantum is distinguished from the standard Euclidean line which is composed of a continuum of points. There is no space of separation between adjacent points along a linear continuum. To impose such a space of separation would, in fact, compose a quantum. The difference between a quantum length and an Euclidean linear length is the number of points contained within the length. There are an infinite number of points contained along an Euclidean line, regardless of the length of the line. There are only two points defining a quantum line and therefore there are only two points contained by the line, regardless of the length of that line.

A quantum cannot be subdivided because it would require multiplication of the quantum to do so. A quantum is defined by only two points and any subdivision would require the addition of points to accomplish, which would make "n" quantums out of one, "n" being the subdivisional factor. An Euclidean line can be subdivided without restriction by identifying a point along the continuum at the subdivisional measure. Any distance measure defined as an Euclidean line can be differentiated by subdivision without restriction.

A quantum can be differentiated only by the negation of subdivision; by a *subtraction* of a subdivisional value from the whole, remaindering a new distance value for the original two quantum points. The quantum can acquire a new distance value based on the subtraction of a subdivisional unit from the original distance. This is called "differentiation by the negation of subdivision."

A quantum harmonic distribution would be differentiated by the negation of subdivision, rather than by Pythagorus' subdivision. Rydberg's light distribution was not Pythagorus' harmonic sound distribution[54] which is built upon geometric subdivision, but it was a geometric distribution nonetheless. It is a harmonic distribution based upon the negation of subdivision.

Rydberg's harmonic distribution was actually the negation of subdivision for the quantum squared. It must be the "quantum squared" rather than the linear quantum because the quantum is a single dimension which cannot exist alone and disconnected from the "system" of Euclidean space. The quantum squared defines a unit of geometric space which includes both a quantum and an Euclidean component: $Q^2 = (Q)(\varepsilon\mu)$

It is true that the quantum geometric principles just outlined were unknown during the early part of the twentieth century, when Rydberg's discovery became important. However, Rydberg's formula itself should have suggested a new form of geometric frequency distribution, especially since radiation output was identified by Niels Bohr as a function of electron orbital distances[55]. Why didn't great minds like Einstein, Bohr, Planck and Millikan identify the Rydberg formula for wavelengths as an harmonic distribution from the highest "root" frequency? The deficiency is not explained by incapacity, but by will.

Rydberg was not in anyway marginal or unimportant to the development of the "revolutionary" physics of the early twentieth century. Indeed the Rydberg discovery was crucial. Bohr used the Rydberg formulation to fashion the key atom-based energy unit which dominated the whole of the last century— the "electron volt." That energy measure is based upon three empirical discoveries; that of Planck; that of Rydberg, and that of the elementary charge of Robert Millikan. Why didn't physicists at least consider the implications of such a root wavelength?

The answer may be simple— as simple as the conversion of Rydberg's Constant to a root frequency is simple. Science had concluded that light could not be pure waveform energy. An harmonic distribution from a root wavelength implies that light might be completely waveform. To pursue the thesis that Rydberg's formula identified a new mathematical form of harmonically related light-energy —a quantum wave harmonics — was unthinkable at the time. The scientific *Zeitgeist* of the period precluded it. The scientific view of light which had emerged in the early part of the twentieth century prevented physicists from recognizing the quantum wave characteristics which Rydberg had actually discovered.

In 1905, Albert Einstein had proposed that light had a dual nature[56]. It was both waveform and "particle-like."; Its energy exchange capacity with matter was via the light-particle— the "photon"— which Einstein had advocated. In other respects it was an "electromagnetic wave."

The "photon" was advanced in defense of Einstein's equation for the energy producing the photoelectric effect. Einstein's equation for the energy required to eject bound electrons into an electric current has proven to be highly accurate and Einstein received the Nobel Prize in 1921 for it.

However, quantum geometry and the research it has generated have shown that the

---

[54] Pythagorus had discovered that the subdivision of a vibrating string by a whole number "n" increases the root frequency of the string by a factor of "n."

[55] *"On the Constitution of Atoms and Molecules"*; Niels Bohr; Philosophical Magazine; Series 6, Volume 26, July 1913, p. 1-25

[56] Ibid.

"photon" is not required to explain Einstein's photoelectric equations. His equations can be resolved to quantum harmonics[57] and is completely compatible with the light-as-waveform hypothesis.

Nonetheless, It was Einstein's mathematics which completely suppressed the theory of "light as waveform energy," the theory which had been experimentally probed 100 years earlier.

In 1803, a physicist at Cambridge University had introduced the field of lightwave research. In that year, the English physicist Thomas Young proved that light possessed wave characteristics[58]. Using a double slit experiment, Young had shown that two beams of light can "interfere" or cancel each other out. In this, he demonstrated that light acted like water waves which he also showed were capable of canceling each other by opposition. Young was also the first to discoverer and measure light wavelengths[59] .

Prior to Young's observation of wave interference and measurement of light wavelengths, the wave theory of light was given little credence. Previous wave theory had fallen under the disapproving scowl of none other than Isaac Newton. In his book *Opticks*[60] Newton had made a strong case for the particle theory of light. Newton was the first to observe light refracted into color bands. He speculated that the bending of light by glass prisms into his "spectrum" was due to the motion characteristics of "corpuscles" (particles) of different colors.

Newton's reasons for postulating the particle theory were to be repeated two centuries later by Einstein. A particle theory of light was the simplest way to explain the differing energy characteristics of light colors (frequencies). Einstein had been confirmed in his particle theory by Max Planck's discovery that light energy was a function of monochromatic frequency *times* his constant, (6.626e-34 joules). Planck had developed his constant from blackbody studies in 1900[61]. The "packets of energy" which Planck's Constant implies suggested a frequency-based light particle.

Nonetheless, Young's 1803 observations require explanation. The "ether" theory emerged to explain the wave characteristics of light.

The proposed "ether" was a mysterious fluid-like substance which was thought to permeate all space. It was a fluid which didn't resist motion since, if it did so, planetary motion would decelerate and orbits would collapse. But the proposed "ether" was given the capacity to transmit light as a compression wave.

On the surface, the proposed "ether" is incompatible with known wave mechanics since fluid resistance provides the energy-exchange potential required for wave impedance. That energy-exchange potential allows the fluid to surrender energy to anything which can impede the wave.

For example, a car crashing into a wall surrenders its energy of motion (kinetic energy) to the wall during the sudden deceleration. A part of this surrendered energy is exchanged with

---

[57] See *Quantum Harmonics by the Negation of Subdivision* in Appendix.

[58] In a paper entitled *Experiments and Calculations Relative to Physical Optics*, published in 1803

[59] John C. D. Brand (1995). *Lines of light: the sources of dispersive spectroscopy, 1800-1930*. CRC Press. p. 30–32. ISBN 9782884491631.

[60] *Opticks or a treatise of the reflections, refractions, inflections and colours of light ;* (1704), Isaac Newton

[61] *on the theory of blackbody radiation* by Max Planck in Annalen der Physik 4, p553 ff 1901

the air and transmitted as the waves which we identify as the sounds of the crash. The wave being conducted through the air is impeded by our eardrums which vibrate at the frequencies of the carrier sound waves. Without the air molecule being a potential partner in the energy exchange, the wave could neither be generated nor impeded. By definition, "ether" lacks the capacity to exchange energy with matter.

The ether theory was supposedly disproved in 1887 by the Michelson-Morley experiment.[62] That experiment used a device invented by Michelson —the interferometer— to test Young's light wave interference patterns relative to something called the "ether wind."

The "ether wind" was supposedly caused by the movement of the earth through the "ether." The interference patterns between light moving in the direction of the earth's travel and at 90° to the earth's travel were supposed to produce interference distortions due to differences in velocity of the light. The velocity of the earth was hypothetically added to the velocity of light in the direction travel.

The premise being put forward by Michelson-Morley seems very strange to people from the age of supersonic jet aircraft. The velocity of a jet aircraft is never added to the velocity of the speed of sound. The jet could never "catch up" with its own sound waves and "break the sound barrier" if the velocity of sound were determined by the jet's own speed.

However, Michelson-Morley "discovered" that a proposed medium for waveform light worked exactly like the medium conducting sound waves. It was curiously concluded that the medium couldn't exist because of this.

The Michelson-Morley finding that the speed of light is independent of the velocity of its source is said to have disproved the "ether" theory and opened the way for alternative light theories such as that proposed by Einstein *cum* Planck.

Light as waveform energy was abandoned because any potential "medium" had been "disproved." It was replaced by the Einsteinian hybrid, the "photon" in cases of energy transfers and as a wave being conducted by unmeasurable "electromagnetic fields" when light needed to be acknowledged as a wave form.

It was, of course, James Maxwell's electromagnetic wave theory[63] which survived Michelson-Morley. Heinrich Hertz[64] had generated electromagnetic radio waves in the same time period as the Michelson-Morley experiments, supposedly proving that radiation was composed of electromagnetic waves being conducted through the vast expanses of the universe.

The contemporary belief in physics that the electromagnetic wave alone can fully explain light as waveform energy is without empirical foundation. If the electromagnetic wave were to conduct light radiation across the vast distances of interstellar space, the presence of measurable capacitance fields would exist in the empty space through which that light transitioned. Maxwell's equation for electric field flux in free space and Planck's Constant for

[62] Michelson, A. A.; Morley, E. W. (1887). *"On the Relative Motion of the Earth and the Luminiferous Ether"*. American Journal of Science 34 (203): 333–345.

[63] James Clerk Maxwell, *A Dynamical Theory of the Electromagnetic Field*, Philosophical Transactions of the Royal Society of London 155, 459-512 (1865)

[64] Hertz, Heinrich, Rudolf (1857-1894), was a German physicist. He opened the way for the development of radio, television, and radar with his discovery of electromagnetic waves between 1886 and 1888.

the energy value in light frequency can be combined to demonstrate that light conducted as an electromagnetic wave would produce measurable capacitance fields in free space[65]. Where are these required nano farad and picofarad field capacitance measurements in the presence of received light? They simply do not exist.

The electromagnetic wave alone cannot explain radiation conduction. Quantum geometry has shown that electromagnetic waves are extremely localized by matter and could never conduct radiation across interstellar space. Further, the free-space permeability/ permittivity formula establishing the electromagnetic wave reveals a direct interaction with quantum space. It is quantum space which conducts radiation across long distances[66]

It is misfortunate that alien and unfamiliar, yet complex concepts from quantum geometry must be rapidly introduced. But such is the requirement for any further discussion of electromagnetic radiation. The electromagnetic wave interacts directly with quantum space but this cannot be discussed without a fuller appreciation of the quantum character of both the vacuous volume of the space through which light travels and of the light itself.

The quantum character of light radiation is demonstrated by the negative radiation frequencies which were discovered in the Rydberg hydrogen distribution by the current author and thermodynamic engineer, David Rule. Negative radiation exists precisely because of the distinction between quantum and Euclidean differentiation of measurement; of "negation of subdivision" versus "subdivision."

The light spectrum itself (meaning all frequencies output or absorbed by atoms) is quantum based and produced by the negation of subdivision, as the Rydberg formulation clearly demonstrates. However, this leaves "gaps" in the spectrum at subdivisional points:

$$\left(\frac{1}{n^2}-\frac{1}{n'^2}\right)\frac{1}{\lambda_r}\neq\left(\frac{1}{n^2}\right)\frac{1}{\lambda_r}$$	*There is always a gap between "subdivision" and "negation of that subdivision." A negation of quantum-squared subdivision can never reach the subdivision.*

The subdivision itself produces "bands" of frequency composed of negations of that subdivision. Thus subdivision "n=1" composes the Lyman Series ultraviolet band. The subdivision "n=2" composes the Balmer Series visible light band. The Subdivision "n=3" composes the Paschen Series high infrared band and so forth.

The Rydberg Hydrogen Frequency Distribution into Bands

Lyman Series (ultraviolet). Formula:
$$\frac{1}{\lambda}=\left(\frac{1}{(1)^2}-\frac{1}{(n')^2}\right)\frac{1}{\lambda_r}\ ;\ 2\leq n'\geq 9$$

Balmer Series (visible light frequencies). Formula:
$$\frac{1}{\lambda}=\left(\frac{1}{(2)^2}-\frac{1}{(n')^2}\right)\frac{1}{\lambda_r}\ ;\ 3\leq n'\geq 8$$

Paschen Series (infrared spectrum). Formula:
$$\frac{1}{\lambda}=\left(\frac{1}{(3)^2}-\frac{1}{(n')^2}\right)\frac{1}{\lambda_r}\ ;\ 4\leq n'\geq 8$$

---

[65] See *The Required Capacitance Measures in the Presence of Light if the Electromagnetic Wave Were the Carrier* ; an Appendix entry in *Quantum/Electromagnetic Field Equations.*
[66] See *Quantum /Electromagnetic Field Equations* in Appendix.

Brackett Series (lower infrared spectrum). Formula: $\frac{1}{\lambda} = \left(\frac{1}{(4)^2} - \frac{1}{(n')^2} \right) \frac{1}{\lambda_r}$; $5 \le n' \ge 8$

Pound Series (far infrared spectrum). Formula: $\frac{1}{\lambda} = \left(\frac{1}{(5)^2} - \frac{1}{(n')^2} \right) \frac{1}{\lambda_r}$; $6 \le n' \ge 8$

In the above formulas, all frequencies are divided by the speed of light "c."

The electron orbital is energized and sustained by radiation pressure against the kinks establishing the internal three-dimensional wave. The maximum light frequency which the orbital kinks can impede is provided by the root wavelength. It is the frequency which provides the maximum terminal velocity for the acceleration/deceleration phase of the orbital wave (see chapter 1). Kink pressure in the root orbital can accelerate the electron to the speed of light.

The Rydberg frequencies represent the quantum structure which is constructed by radiation energy exchanges with the atom. At the high end is the root orbital which is restricted by the speed of light. At the low end is the orbital which can no longer impede radiation frequencies because the required "angle of deflection" for the orbital's accel./decel. phase is greater than available. A set of three-dimensional orbitals in four dimensional space which can impede light is distributed between these two limits. That distribution must be by the quantum principle of the negation of subdivision for the quantum squared with the root orbital holding the quantum position. Electrons can only fall or rise to radiation established orbitals which are proper negations of subdivision. The Rydberg formula identifies this distribution.

Subdivisional Wavelengths are Negative Radiation
In our analysis of atomic structure it was demonstrated that all lower-order negations of subdivision are distributions of the Lyman Series. For example the Balmer series is acquired by subtracting the Lyman negation "n'=2 from all higher negations:

$$\left(1 - \frac{1}{n'^2} \right) - \left(1 - \frac{1}{2^2} \right) = \left(\frac{1}{2^2} - \frac{1}{n'^2} \right) ; \quad n' \ge 3$$

The subdivisions themselves, however, can only be reached by improper negations; that is by their inverses:

$$\left(1 - \frac{1}{n'^2} \right) = \frac{n'^2 - 1}{n'^2} ; \quad \frac{1}{n'^2} = 1 - \left(\frac{n'^2 - 1}{n'^2} \right)$$

None of the inverses, neither "3/ 4" nor "8/9" nor "15/16" nor "24/25" *etc.* are proper subdivisional values. In the case "n=2" the inverse is "3/4" and it gives the following negation value:

$$\frac{3}{4} = \frac{1}{(1.1547005384)^2} ; \quad \frac{1}{2^2} = 1 - \frac{1}{(1.1547005384)^2} \qquad n' \text{ cannot equal } 1.1547005384$$

The same is true of all other inverses
Rydberg-distributed frequencies can never "fall" to the subdivisions because all subdivisions have an improper negation value.

However, the variance in electron voltage (ΔeV) between subdivisions and the highest proper negation of that subdivision is always the same. It is the electron voltage value of the highest negation number, "n'=8" :

$$\Delta eV = \frac{13.6037 \text{ eV}}{n^2} - \left(\frac{1}{n^2} - \frac{1}{8^2}\right)13.6037 \text{ eV} = \frac{13.6037 \text{ eV}}{8^2} = 0.21256 \text{ eV}$$

Energy value for "ΔeV" is 3.4055665859 (10^{-20}) joules (significance shown later)

The atom can interface with subdivisional or "shell" frequencies as negative radiation. Science knows the negative radiation frequencies as those which cause florescence. The two best known are the following:

$$\frac{c / \lambda_r}{2^2} = \frac{c}{\lambda} \; ; \; \lambda = 364.56 \text{ nm} \quad \langle \text{"black light" inducer of visible florescence} \rangle$$

$$\frac{c / \lambda_r}{3^2} = \frac{c}{\lambda} \; ; \; \lambda = 820.26 \text{ nm} \quad \langle \text{infrared florescence - inducer used in medical research} \rangle[67]$$

Science, however, does not generally recognize "365 nm" and "820 nm" florescence-inducers as negative radiation wavelengths. They are unaware that the florescence is caused by energy from the nucleus making up the variance in electron voltage between the proper subshell wavelength and the improper subdivisional wavelength.

The electron cannot naturally "fall" to an orbital position which outputs or impedes "365 nm" or "820 nm" wavelengths since such wavelengths are produced by improper negations. Therefore, the only way the electron can absorb those improper frequencies is by the application of energy from the nucleus making up the difference in electron voltage between proper and improper wavelengths($\Delta eV = 0.21256$ eV).

A study of florescence-induced cotton fibers under 365 nm black light proved the black light to be negative radiation[68]. As the cotton fluoresced it simultaneously dropped in temperature. The drop in temperature was found to be a function of the number of hydrogen bonds in the cotton molecule and of Planck's Constant. The nucleus of the cotton molecule was surrendering energy in the form of lost heat to output the florescence light. The cotton's nuclear kinetic energy was "making up" the 0.21256 eV required to impede 365 nm black light from the nearest proper negation of subdivision orbital which outputs 388.9 nm wavelength.

Subdivisional Wavelength	Closest Proper "Negation" Wavelength
$\frac{1}{\lambda} = \left(\frac{1}{2^2}\right)\frac{1}{(\lambda_r = 91.14 \text{ nm})}$	$\frac{1}{\lambda} = \left(\frac{1}{n^2} - \frac{1}{n'^2}\right)\frac{1}{\lambda_r} = \left(\frac{1}{2^2} - \frac{1}{8^2}\right)\frac{1}{91.14 \text{ nm}}$
$\lambda = 364.56 \text{ nm}$	$\lambda = 388.86 \text{ nm}$

The nucleus was surrendering energy at the rate of Planck's Constant[69]. This energy transferee rate from the nucleus was being applied to the proper negation frequency to achieve the additional 0.21256 eV which were required to reach the higher subdivisional frequency. The energy being output was much less than would be needed to output normal light from the 388.9 nm orbital.

$$\text{Normal energy} = \frac{c}{388.9 \text{ nm}}(h) = 5.107861818 \; 10^{-19} \text{ joules}$$

---

[67] *Near-infrared imaging of injured tissue in living subjects using IR-820;* Prajapati SI, Martinez CO, Bahadur AN, Wu IQ, Zheng W, Lechleiter JD, McManus LM, Chisholm GB, Michalek JE, Shireman PK, Keller C; University of Texas Health Science Center, San Antonio, TX; Mol Imaging. 2009 Jan-Feb;8(1):45-54

[68] *The Drop in Temperature by N-Irradiated Cotton and the Derivation of Planck's Constant* in Appendix

[69] ibid.

The nuclear energy required to make up the 0.21256 eV— at 3.4055665859 (10^{-20}) joules— is 15 times less energetic than the energy required by normal light. This is the reason florescence light is so "ghost like."

$$\frac{5.107861818\ \left(10^{-19}\right)}{3.4055665859\ (10^{-20})} = 14.998$$

Frequency is the inverse of the time factor for the acceleration/deceleration phase of any orbital three-dimensional wave. The acceleration/deceleration time factor is a function of the kink tension for the orbital quantum radius.

Frequency equals the kink-tension constant " $\sqrt{2}\ k^{70}$ " times the quantum radius squared. The constant is the amount of tension generated upon atomic molar volume by "kinking" it into four-dimensional space.

Essentially, the electron is a quantum particle which only achieves mass value by charged attachment to the nucleus[71] . When captured by the atom, the electron's Euclidean bond is rotated 90° to the quantum bond of the proton and kinked into curvature. This forms a three dimensional orbital in four dimensional space. The electron's attachment completes its mass value, the bond-kinks operating much like a tensioned "string" which vibrates at constant frequency. These kink/strings vibrate at Rydberg distributed frequencies because frequency is the kink-tension constant times the radial length squared. The quantum radial lengths (squared) are distributed by the Rydberg formula as negations of the root radial length (squared). Rydberg frequencies are a function of a Rydberg distribution of orbital radii.

Frequency, however, also has an electrodynamic definition. Specifically, frequency is a function of the Piezoelectrical energy of the nucleus and its ability to multiply the particle charge of the electron[72] . The formula for this is very specific:

$$\text{frequency } (f) = \frac{(\text{electron voltage})(\text{Planck capacitance})}{\text{elementary charge}} = \frac{eV(C_P)}{e}$$

"Planck capacitance" is is the maximum electron capacitance which is continuously available to the nucleus. It is determined by the elementary charge squared *divided by* Planck's Constant. "Electron voltage" is the standard Bohr eV which can be assigned to the particular electron shell/subshell orbital.

Since voltage *times* capacitance equals charge, the frequency of light emissions from any electron orbit is the number of times the charge of the electron is multiplied by Piezoelectrical energy from nucleus being transferred to the electron at the rate of Planck's Constant[73].

The Rydberg distribution is the equivalent of the shell/subshell schematic from the periodic table of elements, as defined by three-dimensional orbitals in four-dimensional space. This is proven by the capacity of the Rydberg-based quantum geometric model to derive a

[70] See *Formulations for Light Frequency as the Multiple of the Electron Charge and Basic Quantum to Establish Orbitals* in Appendix , under the " *kink tension constant*" subheading.

[71] See *The Neutrino: Proof that the Electron's Mass is Contained in its Charge* in Appendix.

[72] Op. Cit. See *Formulations for Light Frequency ,,,,,,,,,,,,,,,* in Appendix

[73] *The Drop in Temperature by N-Irradiated Cotton and the Derivation of Planck's Constant* in Appendix

formula which precisely predicts the variance in electron voltage for Zeeman effect doublets. That formula is then used to located sodium electrons within the shell/subshell schematic during D-line doublet emissions (see chapter 1).

The significance of the Rydberg equivalence with the shell/subshell schematic is that it illustrates the two methods by which electron orbital energy states can be acquired[74]. They can be acquired by impedance energy from light-frequency pressure *ala* Rydberg. They can also be acquired by Piezoelectrical energy from the nucleus as with energized sodium nuclei outputting the D-line doublet light emissions.

The energy state of the nucleus, its "heat," is not a quantum value. It is an Euclidean value, representing a point on a continuum. This is why nuclear Piezoelectrical energy can provide the amount of energy required to close the electron voltage gap to negative radiation frequencies, outputting light at the uncharacteristic low intensities known as "florescence."

The Dawson and Rule negative radiation experiments proved that these subdivisional frequency gaps have an unrecognized radiological relationship to the hydrogen atom. Negative radiation is composed of frequencies at or near the predicted subdivisional wavelength which force an hydrogen nucleus to radiate off, or "fluoresce." its energy in the form of heat loss. For this reason, negative radiation can be identified as "cold radiation." David Rule has patented the first, and at the time of this writing, the only cooling device which operates completely by negative radiation[75].

Despite the fact that these restricted negative radiation frequencies have been observed and measured under reproducible experimental conditions, their existence is likely to be resisted by three-dimensional science. On the surface negative frequencies appear to operate contrary to known principles governing light radiation. Light is known to energize atoms[76], not "de-energize" them. However, contemporary physics does not recognize the quantum nature of light, having misunderstood the Rydberg distribution of hydrogen emissions for well over 100 years.

Negative frequencies are, in fact, a mathematical requirement of the quantum differentiation of light frequency. No matter how large a "negation" number, "n' ," is chosen, the negation of any subdivision will never reach the subdivision. There will always be gaps in the radiological spectrum. These gaps exist at all subdivisional frequencies. Any electron impeding (absorbing) the frequency will always do so under "negative radiation pressure."

Under negative radiation pressure quantum electron strings (topographically 1+1 dimensional "kinks") move toward the radiation source, not away from it as under pressure from the impinging wave. This motion towards the radiation source must be provided by the atom's own nuclear energy. Negative radiation pressure had been predicted by soliton mathematics[77] and confirmed by Dawson and Rule.

Negative radiation establishes a radiological heat exchange. The frequencies are negative stimulated emissions of subdivisional frequencies which must be reabsorbed by exchanging heat for return "florescent" emissions at lower frequencies. Negative radiation

---

[74] Once acquired orbitals can be sustained by electron velocity.

[75] A Negative radiation wine cooling system offered by David Rule's Pasco Poly Tank, Inc. under the trade name "Killer Chiller."

[76] By Planck's formula; energy=(number of oscillators)(Planck's Constant)(frequency); $E=n(f)(h)$

[77] See *"Negative radiation pressure exerted on kinks"* : T o m a s z   R o m a D c z u k i e w i c z et. al; Phys.Rev.D77:125012,2008

causes certain hydrogen bonded molecules to "glow" while simultaneously dropping in temperature. The best known example is "black light" with a wavelength at or near 365 nm (soft ultraviolet). Black light is the "predicted subdivisional frequency" for the Balmer Series of the Rydberg distribution; Balmer being the visible band of the Rydberg distribution.

The Balmer Series "n=2" subdivision would predict a frequency of 364.56 nm wavelength. Black light peaks at 365 nm. However, the highest actual frequency in the Rydberg "n=2" Balmer Series has a wavelength of 388.9 nm (n=2; n'=8).

Experimental evidence gathered by the current writer has demonstrated that black-light frequencies between 365 and 371 nm are absorbed by hydrogen as negative radiation. In an environment neutralized for ambient heat exchange, cotton fibers were caused to "glow" in 365-371 nanometer black light. Temperature measurements of the cotton were simultaneously made using an infrared thermometer. Drops in temperature in the range of 3° F in approximately 5 minutes of irradiation were routinely recorded[78].

Because the black light frequencies are related to the Rydberg quantum distribution for hydrogen emissions, the results were calculated by the number of hydrogen bonds in the cotton molecule. Using the measured drop in temperature (in Celsius) and the known nuclear weight of the cotton molecule *per* hydrogen bond, an equation for the loss of energy per bond could be calculated. This equation derived Planck's Constant with an accuracy of six tenths of a percentage point. The equations was also able to predicted the measured cooling performance of David Rule's infrared negative radiation device (Paschen Series infrared).

That equation derived from the cotton/black-light study is the following:[79]

Formula Derived from N-Radiation Study of Cotton Fibers

$$\frac{\Delta \text{Temperature}}{sec.} = -\left(\frac{239(\text{Planck's Constant})(\text{number hydrogen bonds})}{\text{mass of proton}(sec.)} \right)$$

Planck's Constant is given in the energy unit "joules" ($6.6260755 \ 10^{-34}$ joules). The "joule" equates conventional Newtonian energy as a function of mass and velocity to energy as a function of mass and change in temperature. To divide Planck's Constant by a mass value— in this case the mass of the proton— remainders change in temperature. This remaindered "change in temperature" *times* the number of hydrogen bonds, *times* a conversion factor from calories to joules(239) gives the measured drop in temperature (*per* second).

This change in temperature formula for negative irradiation was also shown to be true of David Rule's negative radiation cooling of water[80] Rule's measurement[81] of radiant cooling rates at peak efficiency is consistent with predicted change in temperature ($-1.22°$ F *per* hour) as determined by the above formula developed from the "black light" cooling of cotton. Again, change in temperature was a function of Planck's Constant and the number of

---

[78] See *The Drop in Temperature by N-Irradiation of Cotton a nd Derivation of Planck's Constant* ; Append
[79] Ibid.
[80] Rule's N-radiation is known to be Paschen Series by the established characteristics of water's absorption of infrared (*Segelstein, 1981*). Water's greatest transparency to infrared — and thus its ability to pass out infrared "glow" and not reabsorb it — is for the Paschen Series of the Rydberg distribution.
[81] Proprietorial data from Pasco Poly, Inc. Available on request.

hydrogen bonds[82] with N-radiation frequency being held irrelevant.

The Rule heat exchanges occur by frequencies which are not visible— as black light and its "florescent glow" are visible. Thus, frequencies by which the device operated were not easily identifiable. The operational Rydberg frequency-band was mathematically deduced to be the infrared "Paschen Series" (820 nm medical florescent frequency). The return florescence proved too weak to detect with an off-the-shelf CCD infrared sensitive camera[83].

Data from the frequency absorption coefficients for water provided the identification as Paschen Series IR 820 negative radiation.

Paschen Series (Infrared) Negative Radiation

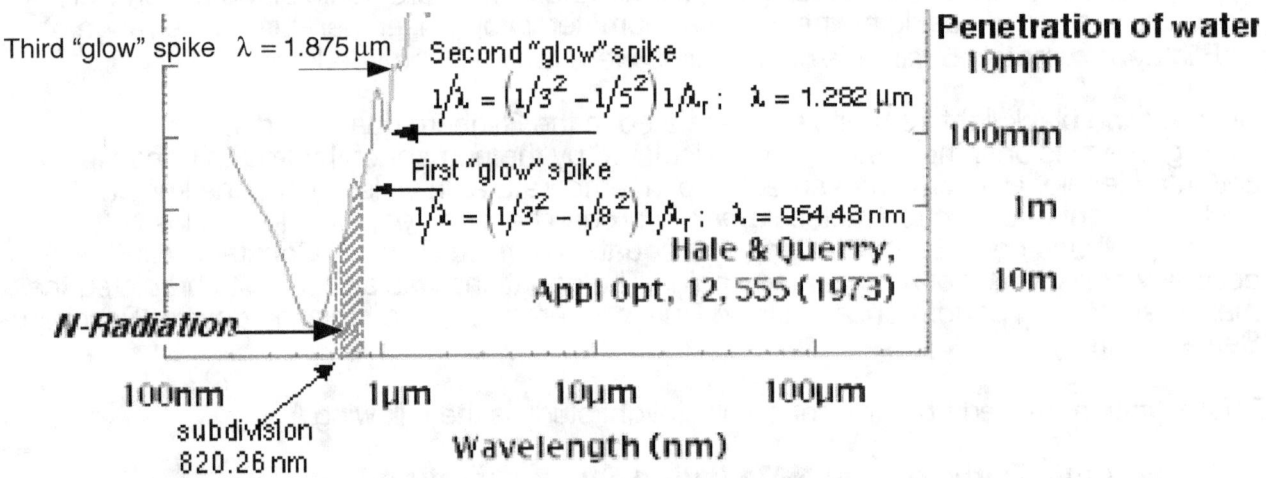

Third "glow" spike $\lambda = 1.875\ \mu m$

Second "glow" spike
$$1/\lambda = \left(1/3^2 - 1/5^2\right)1/\lambda_r \ ; \ \ \lambda = 1.282\ \mu m$$

First "glow" spike
$$1/\lambda = \left(1/3^2 - 1/8^2\right)1/\lambda_r \ ; \ \ \lambda = 954.48\ nm$$
Hale & Querry,
Appl Opt, 12, 555 (1973)

N-Radiation

Penetration of water
10mm
100mm
1m
10m

100nm   1μm   10μm   100μm
subdivision
820.26 nm   Wavelength (nm)

The Paschen Series gives a predicted subdivisional N-radiation wavelength of 820.26 nm:

$$\left(1/n^2\right)1/\lambda_r = \left(1/3^2\right)1/\lambda_r = 1/\lambda \ ; \ \lambda = 820.26\ nm$$

The highest actual quantum wavelength for the Paschen band is 954.5 nm:

$$\left(1/n^2 - 1/n'^2\right)1/\lambda_r = \left(1/3^2 - 1/8^2\right)1/\lambda_r = 1/\lambda \ ; \ \lambda = 954.5\ nm$$

Paschen Series N-radiation is at or near 820.26 nm and the return "glow" is at the Rydberg quantum differentiated frequencies at or below 954.5 nm.

The Paschen Series was identified as Rule's negative radiation band by the infrared absorption characteristics of water. The wavelength absorption coefficient of water must be low enough to allow both the negative radiation and the return "glow" to pass in and out of

---

[82] Water is assumed to be covalently bonded . An oxygen proton and the hydrogen nucleus share an electron. Citation: *"Water is polar covalently bonded within the molecule. This unequal sharing of the electrons results in a slightly positive and a slightly negative side of the molecule."*
http://www.emc.maricopa.edu/faculty/farabee/BIOBK/BioBookCHEM2.html

[83] The low intensity levels of 820 nm florescence has also proved a problem for medical research. See *"Principle and clinical usefulness of the infrared fluorescence endoscopy. J Med Invest. 2006 Feb;53(1-2):1-8"* which discusses the difficulties of locating, let alone "seeing," small cancer tumors with IR 820 florescence. Intensity is a function of hydrogen bonds and and calculations show that the florescent wattage for the two hydrogen bonds of water is well below all CCD camera thresholds.

the Rule cooling tanks which have radii of ± 1.24 meters . The Paschen Series met this criteria. Unexpectedly, the frequency absorption coefficients for water in the Paschen Series range of infrared produced a direct correlation with Rydberg quantum mathematics[84] .

The standard data for water absorption of radiation, as published by Hale and Querry, shows that the line illustrating wavelength absorption coefficients in the above graph almost "flattens" in the slope of the line at the Paschen Series "predicted subdivisional N-radiation" wavelength of 820.26 nm. The penetration of low intensity infrared at this negative radiation frequency is approximately 6 meters. The line then resumes its steep slope until it reaches the highest actual Paschen Series wavelength (n'=8; λ =954.48 nm).

Near the 954.8 nm wavelength, the absorption coefficient "spikes" downward. The downward spike is likely greater than reported by Hale and Querry because the single data point used to establish the spike measures as approximately 940 nm rather than the Paschen Series 954.48 nm, although 954.48 nm is within the width of the spike. The spike likely continues downward from the measured Hale and Querry data point of 940 nm until it reaches the Paschen Series wavelength of 954.48 nm.

The probability that the spike is larger than reported on the Hale and Querry graph is confirmed by the second "glow spike" which uses three data points instead of one (designated "Second 'glow' spike" in above graph). The lowest absorption coefficient for this spike (at the second data point) falls almost exactly upon the Paschen Series wavelength of 1.282 micrometers (n'=5). If the first data point of this spike— which has a wavelength greater than 1.282 μm— were used exclusively as the only data point, both the first and the second spikes would look identical. Instead, two additional data points are used to establish the second spike, showing that the absorption coefficient continues to fall from the first data point until it reaches the actual Paschen Series wavelength of 1.282 μm. This is likely also true of the 954.48 nm spike.

A secondary analysis of the Hale and Querry data shows a direct correlation between water frequency absorption coefficients and Paschen Series N-radiation characteristics of the Rydberg quantum distribution. Coupled with the fact that the penetration of water for the highest actual Paschen Series wavelength fits the radius of the Rule tanks being cooled by negative radiation.

Water is cooled in the Rule device by Paschen Series N-radiation (820 nm) forcing the water molecules to "glow" infrared (954 nm) while simultaneously dropping in temperature as a function of Planck's Constant and the number of hydrogen bonds. The measured drop in the temperature of the water approximately equals the loss of temperature predicted by the N-radiation formula derived from the black-light cotton study.

The actual frequency of the negative radiation is irrelevant to drop in temperature. Only the number of hydrogen bonds determines the change. Both the cotton under black light irradiation and water under Paschen Series negative irradiation changed in temperature as a function of the number of hydrogen bonds. The Paschen Series N-radiation wavelength ($\leq$ 820.26 nm) is over 2.25 times longer than black light ($\leq$ 364.56 nm). Yet temperature drops induced by both frequencies were strictly determined by the number of hydrogen bonds.

Time is a Quantum Force Which Establishes Space and Conducts Radiation
It is currently believed that when light is not a "photon" it is an "electromagnetic wave." Thus

---

[84] Low intensity, room-temperature absorption spectrum data for water published by Hale and Querry.

stated is the consensus catechism of contemporary physics.

The electromagnetic wave as mathematically established by Oliver Heaviside, however, is not accurately understood by conventional physics. The Heaviside formulation of the Maxwell field equation —which is the contemporary SI standard— is actually "meaningless" without quantum geometry. Specifically, the electromagnetic wave is irrational without the time quantum. The electromagnetic wave cannot really be discussed without first addressing an area of reality completely unknown to contemporary physics —the quantum dimension and its time substructure.

Quantum geometry proposes that all vacuous space is sustained by a "quantum time-force."[85] This time force is established by a small variation in time between two spacial points. It is proposed that one time value is slightly behind the other and yet both time values are correct for their perspective positions in space. The variance between them is a force which sustains spacial separation. It is the force establishing the fundamental quantum "α" and is the force which interacts directly with the electromagnetic wave.

The Time-Quantum "ΔT[86]"

Potential time-energy is conserved as the variance between differentiated time values located in quantum points. That potential energy sustains spacial separation between the quantum points. This fits the universal formula for energy: ***"Energy=(force)(distance)."*** ΔT^2 = **Potential Time Energy** = $F_t^2 \alpha^2$	$\text{Separation}^2 = \alpha^2$; F_t^2 = time force $\text{Time-Quantum} = \Delta T$ $(T_2 - T_1 = \Delta T)^2$ = Potential Time Energy $c = \alpha / \Delta T$ $\Delta T^2 = F_t^2 \alpha^2$ $\quad \Delta T^2$ = potential energy $F_t^2 = \dfrac{\Delta T^2}{\alpha^2}$ $F_t^2 = \dfrac{1}{c^2}$

The spacial quantum simply cannot exist without the time quantum. The spacial quantum is composed of only two points separated by "some" distance. The distance of separation is space without geometric definition. What, then, sustains that space and what composes the points? Non-geometric space —quantum separation— can only be produced by a projection of force similar to the force from opposite poles of magnets in proximity to one another. The points must contain a type of charge projecting the force of separation.

Differentiated or "alien" time values cannot exist in the same place and would project a force of opposition separating them. Time as a continuum would only exist in the points. For anything with "dimensionality" or length, time would be a quantum. Time must move in "ticks." Each "tick" representing the time differential "$T_2 - T_1$."

The time quantum "ΔT" would determine the maximum velocity with which anything could move across the space of separation. Velocities taking less time than "ΔT" could arrive before time itself arrived. This restriction on velocity is the speed of light which equals the space of separation divided by the time quantum or "$c = \alpha / \Delta T$."

---

[85] See *"The Theory of Time-Enforced Quantum Space"* . Chapter 5

[86] The time quantum can only exist as "ΔT^2" since the quantum exists only within the "quantum-squared."

68

The value of both the space of separation and the time quantum can be closely estimated by known scientific measurements. Nothing could exist with less length than the space of separation "α." The minimum measured distances are represented by the diameter of a proton and the shortest wavelength. These two measurements are known to be of the same order of magnitude of " ($\approx 10^{-15}$) meters." The alpha space must be the radius of the electron/proton and therefore estimated at "($\approx 10^{-15}$)/ 2= $\approx .5(10^{-15})$."

The radius of the electron/proton are equal because the proton and the electron are actually the same four-dimensional particle with their axii rearranged:

$$\left(\overline{\alpha}^2\right)^2 = \text{particle} \quad ; \quad \left\langle "\overline{\alpha}^2" \text{ is fundamental quantum - squared.}\right\rangle$$

$$\alpha(\text{Euclidean leg}) = \overline{\alpha}(\text{quantum leg}); \quad \left\langle "\alpha^2" \text{ must have quantum and Euclidean leg.}\right\rangle$$

$$\left(\alpha^2\right)^2 = \text{charge(volume)} = \overline{\alpha}\left(\alpha^3\right) = \alpha\left(\overline{\alpha}\,\alpha^2\right); \quad \overline{\alpha}\left(\alpha^3\right) = \text{proton}; \quad \alpha\left(\overline{\alpha}\,\alpha^2\right) = \text{electron}$$

Quantum geometry is always adequate to measure our space as shown by its derivation of the alpha space (radius of proton). The radius of the proton can be derived from the kink-tension constant which relates light frequency to the orbital radius of the electron:

$$f = \sqrt{2}\,k\left(r_Q^2\right) \quad \langle \text{from Dawson's Tensor; see Appendix}\rangle$$

$$\sqrt{2}\,k = \text{kink tension constant} = \frac{1^2}{\alpha^2} \quad \left\langle \text{units of kinked } "\alpha^2" \text{ quantums per meter}^2\right\rangle$$

The constant can be applied to the root orbital with a known frequency, orbital velocity, and a radius which can be calculated by an acceleration/deceleration terminal velocity which is equal to the speed of light.

$$f_r = \text{root frequency} = \frac{c}{\lambda_r} \quad ; \quad Q_r = \text{root radius} = \left(1 - \frac{v}{c}\right)\frac{\lambda_r}{\pi}$$

$$v = \text{electron velocity} = \sqrt{\frac{2eV_r(e)}{m}}$$

$$f_r = \frac{Q_r^2}{\alpha^2}$$

$$3.2893620584\ 10^{15}\ \text{hz} = \frac{8.2938681797\ 10^{-16}\ \text{m}^2}{\alpha^2}$$

$$\alpha^2 = 2.5214214892\ 10^{-31}\,\text{m}^2$$

$$\alpha = 0.50213757967\ 10^{-15}\text{m} = radius\ proton \quad \left(\text{measured value} \approx 0.5\left(10^{-15}\right)\text{m}\right)$$

The fact that quantum geometry accurately derives physical measurements of our space is definitive proof that the geometric system is descriptively true. The first case was the derivation of the Zeeman effect doublets as measured by changes in electron voltage. The second case in point is the above derivation of the radius of the proton. The third example is the prediction of a loss of temperature for fluorescing hydrocarbons in black light as a function of electron bonds and the resultant derivation of Planck's Constant. Anyone attempting to discount these derivations and describe them as coincidental must provide some evidentiary support for this contention. Otherwise they should simply be dismissed as practitioners of artificial science and mathematics.

By the quantum geometric hypothesis, the alpha space is equal to the radius of the proton. It is the minimal distance of separation which must be sustained between the hydrogen proton and an enveloping field from an attached electron(see chapter 1). Using the value "0.50213757967 (10^{-15}) meters" for "α" in the speed of light equation, the time quantum "ΔT" would equal "1.6749506743 10^{-24} seconds." This is the "tick" by which time must proceed.

Proof that Time Is a Quantum

The negative radiation study of cotton determined that originating frequency is eliminated as a factor in determining temperature drop. The cooling of the water by Paschen Series "820 nm" infrared was the exact equivalent of the cooling of the cotton by Balmer Series "365 nm" black light. The "per bond" formula derived from the cotton study correctly predicted the drop in temperature for the water as measured over a great number of experiments. Frequency of the negative radiation was irrelevant. Only the number of bond oscillators determined temperature change. Frequency was removed as a factor in energy determination. The energy equation "$f(h)$" became "$1(h)$."

Frequency is determined by the kink-tension for any quantum squared orbital distance. Planck's Constant is the rate at which energy is exchanged with the nucleus per unit of time. Total energy exchanged with the nucleus is Planck's Constant *times* units of cycle time:

$$t = \text{time of one cycle}; \quad f = \frac{1}{t} = \text{number of units of } "t" \text{ in one second.}$$

$$\text{Energy} = f(h) = \frac{h}{t} \quad \langle \textit{Energy exchange} = \textit{Planck' s Constant times units of "t" in sec.} \rangle$$

For the N-radiation studies, the value of "Planck's Constant per unit of time" changed from "$t = 1/f$ sec." to "$t = 1$ sec.":

$$E = t\frac{(h)}{t} = 1\frac{(h)}{1} \quad ; \qquad t = \frac{1}{f} \quad \textit{The time factor has shifted from "1/f sec." to "1 sec."}$$

The "units of time *per* second" is equal to frequency and, therefore, is a quantum value:

$$\text{units of time } per \text{ second} = f = \frac{\text{second}}{t} \quad ; \quad t = \text{cycle time}$$

$$\frac{second}{t} = \left(\frac{1}{n^2} - \frac{1}{n'^2}\right)\frac{c}{\lambda_r} = \left(\frac{1}{n^2} - \frac{1}{n'^2}\right)\frac{second}{t_r}$$

$$\frac{second}{1} = \left(\frac{1}{n^2} - \frac{1}{n'^2}\right)(second)\frac{t}{t_r} = \left(\frac{t}{t_r n^2} - \frac{t}{t_r n'^2}\right)(second)$$

$$\frac{t}{t_r} = \langle \textit{orbital cycle -time as multiple of root cycle time} \rangle$$

Frequency/ cycle-time can be brought to unity ($f = t = 1$) for any orbital by the negation of subdivision. The orbital's cycle-time as a multiple of the shorter "root cycle-time" can be differentiated by orbital negation of subdivision to produce unity. Time can be brought to quantum unity by the rules governing negation of subdivision and, therefore, time is a quantum. The time quantum "ΔT" represents a time- differentiation of "1.6749506743 (10^{-24}) sec." and provides a force which separates the quantum alpha space.

The Electromagnetic Wave and Quantum Time Force

Returning to the electromagnetic wave. The electromagnetic wave is established by the flux in electric permittivity across the constant magnetic permeability of free space. The electromagnetic field equation for free space can be used to verify that the electromagnetic wave cannot exist in vacuum at great distances from matter or mass. Further, the EM flux in free space can be demonstrated to be a mathematical function of the quantum time force. The flux in intersecting magnetic and electric fields generates an equivalent "pulse" in the quantum time-enforced spacial field.

The EM Wave is Local in Character

The "field" establishing magnetic permeability for free space incorporates mass and, therefore, only vacuous volume (free space) in near proximity to mass possesses magnetic permeability. This fact is disguised by the Heaviside formulation of Maxwell's electromagnetic field equations[87] . The Heaviside formulation of magnetic permeability[88] disguises the field equation by converting it to an electrodynamic unit— to "amps":

Heaviside's Formulation of Magnetic Permeability of Free Space

$$\mu_0 = \text{magnetic permeability of free space} = 4\pi\left(10^{-7}\right)\frac{\text{Newtons}}{\left(\text{amp}\right)^2}$$

Amps, however, are actually determined by a measure of electric field strength. An amp is defined as the amount of current flowing through two parallel wires set one meter apart which establish a field equal to 2 (10^{-7}) Newtons of force over one meter of distance along the wire[88] . The amp is actually an electric field value equal to 2 (10^{-7}) Newtons *per* meter squared. Heaviside's magnetic permeability unit of measure is achieved by squaring the amperage field which produces a four-dimensional field:

$\text{amp} = \dfrac{2\left(10^{-7}\right)\text{Newton}}{\left(\text{meter}\right)^2}$ $\mu_0 = 4\pi\left(10^{-7}\right)\text{Newton}/\text{amp}^2$	$\mu_0 = 4\pi\left(10^{-7}\right)\dfrac{\text{Newton}}{\left(\dfrac{2\left(10^{-7}\right)\text{Newton}}{\left(\text{meter}\right)^2}\right)^2} = \pi\dfrac{\text{meters}^4}{\left(10^{-7}\right)\text{Newton}}$

Excluding the circularity factor "π," free-space magnetic permeability is a measure of meters to the fourth power. Since free-space permeability must be a field value— force per unit of space— the correct unit of measure must be geometric, not electrodynamic. By squaring the amp Heaviside has squared the amperage field and cannot escape from this fact. He retains the force value in Newtons and this force must be expressed into space. The squaring of amperage also squares the amperage field, resulting a four-dimensional geometric topography.

But what can "meters to the fourth power"— Heaviside's four-dimensional spatial topography— actually mean? It can be given definition only by quantum geometry.

The amperage unit of area is achieved by multiplying a length of wire (mass) by a distance of space (vacuum). The two nonequivalent distance legs influence force differently. If the

---

[87] The modern formulation of Maxwell.
[88] *Electromagnetic Theory*, Vols. I, II,. and. III. Reprint. Oliver Heaviside. New York:. Dover,. 1950. P. 21
[89] See "Dictionary.Com"

length of wire is increased, force goes up. If the distance of separation is increased, force decreases. The two legs influence "force" in exactly the opposite manner.

This duplicates the *soon-to-be-introduced* quantum-squared spatial field. By increasing the Euclidean component of the quantum-squared field(equivalent of the mass leg) the force value is increased. By increasing the quantum component of the Q^2 field (equivalent of the spatial leg), force is reduced. The amperage field mirrors the quantum squared field.

Heaviside's squaring of the amperage field is the equivalent of the squaring of the quantum-squared field. By quantum geometry, the squaring of the quantum-squared field produces a unit of mass *times* a unit of vacuum[90]. That is, four dimensional space is achieved by the squaring of the quantum-squared field and that squaring produces a mass which multiplies vacuum outside itself. Only in quantum geometry can field area be squared to produce rational four-dimensional geometric space and that space must incorporate both mass as well as vacuum[91].

This means that the four-dimensional field of free-space magnetic permeability must incorporate mass within the free-space. Free space permeability is a function of mass interacting with the free-space. Free space magnetic permeability simply does not exist outside the presence of matter.

If the quantum geometric resolution of the field squared cannot be accepted, a strict Euclidean interpretation provides the same results. In the most limited sense, the squaring of the amperage field must mean the squaring of the current generating the field. The squaring of current means the conversion of the linear conduction wire into a plane or conduction plate. Free-space permeability is a characteristic of the space proximate to plates generating the magnetic field.

Free space can only conduct the electromagnetic wave in near approximation to the matter which is generating the magnetic field. The projection of the electromagnetic wave will be determined by the strength of the electric field (electric permittivity defined as capacitance in Farads *per* meter of distance). However, projection is irrelevant because the interface between the electromagnetic field and the quantum-squared spatial field— the field which actually conducts waveform radiation over long distance— is immediate and direct.

The Heaviside formulation of free-space permeability/permittivity identifies the interface between the electromagnetic wave and the spacial field sustained by the time-quantum. By the theory of four dimensional quantum geometry, vacuum is sustained by the force of the time-quantum[92]. Time force is a force generated by two incompatible time values which must sustain a space of separation between them in order to coexist. These incompatible time values establish the time quantum " ΔT ." The space of separation is the fundamental distance quantum " α ."

The Heaviside formulation can be use to calculate a standard time-force value in amp-Newtons *per* the alpha-squared field of space *per* the time-quantum squared of force which sustains the field. Time force is "pushed" directly by electromagnetic flux in free space and provides the transmission mechanism by which waveforms may be conducted by vacuous space itself :

---

[90] See *"Quantum/Electromagnetic Field Equations"* in the Appendix.
[91] Ibid.
[92] See *"The Theory of Time-Enforced Quantum Space"* Chapter 5

The Heaviside Formulation of Permeability/Permittivity of Free Space[93]

c = speed of light ; μ_0 = magnetic permeability of free space = $4\pi(10^{-7})\dfrac{\text{Newtons}}{(\text{amp})^2}$

ε_0 = electric permittivity of free space ; F_t^2 = time force squared = $\dfrac{1}{c^2}$

$$c = \frac{1}{\sqrt{\mu_0\varepsilon_0}} \quad ; \quad \mu_0\varepsilon_0 = \frac{1}{c^2} = F_t^2$$

Using the Heaviside Formulation to Calculate Time Force

By converting Heaviside's electrodynamic units (amps and Farads) back to the original field values of Newtons per unit of space, we can express Newtons in spacial field and time values :

$$\text{The electric permittivity "}\varepsilon\text{" unit of measure} = \frac{\text{Farad}}{\text{m}} = \frac{\left[2(10^{-7})\right]^2\text{Newton}(s^2)}{\text{meter}^6}$$

The Conversion of $\mu_0\varepsilon_0$ to Field Force Unit of Measure:

$$\mu_0\varepsilon_0 = \varepsilon_0 \frac{\left[2(10^{-7})\right]^2\text{Newton}(s^2)}{\text{meter}^6}\left(\frac{\pi m^4}{(10^{-7})\text{Newton}}\right) = 4\pi\varepsilon_0\left((10^{-7})\text{amp}-\text{Newtons}\frac{sec.^2}{m^2}\right)$$

$$\frac{4\pi(10^{-7})^2\text{Newton}}{(10^{-7})\text{Newton}} = 4\pi(10^{-7})\text{ amp - Newtons}$$

$$\text{amp - Newtons} = (10^{-7})\text{ Newton}$$

$$F_t^2 = \frac{1}{c^2} = \frac{\Delta T^2}{\alpha^2} = \frac{\left(1.67495\ 10^{-24}\,sec.\right)^2}{\left(0.50214\ 10^{-15}\,m\right)^2} = 1.11264(10^{-17})\frac{sec^2}{m^2}$$

and

$$F_t^2 = 4\pi\varepsilon_0\left((10^{-7})\text{amp}-\text{Newtons}\frac{sec.^2}{m^2}\right) = 1.11264(10^{-17})\text{ amp- Newtons of }\frac{\Delta sec^2}{m^2}$$

$$1.11264(10^{-17})\frac{sec^2}{m^2} = 1.11264(10^{-17})\text{ amp - Newtons }\frac{\Delta sec^2}{m^2}\text{ of time variance}$$

This is an important equation because it equates amp-Newtons and time variance (squared) with the straightforward time per meter of space from the inverse of the speed of light (squared).

Because Heaviside's electromagnetic permeability/permittivity formula can be calculated in amp-Newtons of force in a field and because that formula is exactly equal to the "time force" establishing the time quantum, this time force can be given a value in amp-Newtons.

Specifically, the quantum spacial unit is sustained by a force generated by the quantum

---

[93] Standard SI formula in physics.

time differential. This force is equal to 1.11265e-7 amp-Newtons, the same value as the "seconds squared per meter squared" from the inverse of the speed of light (squared).

The Heaviside formula shows that there is a direct relationship between the flux of the electromagnetic wave and the force which sustains vacuous space. An electromagnetic wave fluxes against the time-force sustaining vacuous space and must reach the limit of that time force in order for the wave to be conducted.

Magnetic permeability of vacuum is a constant[94] and the electric field must reach the permittivity threshold (in capacitance *per* meter) in order for vacuum to conduct the flux as a wave. This threshold, which the electric field must reach, is exactly determined by the quantum time force sustaining vacuous volume. In the Heaviside formulation, the permeability constant *times* the permittivity threshold is equal to the inverse of the square of the speed of light. The inverse of the square of the speed of light is, itself, the quantum time force required to sustain vacuous volume (to be demonstrated below). The electric field permittivity must exceed the quantum time force sustaining volume in order for an electromagnetic wave to be conducted across free space.

The Heaviside formulation is typically taught as a two-dimensional field equation since it reduces, geometrically, to "meters squared." How then, can it be claimed that the Heaviside formulation identifies the quantum time force which sustains the "volume" of vacuum? The answer is that vacuous volume is defined in a "counter intuitive" way. The volume of space is actually two dimensional, as defined by the quantum squared.

The fact that the electromagnetic waves interact with or "push" vacuous space is proven in millions of kitchens by microwave ovens. The electromagnetic wave can interact with nonlinear vacuum contained within matter (nonaligned "soliton vacuoles" as described below). Electromagnetic waves interact with vacuum in ways that light (which is not an electromagnetic wave despite common belief) cannot. Electromagnetic Microwave penetrates food and heats it from within by generating volume pulses in the nonaligned quantum vacuoles contained within the food.

Physics, of course, knew nothing of this in the early twentieth century. With light as sometimes a misunderstood electromagnetic wave and sometimes an unproved photon particle, light theory became *ad hoc* and completely devoid of systematic physics. Most certainly, Rydberg as wave harmonics would never have been recognized.

Now enter four-dimensional quantum geometry to provide that which was missed over a hundred years ago. Light as a pure waveform energy can be conducted by a completely unanticipated source. Quantum defined space itself— the vacuous volume separating islands of matter— can conduct waveform energy. Further, this waveform is of a known type. It is a transitional wave consisting of pulses of increasing and decreasing volume in transverse motion to the direction of wave travel. Soliton-like quantum units compose vacuous space. These "soliton vacuoles" can be made to expand and contract in volume.

How can this be true if conventional physics knows nothing of it? It is true because quantum geometry proves it is true. And it is true because science has observed and measured the pulsing of quantum vacuous space in soliton water waves. Nineteenth century science was mystified, however, by the soliton observation. Fully understanding the soliton water wave would also have to await the arrival of quantum geometry.

---

[94] *Electromagnetic Theory*, op. cit.

Elsewhere I have proven that three dimensional Euclidean space cannot exist without a quantum dimensional substructure.[95] Let us briefly review that proof.

Modern scientific mathematics are built upon calculus. Yet, the impossibility of differentiating adjacent points along the Euclidean continuum to "0" by subdivision has been known since Newton's time.[96] Even to exist, differentiating a Euclidean distance to "0" — as scientific calculus necessitates — requires the existence of the quantum negative integral.[97]

In the following pages I will demonstrate that knowledge of the quantum dimension changes our understanding of many of the so-called "factual foundations" of the physical sciences: what Rydberg actually discovered; what Planck's Constant actually is and how it works; what radiation is; what the electromagnetic wave is and its resonance with quantum space; how the atom is structured and composed; what a soliton wave is and why its emerging mathematical descriptions —currently on the periphery of physics— may become central to the future of science.

One of the more important of these misunderstood "factual foundations of science" is our concept of "vacuum." In agreement with "1+1 ' kink' mathematics," quantum geometry identifies vacuum as distorted by tension from an intersecting quantum dimension. More accurately, vacuous space is the forced curvature of planes by quantum dimensional intersections of Euclidean lines.

"1+1 dimensional" soliton geometry describes vacuum as tensioned by "kinks" in a single dimensional axis. In this concept, there is a merging of soliton and quantum mathematics. Four-dimensional quantum geometry holds that a straight Euclidean line in vacuous space is "kinked" upward to become a curved line.

Quantum geometry also reveals that two dimensional planes can be "kinked." In the vernacular of soliton geometry, "2+1 dimensional space" is also possible and it is this "kink" which defines volume. The linear quantum plane defined by a straight Euclidean line and a quantum point is "kinked" upward to become a curved plane which "contains" volume. Vacuum is the volume contained by the forcible curvature of quantum planes.

Soliton geometry is the only off-the-shelf strict Euclidean-based geometry which applies to the quantum dimension.

Curiously, soliton geometry emerged contemporaneously with Rydberg and Planck to explain an earlier mystery. The soliton is a fast moving solitary water wave without leading or trailing troughs. It was first observed in 1834, but was denied as a scientific possibility for well over 50 years until nonlinear mathematics were applied near the turn of the twentieth century. Later, in the 1950's, Enrico Fermi used a nonlinear soliton model for an array of weakly linked harmonic oscillators which can be used to explain how quantum vacuous space can conduct wave form energy within matter itself.

In his 1997 review of soliton wave geometry, Sascha Vongehr [98] writes,"For any nonlinear theory the soliton is at least as fundamental a solution as the sine wave. Recently, there have been profound advances in finding solitons in higher dimensional theories and in

---

[95] See *"Dawson's Theorem: Proof of the Quantum Dimension"* in Appendix
[96] Implicit in the Maclaurin Derivative Series.
[97] *Ibid.* *"Dawson's Theorem."* The derivative is the "anti-integral." The quantum negative integral is the "anti-anti-integral" or the Euclidean derivative. Two negatives make a positive.
[98] *"Solitons"* ; Sascha Vongehr; 1997 available on internet.

quantizing them. Doing quantum mechanics one finds relations between solitons that go very deep and are entirely unexpected from a classical viewpoint." It was one of these "unexpected relationships" which correctly predicted the existence of negative radiation several months before its empirical confirmation by the current writer[99].

The value of soliton geometry to four dimensional quantum geometry is that it explains how the quantum dimension can interface with strict Euclidean space. The quantum dimensional axis cannot be Cartesian. One cannot plot the quantum dimension intersecting an unrestricted Euclidean dimension on a standard Cartesian graph. If the "y" axis were the quantum dimension, the distributions "y=1" and "y=2" along the x axis would be parallel lines separated by the quantum value of "1." However, every point along "y=1" would establish a quantum value with every point along "y=2." That is, the quantum could not be restrained to "Q=1" between the two parallel x axis lines. The quantum unit of distance requires a single point interacting with a line, not two lines interacting from every point along their respective lengths.

A quantum is defined as two points separated by "some" distance. The only way a quantum can be arrayed with an Euclidean unit of distance "x" such that the quantum is restricted to the distance "x" is the following: a single point must be set off the unit distance "x" such that the distance from the external point to the endpoints of "x" are equal. That is the quantum must be interfaced with the Euclidean distance unit "x" as a "stand-alone" equilateral triangle (see the illustration below).

This "stand alone" unit of space will be defined as a soliton. The "stand alone unit of space" is being put forward as the precise definition of a soliton. At present, the soliton has no such scientifically agreed upon definition. Speaking to the issue, Vongehr writes "*definitions change from author to author.*"[100]

The soliton illustrated below is the quantum squared unit of area. The quantum squared interfaces with an Euclidean distance by making that distance a component of itself. the Euclidean unit "x" is a linear component of the area circumscribed by the triangle, a linear component which can be differentiated by conventional subdivision (1/2 a unit, 1/3 a unit etc.). The quantum component of area (designated "Q" in the below illustration) can also be differentiated — but by the only method of distance differentiation available to the quantum; by the negation of subdivision for the quantum squared.

A quantum cannot be differentiated by subdivision because the quantum is composed of only two points. To subdivide the space of separation between these two quantum points would require adding more points to the space of separation. The quantum would be multiplied by the addition of these points. It would create more quantums all of which would have "some" space of separation between two points.

The only way a specific quantum distance can be differentiated is by subtraction of a unit of subdivision. This is called "the negation of subdivision." If one subtracts a subdivisional distance from the original quantum distance, the original two quantum points are remaindered, but at a new space of separation. This new space of separation is a mathematically rational differentiation of the original quantum. The quantum must be differentiated by negation of subdivision. The Euclidean line composed of a continuum of points can be differentiated by subdivision. The essential difference between a quantum

---

[99] *"Negative radiation pressure exerted on kinks"* : T o m a s z R o m a D c z u k i e w i c z et. al; Phys.Rev.D77:125012,2008

[100] ibid.

distance and an Euclidean distance is the manner by which the distance must be differentiated.

The quantum squared unit integrates a quantum unit and an Euclidean unit of the same measure. The quantum squared must be a "stand alone" unit of distance derived from a single point relative to the x axis. The quantum squared unit of measure must be a soliton, as we have defined the term, or a "stand alone" unit of space.

Any discrete Euclidean distance along the x axis must establish the quantum as a soliton, or a discrete unit of area converging upon a single point in "equidistant" quantum space.

The Quantum as Soliton: Soliton = "Stand-Alone" Unit of Space

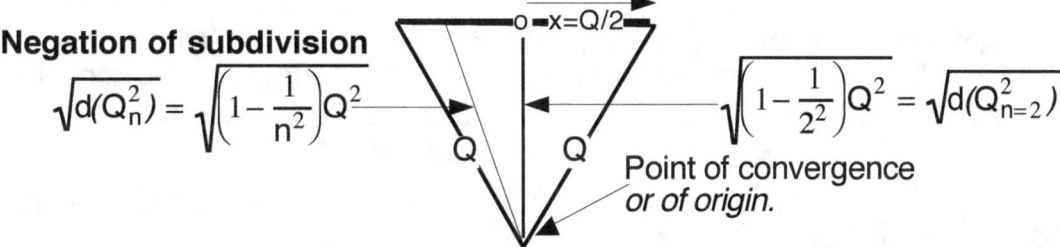

All quantum values intersecting "x" can be restricted to the square root of the negation of subdivision for the quantum squared, (designated " $d(Q_n^2)$ " above)[101] with a lower limit of "n=2" as shown. Quantum space can be integrated with Euclidean space only by means of this "differentiated quantum squared" soliton.

However, this initial quantum squared soliton is a strict Euclidean definition. It is not adequate to the factual quantum. The quantum forces a second soliton definition, a "kink" in the x axis. This is true because the basic quantum, designated "the Alpha space," is a time-enforced distance[102] which cannot be compressed to the quantum differential, " $d(Q_n^2)$," without violating time restraints.

The differentiated quantum, designated " $\sqrt{(1-1/n^2)Q^2}$ " above, must be converted to "Q." To achieve this conversion, the distance "x" must be "kinked" upward and become curved relative to the linear definition of "x." Thus, the distance "x=Quantum (Q)" of strict Euclidean definition becomes "Kx=Qπ" of quantum definition. The Euclidean distance "x" becomes the quantum circumference of a circle of diameter "x."

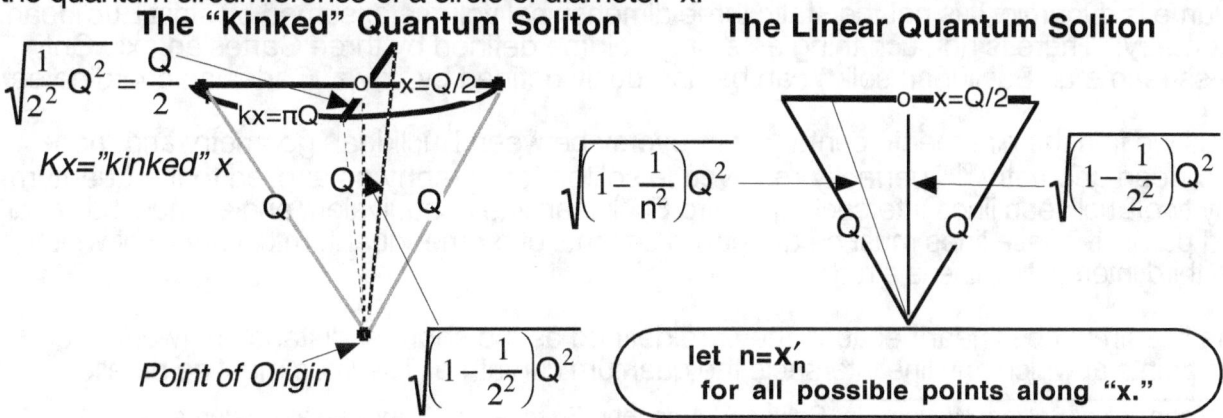

[101] See "The Quantum Dimension and the Definition of the Quantum Squared" in Appendix
[102] "The Theory of Time Enforced Space" Chapter 5

Each and every quantum value between the "Point of Origin" and the circumference "Kx" is equal to "Q." By the Pythagorean Theorem the quantum squared is always equal to the negation of subdivision for the quantum squared at "x" and the square of the "kink" value for "x":

$$Q^2 = \left(\frac{Q}{n}\right)^2 + \left(1 - \frac{1}{n^2}\right)Q^2$$

Kink of $x = \dfrac{Q}{n}$

Negation of subdivision for quantum squared $= d(Q_n^2) = \left(1 - \dfrac{1}{n^2}\right)Q^2$

The linear quantum plane of the original soliton formed by the "Point of Origin" and "x" has also been forcibly "kinked" to become the curved plane of the surface of the cone so formed. The surface area of the "kinked" quantum plane is equal to "$\pi Q^2 / 2$":

Surface area of cone = π(radius of base)(slant height)

Standard solid geometry

$$\text{S.A.} = \pi\left(\frac{Q}{2}\right)(Q) = \pi\frac{Q^2}{2}$$

In summary, the linear distance "x=Q" has been "kinked" to become the curved distance "Kx=πQ" and the original linear soliton plane has been "kinked" to become the curved plane with a conical surface area equal to "$\pi Q^2 / 2$."

The kinked soliton is a function of the quantum squared. Volume is defined only by the "kinking" of a linear plane. Vacuous space is defined completely by the quantum squared and is achieved only by the distortion or "kinking" of the linear quantum plane.

Soliton geometry identifies the pushing of the quantum off the secant as a "kink" in the line "x" or an assertion of "x" into another dimension (1+1 dimension)[103]. The length of the "kink" is identified by quantum geometry as the subdivision by which the quantum is negated (for the quantum squared). The length of the kink is "(Q)/ n," where "n" is the subdivisional value by which the quantum squared is negated.

All vacuous space is composed of scaled multiples of the kinked soliton from the fundamental quantum, Alpha. Vacuous space is pushed, distorted and expanded by attempted Euclidean definitions imposed by the existence of matter. Quantum squared volume is dynamic. It is not the static three dimensional volume assumed by strict Euclidean geometry. There is no such thing as empty volume defined by three Cartesian axii. Only mass (same as Euclidean solid) can be rigorously defined by "x,y,z" Cartesian coordinates.

Thus we find the nineteenth century controversy between Euclidean geometry and "non-Euclidean geometry[104]" perfectly resolved in soliton topography as applied to the quantum. Any two Euclidean lines intersecting a third Euclidean line at equivalent angles must be linear and parallel. These lines must be quantized secants of "some" quantum curvature of which the third intersecting line is a radius.

The lines must be linear because they are defined as the shortest distance between the two points at which any line intersects the quantum curvature. The lines must be parallel

---

[103] *"Topological Solitary Waves"* in "Solitons;" Vongehr, Sascha; 1997; available on internet.
[104] Primarily the work of Carl Friedrich Gauss and Janos Bolyai to disprove Euclid's 5th Postulate which establishes the possibility of parallel lines.

since the curvature distances between any two points of intersection by different lines is always equal. That is, the distance between the lines, *as defined by curvature,* is always equal.

Einstein was correct in his supposition that vacuum has "physical reality," though his specifics are wrong The volume of vacuum is the forced curvature of linear quantum planes. There is a physical reason that vacuous space is defined as the forcible curvatures of linear planes. Quantum space is structured by potential time energy[105]. The curvature of space is forced by time.

Vacuous space is composed by vectors of time-force "pushing" linear planes into curvature.[106] Space can acquire wave "compression" by shortening the quantum vector in the distance of travel, increasing "kink" vectors which are set at 90° angles to the direction of travel. Just as a water waves are increases of height relative to the direction of wave motion, so these compressions of quantum space compose pulses of volume; the shortening and lengthening of quantum distances accompanied by related up and down extensions of the "kinks." The volume pulses compose waves moving at the speed of light through vacuous space, conducting energy.

The point of this brief review is not to critique Einstein *et. al.* but to document the inadequacy of non-quantum wave theory at the time when adequate theory was needed most. The year following Michelson-Morley, Janne Rydberg detected quantum light characteristics. The Rydberg formula could not be recognized for what it is, as quantum geometric wave - harmonics; as a harmonic pattern which varies systematically from the Euclidean geometric harmonic pattern which governs musical sound waves. Science was unprepared for Rydberg and has continued in that unprepared state for the last 100 years.

Science suffers a deficiency in geometry, a failure to recognize that our geometry is four dimensional not three. It is a mathematical unpreparedness for the quantum dimension, the quantum dimension which operates by a different set of mathematical principles than does the geometry of our known three dimensions. The Rydberg formula tapped into one of those differences. The substance of this work is the revelation of those missing mathematical principles.

---

[105] See *"The Theory of Time-Enforced Space"* Chapter 5
[106] *Ibid.*

The Vacuum Soliton
and its Volume Pulse
vacuous space as radiation-conduction medium

Planck's discovery that light energy is frequency-only was startling, so much so that Planck himself originally thought it would be overthrown or modified by subsequent research[107]. It was startling because it operates contrary to known principles of wave mechanics. Wave energy has amplitude and frequency. Planck had discovered the energy in light was "frequency only." Amplitude seemed irrelevant.

Geometrically, wave energy is two dimensional, not one. Frequency is only a single dimensional component. Technically, frequency is defined as the number of high pressure "hits" or energy peaks per second which pass a single point in space as produced by a pressure differentiated wave in motion through a medium. Amplitude is the amount of energy contained in any one high pressure peak. This can be graphically represented as a scaled sine wave.

The Pressure Differentiated Scaled Sine Wave
The number of density peaks and density troughs passing a fixed point over time is frequency

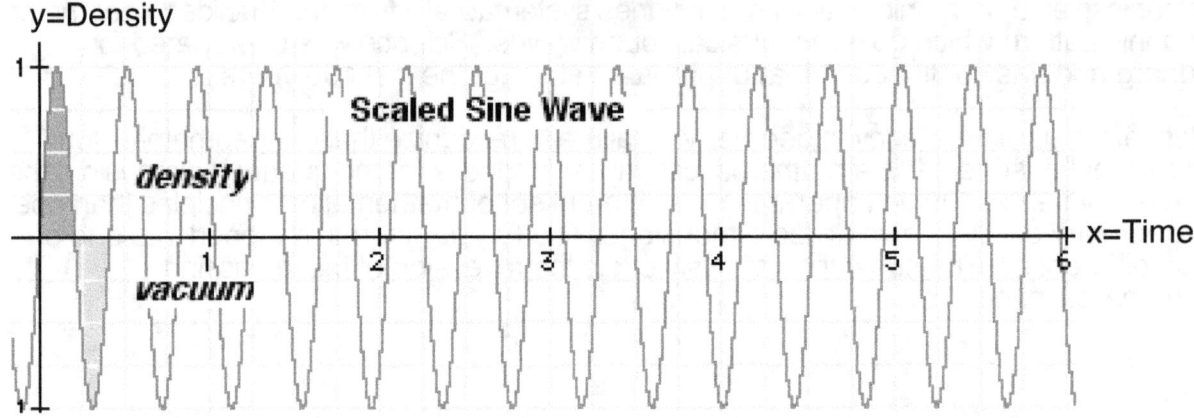

Cycles of high energy density "peaks" and low energy vacuum "valleys" are in motion through the fluid medium. The valleys and peaks are densities and the motion is due to the pressure differential between them. The peaks are increases in the medium's density and the valleys are akin to "vacuum." Waves are changes in densities in motion and the higher density peaks exchange energy because they have a greater mass. .

The particles composing the medium, however, may or may not be in motion. Density can be increased by compression as with a sound wave. The particles are compressed during the peak into a higher density which is followed by a dispersion or partial "vacuum."

Density can also be increased by transverse motion[108] as with a water wave. Density, in this case is converted into height and is relative to the "vacuum" outside the wave. This

---

[107] Planck said, *"My unavailing attempts to somehow reintegrate the action quantum (Planck's Constant) into classical theory extended over several years and caused me much trouble."*

[108] Transverse motion is motion of media particles which is 90° to the motion of the wave.

"peak" is followed by the "trough" in the medium. Negative wave density is composed of a "vacuum hole" in the medium (water).

Light waves are not pressure differentiated sine waves as both sound and most water waves are. They are "pulses" in vacuous space which were loosely modeled by "soliton" geometry in 1955[109] . . Non-linearly connected harmonically related "soliton" oscillators produced a wave- like sequence when a single oscillator was stimulated.

The Single or Soliton waveform (Modeled on a Gaussian Distribution)
There are no negative densities. Density is only a probability distribution.

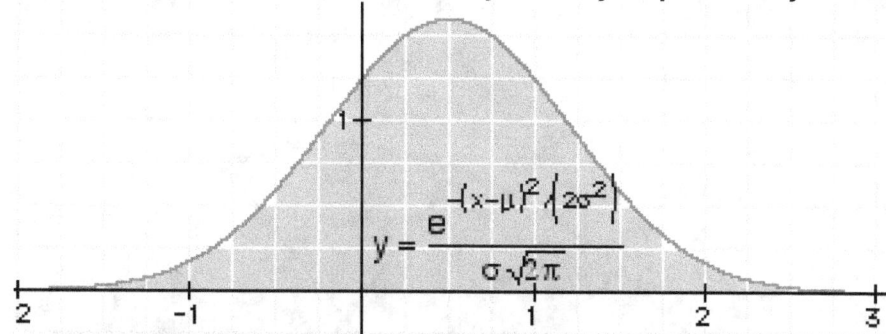

$$y = \frac{e^{-(x-\mu)^2/(2\sigma^2)}}{\sigma\sqrt{2\pi}}$$

The soliton is a single wave without trailing or leading vacuum troughs. It proved very difficult for scientific geometry to accept. The wave form was first reported in 1845, but it took science over 50 years to fully concede that they could exist. Despite strong empirical evidence that solitons existed, acceptance didn't come until the Korteweg de Vries equation of 1898 which identified the soliton as a nonlinear wave form.

The problem was this. The soliton produced no corresponding negative density. If it did, the single wave would disappear, distributing itself into a conventional sine wave. Where did the wave come from? Where did the "peak" come from if there were no corresponding trough? Sascha Vongehr, puts it this way: "*This solitary wave looks similar to a Gaussian bell in its smoothness and localization and in that it does not go below zero displacement before or after the very wave body.*"[110] It is simply a "hump" moving across the surface.

Even though the Korteweg de Vries equation mathematically described a solitary wave, it did not solve the geometric problem. The wave form is loosely based upon a Gaussian distribution and therefore describes "probable densities." It doesn't describe actual density.

If density is to be distributed unequally through a fluid medium, it must be distributed into one full wavelength. If there is a "field" of high density, there must be a corresponding "field" of low density. This is a mathematical requirement of Boyle's Law which states that pressure times volume is a constant and if you increase pressure (density) in one part of the volume, you must decrease pressure (density) in an equivalent part of the volume to retain constancy.

The Quantum-Squared Soliton

The soliton must be a quantum construction. The waveform is produced by a change in the "shape" of space being conducted through the volume of water. Quantum distance is being exchanged for an increase in volume. How this might work can be illustrated by the quantum squared unit of vacuous space shown below.

[109] '*Studies of Non Linear Problems in: Collected Papers of Enrico Fermi*, general editor E.Segre, *University of Chicago Press*, 1965, Vol. II, p. 978 as reported in "Solitons" op. cit.
[110] *"Solitons"* op. cit. p.3

The quantum lines between the point of origin and the x axis, are under tension because they are only partials of the full quantum. The quantum distance is sustained by a time differential "T_2-T_1."

The time variance is a pressure sustaining the distance between the two time values. Since most quantum distance between the point of origin and the x axis are less than the required distance they are forcibly vectored or "kinked" upward to "T1."

The Quantum Squared Unit of Vacuous Space

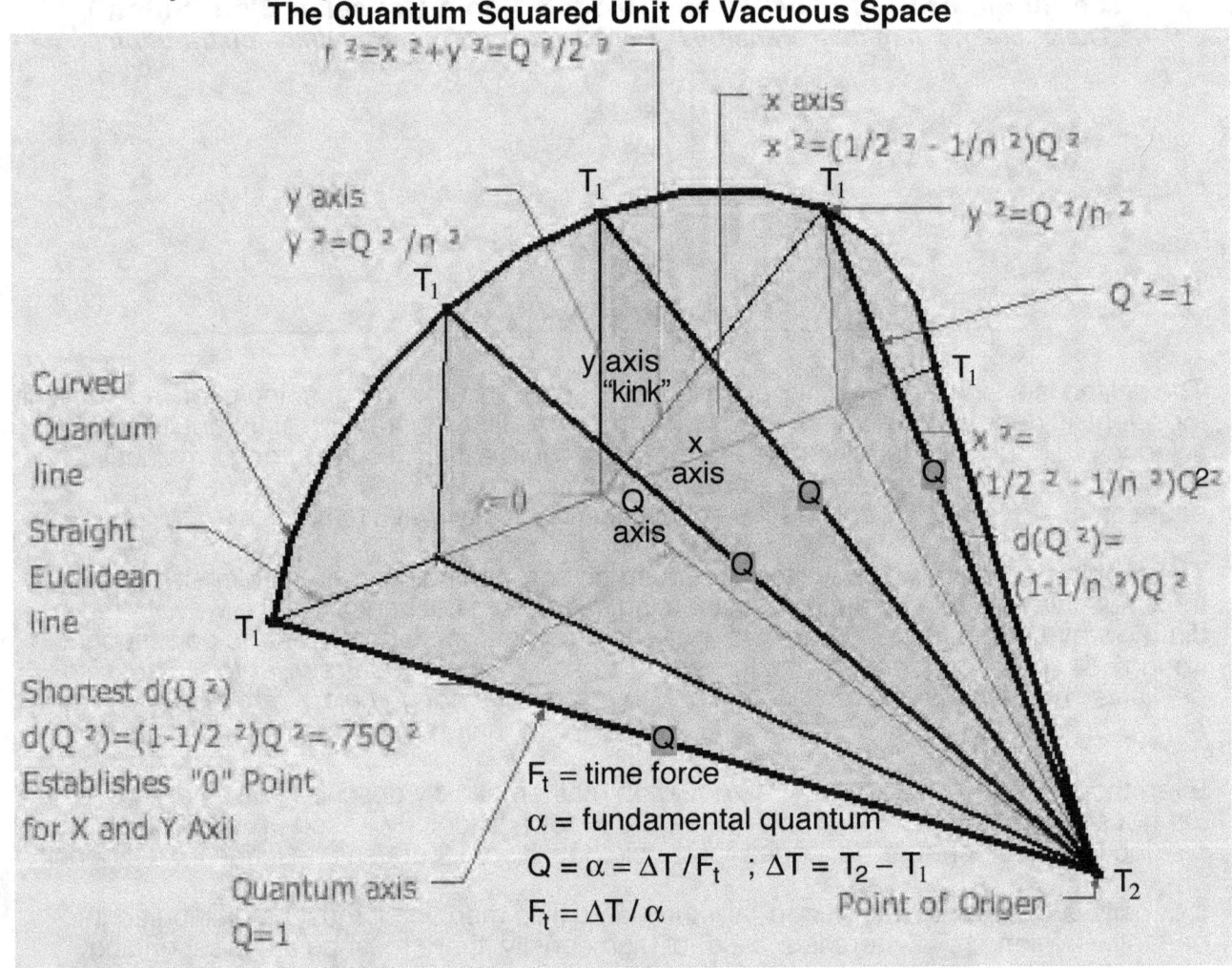

if you forcibly shorten the height of the cone (identified as the "Q axis"), all upward "kinks" must be lengthened. This mathematically increases the volume subscribed by the cone.

Such an exchange between quantum distance along the Q axis and cone volume is the only way a soliton can be produced without violating Boyle's Law.

A review of the quantum mathematics will reveal how this operates. All quantums lying on the linear plane made by the point of origin and the x axis are quantum differentials. They are negations of subdivision for the quantum squared equal to:

$$d(Q^2)=(1-1/n_x^2)\, Q^2.$$

These partials of the original quantum squared, however, cannot exist independently. They are under tension.[111] They are only a dependent vector of the force established by the complete quantum squared. The x axis must be "kinked" at 90° to itself to establish a second vector of force equal to the subdivision by which the quantum squared has been negated:

subdivision$=Q^2/n_x^2$; negation of subdivision$=(1-1/n_x^2)Q^2$

By the Pythagorean Theorem:

$$(1-1/n_x^2)Q^2+Q^2/n_x^2=Q^2$$

This re-vectoring of the differentiated quantum squared produces the semicircle which is the base of the illustrated cone. The x axis has been "kinked" upward to produce a circular plane with a continuum of points composing all possible lines lying within this plane. This Euclidean plane has "laminar density" in the sense that it is a plane against which Euclidean-defined mass can push.

In contrast, the differentiated quantums intersecting the x axis from the point of origin also produce a plane composed completely of quantum lines which have a beginning and an end point— but no points in-between. The quantum plane is a "vacuum" and has no "laminar density."

The differentiation of the quantum squared across the x axis exchanges quantum "vacuum" for Euclidean "laminar density." The quantum differentiations across the x axis get progressively smaller to the limit of the mean :

$$\sqrt{1-1/2^2}\,Q.$$

The Euclidean "kinks" get progressively larger to the limit of the radius of the constructed circle $(Q/2)$. Vacuum is being exchanged for laminar density with the end result being the construction of a vacuous volume with the ability to exchange energy with mass.

Density is defined as mass per volume of space. It is an established fact of physics that vacuous space exists between the nucleuses of water molecules. That vacuous space is defined by the quantum squared as just outlined.

It now can be demonstrated that the solitary wave of 1/2 wavelength —the "soliton"— results from an energy exchange between the mass of water and the quantum-squared vacuum separating molecular nucleuses. This energy is then conducted as a nonlinear distribution of quantum-squared pulses "bulging" overall volume at velocity— that is, as a soliton wave.

One of the best and most important description of the natural soliton may be that given by its discoverer, J. Scott Russell[112] :

> "I was observing the motion of a boat which was rapidly drawn along a narrow channel by a pair of horses, when the boat suddenly stopped - not so the mass of water in the channel which it had put in motion; it accumulated round the prow of the vessel in a state of violent agitation, then suddenly leaving it behind, rolled forward with great velocity, assuming the form of a large solitary elevation, a rounded, smooth and well-defined heap of water, which continued its course along the channel apparently without change of form or diminution of speed. I followed it on horseback, and overtook it still rolling on at a rate of some eight or nine miles (14 km) an hour, preserving its original figure some thirty feet long and a foot to a foot and a half in

---

[111] See "The Theory of Time-Enforced Quantum Space" in Appendix.
[112] J. Scott Russell in the "Report of the British Association for the Advancement of Science".

*height. Its height gradually diminished, and after a chase of one or two miles (3 km) I
lost it in the windings of the channel."*

The energy the boat surrendered when it decelerated to a stop generated a "density-only"
wave moving at 14 km/ h. It's amplitude but not its velocity "gradually diminished" over 3
km.

The wave was not a sine wave with a trailing trough. Russell called it a "wave of translation,"
meaning he thought the wave was moving water along the channel. The "heap" of water
without a trailing trough suggested to Russell that the wave was "carrying" a volume of water
in the peak as it moved.

His own subsequent research showed this to be wrong, however. Soliton velocity was
shown to be a function of wave amplitude. When a large soliton overtook a smaller soliton,
the waves did not merge, as conventional sine waves tend to do[113] , but again separated
into the two waves without apparent interference. The "translation" thesis does not hold
because any theoretical density momentum from the smaller soliton was not increasing the
density momentum from larger soliton, as it would with a pressure differential sine wave.

Russell believed that the soliton contained a larger volume of water over its length than
would a similar length at "water level." But he was incorrect in this assumption. It would mean
there would be greater water pressure in the wave than in the surrounding water. There is
not counterbalancing negative or "vacuum" pressure differential to sustain Boyle's
constancy, as with the sine wave.

The wave does not actually contain more water mass than the surrounding water. Wave
pressure is not greater because the vacuous volume between the water molecules has
expanded but the mass in the wave is no greater than the surrounding water. Only the
volume has increased, not the mass. Velocity is a pulse of quantum volume increase in the
direction of travel.

How do I know this? Because the soliton is "nonlinear" with respect to conventional water
density/pressure waves. This can determined by the variance in characteristics between the
soliton and the conventional sine water wave.

Velocity for the sine wave is a function of the square root of wavelength; longer
wavelengths, higher velocities:

$$v \approx \sqrt{\frac{g\lambda}{2\pi}} \text{ for deep water, depth} > \frac{\lambda}{2} \; ; \; \lambda = \text{wavelength} \; ; \; g = \text{gravity constant} [114]$$

This is exactly as expected for waves accelerated by water pressure differentials. The force
is vectored from high pressure towards relative vacuum. The high pressure peaks move in
the direction of the initiating force into a leading low pressure trough, leaving behind another
low pressure trough for the following peak. For longer wavelengths, there is more distance
over which the wave motion can accelerate (via gravity constant) and velocity will be
greater.

_____

[113] Vongehr describes this as *"Seegang "* or swell *"Bigger waves gain energy from smaller ones - they do
not go through each other and reappear again undisturbed."* op. cit.
[114] *Waves and Beaches*, Willard Bascom, Doubleday, 1964 p58-59; *Ocean Waves-Their Energy and
Power*, Ned Mayo, *Physics Teacher* , 35, September 1997 p352

For solitons, however, velocity is a nonlinear function of wave height:

The Korteweg de Vries Equation for Velocity of Soliton

$$v = c \left(1 + \frac{b}{3}\Psi_{max}\right) ; \quad \Psi_{max} = \text{wave peak}$$

c=phase speed ; b=non-linearity function

With increased "non-linearity" or "Ψ_{max}," the speed increases.[115]

NOTE: *Equation does not really determine velocity, since phase speed is an independent variable and velocity is dependent upon phase speed. The same nonlinear function and wave peak would increase phase speed by the same proportion to equal velocity, regardless of the value of phase speed. Phase speed is the independent variable. Velocity is the dependent variable.*

Can variance in water pressure explain the velocity of the soliton? It is true that— if *solitons contained a pressure differential* — a higher wave height would increase the pressure differential with the water level. However, the velocity of the wave would still be determined by the time of acceleration down the slope (*tanh*) of the wave and therefore by wavelength.

A fixed angle of deflection for the soliton would allow distance down the slope to be directly proportional to wave height. But, the Korteweg de Vries equation for the soliton will not allow this solution. The hyperbolic trigonometric function "slope over wave length" (*sech*) is an independent variable in the equation. Slope of the wave must change independently of the wavelength[116] :

$$\Psi = 3(v-c)/(b-c)\ \text{sech}^2[\ z\ ((v-c)/4d)^{1/2}\ ] \ ; \ d=\text{dispersion}$$
$$z=(vt-x) \ (\textit{i.e. a point moving along the x axis at velocity "v."})$$

The velocity of the soliton, as empirically observed and determined, cannot be explained by a pressure differential with the surrounding water. Neither can the peak without a trailing trough be explained. To build a mound on flat earth, one has to dig a hole. It appears the soliton leaves no such "hole" in the water behind.

The second characteristic variation between solitons and sine water waves leaves no doubt that the soliton is not a density/pressure wave. Solitons are transparent to one another. Sine waves are not. Solitons can pass through one another without interference. Depending upon the vectors of motion water sine waves will either energize or interfere with one another.

The only explanation for this variation between solitons and sine waves is that the latter are pressure differentials in motion and the former are not. If there are no pressure differentials between solitons, they will be invisible to one another when they intersect. They are neither adding nor subtracting pressure when they intersect. They will behave exactly as Russell's data showed. Larger, faster solitons will pass through smaller slower solitons without effect.

Since quantums must be distorted into curvature by Euclidean dimensions, the soliton wave can be explained by expansion of the quantum space between and within water molecules (*the reader should now know that vacuum is defined by the quantum squared*). In order to do this we need to know the absolute density of water.

The concept of "density" is premised on the fact that mass does not completely fill volume:

---

[115] From the *Korteweg de Vries equation.* See Vongehr , op. cit.
[116] Vongehr, op. cit.

$$\text{density} = \rho = \frac{\text{mass}}{\text{volume}} = \frac{m}{x^3}$$

The conventional density measure "ρ," however, is inadequate to our purposes. It is not an absolute measure of mass as a percentage of volume. It is only a relative measure comparing the weight of a material's mass with the volume it occupies. It will tell us how different materials vary in density, but it cannot tell us what portion of the volume is actually occupied by solid mass and, therefore, cannot give us the portion of volume which is occupied by vacuum. We need an "absolute density" (ρ_{ab}) to compare vacuum with mass. Absolute density=(mass volume)/ (total volume)

Absolute density can be found if we have a known material the density of which is 100%— that is, volume is 100% composed of mass. A comparison of any material's measured density "ρ" with a known 100% density will calculate the material's actual mass volume as a percentage of total volume.

We know of such a material. The volume of the proton contains no vacuum. Both the proton's weight in kilograms and its radius in meters are known or reliably estimated.

The density of a proton[117] is determined to be 3.9931 10^{17} kg/m^3. This figure can be used to calculate an absolute density for any material and the percentage of vacuum in its volume:

$$\rho_{ab} = \frac{\rho}{3.9931 \ 10^{17}} = \%\text{mass to volume} \ ; \ \text{vacuum} = Q^2 = 1 - \rho_{ab}$$

Absolute density can be used on the standard water pressure formula.
Water pressure is a function of density as defined by the following:

Pressure $= P = \rho xg$; x = water height ; g = gravity constant $\left(9.80665 \text{ m/s}^2 \text{ at sea level}\right)$
Water pressure can also be defined by the vacuum in the water's volume:

$$P = \rho xg \ \ ; \ \ \rho_{ab} = \frac{\rho}{3.9931 \ 10^{17}} = 1 - Q^2 \ \ ; \ \rho = 3.9931 \ 10^{17}\left(1 - Q^2\right)$$

$$P = \left(3.9931 \ 10^{17}\right)\left(1 - Q^2\right)xg$$

Define water pressure for the soliton and pressure for the water level, making them equal:

let Δx = height of wave ; let ΔQ^2 = change in vacuum volume

let $P_{wl} = \left(3.9931 \ 10^{17}\right)\left(1 - Q^2\right)xg$ = water level pressure ;

let $P_s = \left(3.9931 \ 10^{17}\right)\left[1 - \left(Q^2 + \Delta Q^2\right)\right](x + \Delta x)g$ = soliton wave pressure

$$P_{wl} = P_s$$

This equality reduces to the following[118] :

$$\Delta Q^2 = \rho_{ab}\frac{\Delta x}{x + \Delta x} = \rho_{ab}\frac{\text{(wave height)}}{\text{(water depth)}}$$

This equation shows that an increase in vacuum (ΔQ^2) is being exchanged for absolute water density (ρ_{ab}) in the wave "swell." Change in the quantum squared is equal to absolute density *times* wave height as a proportion of the total water depth. The wave or "swell" is being generated by an increase in vacuum between the actual water mass. This

[117] See "*The Soliton; Absolute Density vs. Quantum Vacuum* " in the Appendix
[118] Ibid.

produces no pressure differential because the volume is being increased without any increase in the mass component of the volume. The wave and its motion are no longer explained by water pressure differential.

The actual absolute density for water is miniscule. At 4° C, the density of water is 1000 kg/m³. The absolute density is 2.50 (10⁻¹⁵) kg/m³. Wave height would be a function of 3.3391(10¹⁴) *times* change in volume. Because the absolute density of mass is so low, very small changes in vacuum volume would produce great changes in wave height. The changes in quantum vacuum are miniscule.

How could such changes in vacuum volume be made at all? In the case of Russell's initial observation of a soliton, the wave was generated by the sudden stopping of a canal boat. The deceleration of mass over a distance releases energy:
Energy=(mass)(deceleration)(distance).=(Force)(distance)=Fd

This energy of deceleration was absorbed by the quantum vacuum composing the water. The force of the mass of the boat is exchanged with quantum vacuum by the laminar density plane in the quantum squared unit. The volume of water was sequentially "bulged" by units of the quantum squared in pulses sent along the vector of force.

The Quantum Squared Unit of Volume

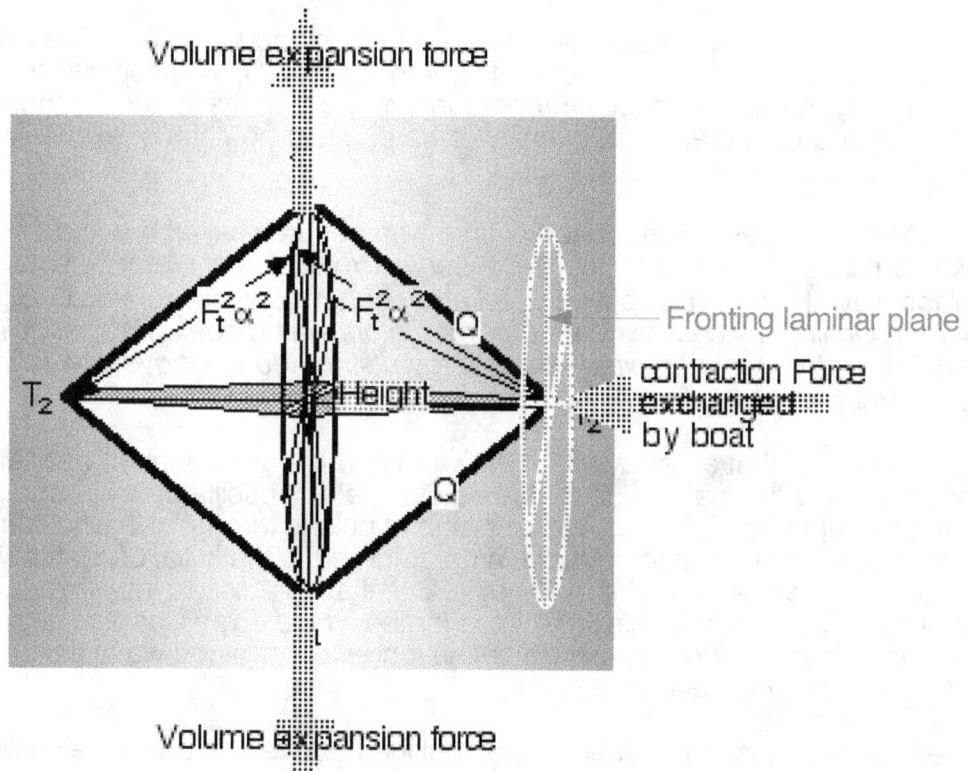

It can be mathematically demonstrated that, if you push on the peak of the above cone compressing the height while keeping the slant height "Q" constant, the volume will increase. The actual static Pythagorean values of the base radius and cone height are the following:

$$Q^2 = \frac{Q^2}{2^2} + \left(1 - \frac{1}{2^2}\right)Q^2 \; ; \; radius^2 = \frac{Q^2}{2^2} \; ; \; height^2 = \left(1 - \frac{1}{2^2}\right)Q^2; \; Volume = 0.227\ Q^3$$

The cone height is the negation of subdivision for the quantum squared. The circle radius is the subdivision by which the quantum squared is negated.

The quantum squared is under tension because the quantum "Q" is composed of two points separated by time potential energy (designated "$F_t^2 \alpha^2$[119]"). Force applied against this tension shortens the height of the cone and increases the radius of the base. This increases volume. An example of shortened height increasing volume is the following:

$$Q^2 = \frac{Q^2}{1.5^2} + \left(1 - \frac{1}{1.5^2}\right)Q^2 \;;\; radius^2 = \frac{Q^2}{1.5^2} \;;\; height^2 = \left(1 - \frac{1}{1.5^2}\right)Q^2 \;;\; Volume = 0.347\, Q^3$$

The cone height has been compressed from ".866 Q" (radius Q/ 2) to ".745 Q" (radius Q/ 1.5). The volume has been increased from ".227" to ".347." The reduction in height has been ".121 Q" or -14%. This results in an increase in volume of ".12 Q^3" or $+52.9\%$. A small reduction in height produces a relative large increase in volume.

The Major Difficulty with the Quantum-Pulse Theory is Time

A pulse in vacuous volume is being argued as the motivational source of Russell's single water wave. It is the same pulse which would conduct light through the vacuum of space. The problem is obvious. The pulse which conducts radiation at the speed of light is said to conduct a water wave at Russell's 14 kilometers per hour.

It is true that the speed of light through water is less than in vacuum. Water has an index of refraction of 1.33 which restricts the speed of light to 75.19% C. The suppression of the speed of light in any material is loosely proportional to the material's density, but the relationship is nonlinear. The absolute density of water could not, alone, explain the observed velocities of soliton water waves.

Before I am accused of wasting the reader's time with absurdities, let me show how quantum vacuum pulses could produce the observed velocities of water solitons. In the 1950's, soliton wave theory was applied to an array of harmonically related oscillators.[120] *"64 weakly and non-linearly coupled harmonic oscillators was modeled numerically. Starting with only one oscillator [being] excited, the energy distributed itself over the whole mode system but returned almost completely to the first excited one.*[121]*"*

The experiment proved that an array of "weakly and non-linearly coupled" oscillators superimposed a new wave pattern upon that established by oscillator in linear alignment. The pulse from a single oscillator initiated a spread of pulses through the whole array of oscillators. The "wave" of pulses through the array returned to the initial oscillator which began the sequence over again. The time between initiating pulse to initiating pulse was a "wave time" for the whole array. This time value for the whole array was a multiple of individual oscillator times. Individual oscillators no longer determined wave time. The array determined a much slower wave time value.

Consider the individual quantum squared "cell" outlined above to be a single oscillator in an array of "non-linearly coupled" oscillator cells. These vacuum oscillator cells must act differently when interacting with the water than when interacting with radiation forces impinging upon the water from outside.

---

[119] See *The Theory of Time Enforced Quantum Space* in Appendix

[120] *Studies of Non Linear Problems* in: *Collected Papers of Enrico Fermi*, general editor E.Segre, *University of Chicago Press*, 1965, Vol. II, p. 978

[121] Vongehr, op. cit.

Radiation is always linear with respect to water. Light can only invade when the water aligns quantum cells to allow such penetration. The effect can be seen when small waves on the surface of water alternately reflects light and allows the penetration of the light, depending upon the angle of surface at any moment relative to the source of the light. Invading light is linear and only linear as defraction — or the straight-line bending of light — so eloquently testifies.

As components of the water, quantum cells are aligned by the motion of the water. They must meet the condition of being *"weakly and non-linearly coupled harmonic oscillators"* when aligned by turbulent flow. Quantum cells composing the space separating water mass can never be linearly aligned while the water is non-laminar and turbulent. By definition, a water wave is turbulent and, therefore, must be composed of non-linearly aligned quantum cells.

Because the dimensions[122] of the individual quantum squared cell are known, the number contained in a cubic meter of water can be calculated:

$$1\,m^3 = \left(4.4106 \ 10^{45}\right) Q^2 \text{ units}$$

$$1\,Q^2 \text{ unit} = \left(2.2672 \ 10^{-46}\right) m^3$$

To transverse a quantum squared unit at the speed of light in water takes 2.508 (10-24) seconds. If these 4.4106 (1046) quantum squared units in a cubic meter were pulsed sequentially— one right after another— it would take 1.106 (1022) seconds (about 356 trillion years). However, such a single dimensional nonlinear sequencing is unrealistic. A three dimensional sequencing would require the cube root of this number or 22,281,364 seconds (about 8.5 months). This assumes the sequencing is occurring in three dimensions simultaneously.

The time for Russell's observed velocity was .257 seconds per meter. The three dimensional nonlinear model would take 8.66 (107) times as long as Russell's observed wave time-per-meter. Russell's observed time was 5.793 (107) times the time-per-meter of the speed of light in water. These variances between the nonlinear model time and Russell and between Russell and the speed of light are of the same order of magnitude (107). Russell's observed times are in the mean range between the nonlinear model and the linear speed of light. They are in a high probability range if velocities are distributed by the quantum squared unit of vacuous space as described.

The Statistical Regularity of Soliton Velocities Relative to Quantum Space
We have described two opposing modes by which quantum squared units can be pulsed in volume by applying force to quantum height. Those modes are the *absolute nonlinear* whereby the units are sequenced one-by-one in three dimensional directions simultaneously. The second mode was the *absolute linear* whereby quantum units are parallel pulsed along one dimensional axis of the volume or the method by which light is conducted.

The velocity of motion through a cubical meter of water could be calculated for these two modes since the volume value of the quantum squared unit is known and the number contained in a meter cubed could be determined. The velocity of the *absolute nonlinear* was determined to be 4.488 10-8 m/ sec. The velocity of the *absolute linear* is the speed of light in water or 225407863 m/sec.

_____
[122] See *"Calculating Alpha and Delta T"* in the Appendix

An interesting statistic can be developed for measured soliton velocities, using the two "polar'" velocities for *absolute nonlinear* and *absolute linear* quantum-squared pulse motion. The statistic uses the soliton as a division point along a time-of-transition scale separating *absolute nonlinear* and *absolute linear* velocities.

v = soliton velocity ; v_n = absolute nonlinear velocity ; v_l = absolute linear velocity

$$\frac{(v_n)(v_l)}{v^2} = \text{ratio [soliton time / linear time] to [nonlinear time / soliton time]}$$

$$(v_n)(v_l) = 10.116$$

"*Soliton time / linear time*" is a multiple of the time it takes *absolute linear* light to cross a meter of space. That multiple produces the time it takes a soliton wave to cross a meter of space.

"*Nonlinear time / soliton time*" is a multiple of the time it takes the soliton to cross a meter of space. That multiple produces the time it takes *absolute nonlinear* velocity to cross a meter of space.

Soliton Linearity Variance by Quantum Negative Calculus

$$D(1 - \frac{10.116}{v^2})$$

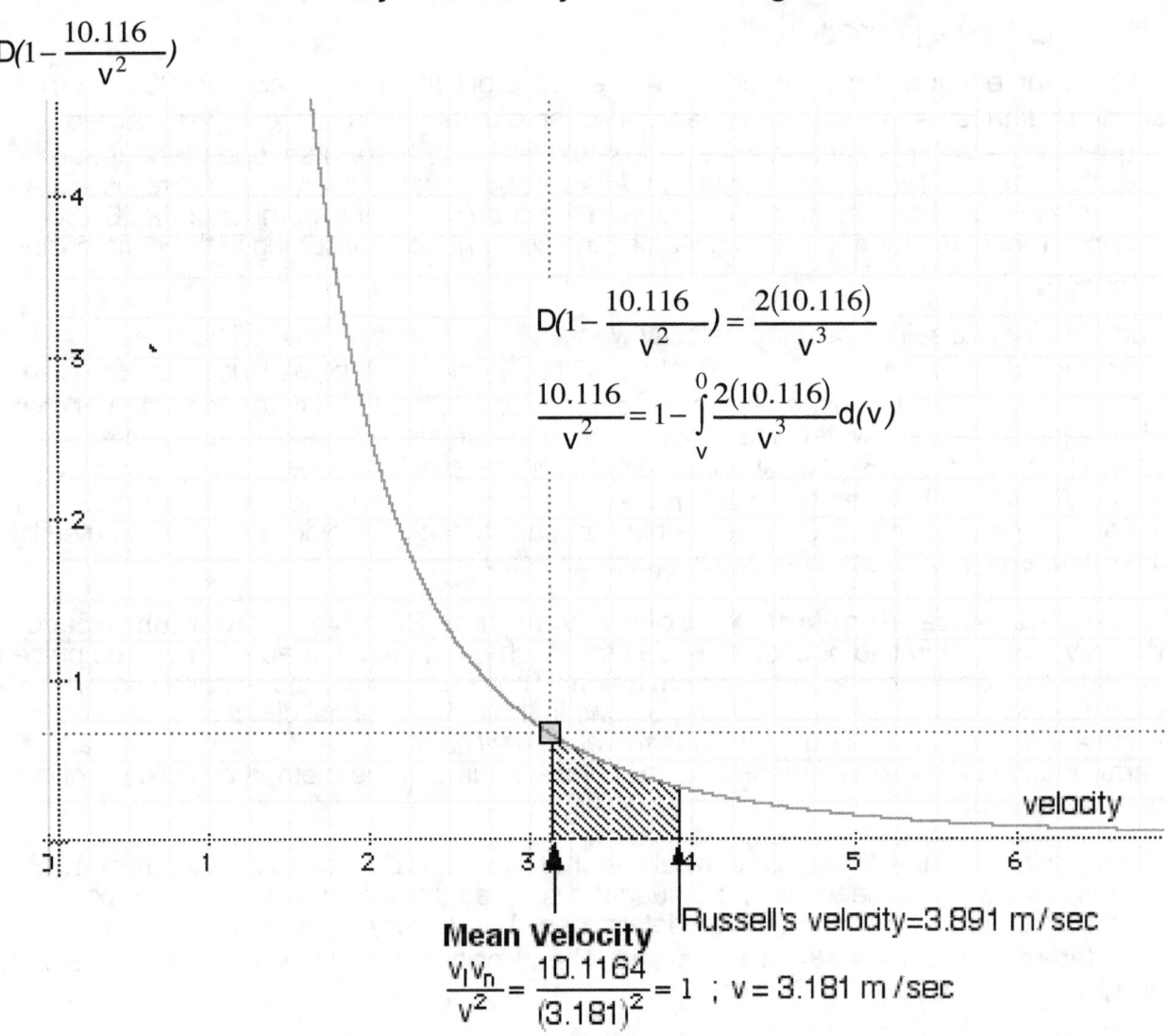

$$D(1 - \frac{10.116}{v^2}) = \frac{2(10.116)}{v^3}$$

$$\frac{10.116}{v^2} = 1 - \int\limits_{v}^{0} \frac{2(10.116)}{v^3} d(v)$$

velocity

Russell's velocity=3.891 m/sec

Mean Velocity

$$\frac{v_l v_n}{v^2} = \frac{10.1164}{(3.181)^2} = 1 \ ; \ v = 3.181 \text{ m/sec}$$

In the above graph, the velocity designated "mean velocity" is the dividing point between linearity and non-linearity. This means that *absolute linear velocity* divided by mean velocity times *absolute nonlinear velocity* divided by mean equals "1." That is: "$v_n / v_m = v_m / v_l$" The speed of light (in water) as a multiple of mean velocity is equal to *absolute nonlinear velocity* as a multiple of mean velocity.

The mean velocity of "3.181 m/ s" is the dividing point between linearity and non-linearity in the sense that velocities to the left of the mean are tending towards non-linearity and velocities to the right of the mean are tending towards linearity.

We can use quantum negative integration to develop a "linearity probability coefficient" for Russell's observed soliton velocity. The coefficient will tell us how probable the observed velocity is relative to the quantum determined mean.

Probability is determined by the shaded area in the above graph. There is a variance of only ".71 m/ sec" between mean velocity and Russell's observed soliton velocity. The shaded area between Russell's and the mean is 33.2% of the whole area under the graph to the right of mean velocity[123].

The shaded area is a probability density for "linear tending velocities." 33.2% of such "linear tending velocities" will reside within a variance of .71 m/ sec from the mean velocity. 66.8% of linear tending velocities will reside between 3.891 m/sec and the speed of light (in water) at 225,407,863. m/sec (225 million). A variance of ."71 m/ sec " accounts for 33.2% of linear tending velocities. A variance of "225,407,859 m/ sec" accounts for the remaining 66.8%.

Just to demonstrate how rapidly probability densities are falling, a velocity of "5 m/ sec" with a variance from the mean of "1.819 m/ sec" will account for 60% of all linear tending velocities. A velocity of "6 m/ sec" (21.6 km/ hour) will account for 72% of all linear tending velocities. The probability densities are the following:

Probability Density=(area under graph between mean and velocity)/ (variance)

Velocity m/ sec	% area under graph	variance m/ sec	probability density= % area graph / variance
Russell (3.891)	33.2	.71	.4676
5	60	1.819	.3299
6	72	2.819	.2554

The point is that Russell's initial observation of a fast-moving soliton wave has a high probability density relative to the mean of a linearity/non-linearity scale. This scale was developed from the theory that solitons are composed by pulses in the quantum space separating the mass of water. The velocities of all water solitons would be within the high probability density range of a quantum attained statistical mean. This would not be true of pressure differentiated sine waves, the velocities of which are a function of wavelength and the gravitational constant.

The relationship between the quantum-based soliton mathematics just outlined and the Korteweg de Vries nonlinear hydro dynamical equations[124] is rather nebulous. In the first place, Korteweg de Vries "non-linearity" is not the equivalent of the quantum linearity/non-

[123] See *"Dawson's Theorem"* in Appendix for explanation of quantum mathematics involved.
[124] D.J.Korteweg and G. de Vries, *Phil Mag.* 39, 422 (1895)

linearity scale.
The scale is premised on the dispersion of quantum-squared volume pulses. The Korteweg de Vries nonlinear operator "b'" is actually an unknown scalar.

In physics, a scalar is a simple physical quantity that is not changed by coordinate system rotations or translations.[125] Since Korteweg de Vries describes wave velocity as motion along the x axis (z=vt-x), the x axis must be "curved" to accommodate the hump of the wave. The nonlinear "b" describes this translation.

Korteweg de Vries nonlinear Velocity of Soliton

v = wave velocity; c = phase velocity ;b = non-linear operator ;ψ_{max} = wave height

$$v = c\left(1 + \frac{b}{3}\,\psi_{max}\right)$$

Korteweg de Vries holds that the wave velocity, that is the velocity "over the hump," is equal to the phase velocity (linear velocity of wave) plus a nonlinear function of the wave height. The nonlinear function gives a fraction of phase velocity which, when added back to phase velocity, gives wave speed "over the hump."

A "nonlinear operator" is an unknown or unrecognized algebraic factor which prevents direct equality manipulation of the equation. Wikipedia defines it as follows: *"a nonlinear system is any problem where the variable(s) to be solved ... cannot be written as a linear combination of independent components."*

While the nonlinear operator must be left with Korteweg de Vries, quantum soliton mathematics can be integrated with respect to wave height:

The change in vacuous volume which constructs the soliton has been shown to have wave height as one of its factors. Therefor, wave height "Δx" can be written as a function of change in volume, water depth and the constant absolute density:

$$\Delta Q^2 = \rho_{ab}\,\frac{\Delta x}{x + \Delta x} \quad ; \quad \Delta x = \Psi_{max}$$

$$\Delta x = \frac{\Delta Q^2 (x + \Delta x)}{\rho_{ab}}$$

This quantum formula for wave height can be applied to Korteweg de Vries:

$$v = c\left(1 + \frac{b}{3}\,\Delta x\right) = c\left(1 + b\,\frac{\Delta Q^2 (x + \Delta x)}{(3)\rho_{ab}}\right)$$

This substitution forces Korteweg de Vries onto the microscopic quantum level. The absolute density for water —used as a divisor — is "$2.50(10^{-15})$ kg/m^3." This is the order of magnitude of the smallest known wavelengths and the diameter of the proton/neutron. . Yet the change in vacuous volume "ΔQ^2" must be smaller still since it must modify water depth to produce wave height. An infinitesimally small "bulging" of quantum squared volume is required to produce wave height.

This is as it should be because the value of the basic quantum "α" measures 10^{-15} meters, the same order of magnitude as the absolute density by which "ΔQ^2" is divided. The mathematics are now operating completely on the quantum level. Wave height is being

[125] Wikipedia

explained by minute changes in the "10^{-15} meter" quantum.

Summary of the Quantum Induced Soliton

The unitary "soliton" water waves first observed and measured by J. Scott Russell have been shown to be the distributions of quantum-squared volume pulses. When linearly aligned these pulses conduct radiation at light speeds. But when incorporated as part of the mass of water, they can become a nonlinear distribution under the turbulence caused by energy exchanges between solid mass and the water.

They become the *"weakly and non-linearly coupled harmonic oscillators"* identified and modeled by Enrico Fermi[126] . This nonlinear coupling produces a wave through the array of oscillators with a time value much slower than the oscillators themselves.

By modeling quantum squared units of volume as Fermi's nonlinear coupled oscillators, a velocity value for *absolute non-linearity* could be calculated. When compared with the speed of light in water as the *absolute linearity* velocity, a statistical mean between the two absolutes could be calculated. This mean was located within the range of observable water soliton velocities. Russell's observed velocity was shown to have a very high probability density in a single-tail comparison of the population of possible velocities as established by the mean.

A formula was also developed which related soliton wave height to changes in the quantum volume separating the mass component of water and with the absolute density of water. This formula was partially integrated with the conventional formula used in soliton geometry, the Korteweg de Vries equations of 1898. The integration forced Korteweg de Vries into a microscopic quantum order of magnitudes.

The quantum mathematical descriptions of the soliton are much more precise and possess predictive powers not shared by Korteweg de Vries. By this fact alone, they must be accepted as the actual descriptors of the soliton phenomenon. The soliton wave is as quantum geometry says it is.

Solitons are the quantum-induced wave forms sent through vacuous space as incapsulated by turbulent water. They are produced by an energy exchange between decelerating matter and the vacuous space within the framework of that water. The first evidence for the quantum structure of vacuous space, the structure which allows it to conduct wave form energy, was acquired by the nautical engineer, J. Scott Russell, in 1834. However, 174 years would have to pass before the significance of Russell's observation could be revealed.

---

[126] *Studies of Non Linear Problems in: Collected Papers of Enrico Fermi* op. cit.

The Quantum Curvature of Space vs. An Expanding Universe
comparisons by Hubble's original redshift data

The best evidence that linear Euclidean distances in space becomes "kinked" upward to become curved by an intersecting quantum dimension may be the redshift in light frequencies reaching us from far galaxies. The linear distance between us and the source of the light is "kinked" into curvature and light follows the greater curved distance rather than the linear distance producing a redshift "Z". Wavelengths are "stretched" when distances are increased in a single dimension.[127] Light follows the curved arc between "E" and "G" in the strictly Euclidean illustration below.

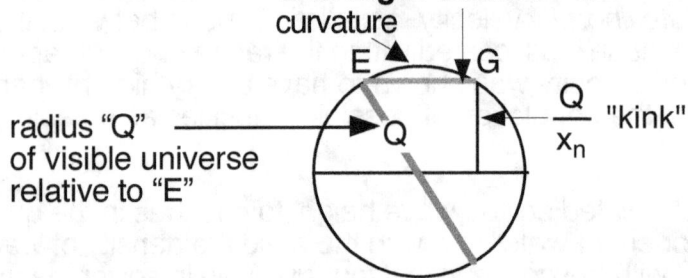

This Euclidean illustration is not an accurate depiction of the quantum curvature of space. The astronomer Edwin Hubble provided the actual mathematical description of curvature, although he never fully understood his contribution. Hubble failed to recognize that the quantum curvature of space gives the universe an appearance of expansion without actually expanding it; that curvature generates an apparent variance in the velocity of light which is the exact equivalent of a recession velocity. Hubble's Constant identifies how the apparent variance in the velocity of light due to curvature varies with distance.

The expanding universe concept is built upon Hubble's discovery that the redshift measured from foreign galaxies is a function of the measured distance to the galaxies and fit a Doppler Effect explanation of the redshift. The Doppler Effect formula[128] for redshift "Z[129]" is the following:

$$Z = \frac{c}{c - v} \quad ; \ c = \text{speed of light} \ ; \ v = \text{velocity of recession}$$

$$cZ - c = vZ$$

$$v = \left(1 - \frac{1}{Z}\right) c$$

---

[127] This is the principle governing trombones and other wind instruments which are lowered in pitch by lengthening the distance of the air passage.

[128] $f_{rs} = f_o \left(c - H_o d\right)/c$; f_{rs} = redshifted frequency ; f_o = original frequency. Taken from the universal formula for the Doppler Effect.

[129] Z=(redshifted wavelength) / (original wavelength)

Hubble's Constant identifies the velocity of recession "v" as a function of the distance to the particular galaxy, hence redshift is also a function of the distance to the galaxy:

Hubble's Constant = H_o ; d = distance to galaxy in mega parsecs (Mpc).

$$v = H_o d$$

$$Z = \frac{c}{c - H_o d}$$

Hubble didn't realize that the apparent change in velocity for the speed of light "c" due to the forced curvature of a linear distance is nearly the same as the Doppler equation for recession velocity, especially for the close distances of his data table. For the curvature hypothesis, redshift "Z=(curvature distance)/(linear distance)." Both curvature distance and linear distance are measured between the observation point and the light-source:

Redshift by Quantum Curvature and its "Apparent Velocity of Recession"

$d = \{linear\ distance\ to\ light\ source\}$; $\chi = \{curved\ distance\ to\ which\ "d"\ is\ kinked\}$; $Z = \frac{\chi}{d}$; $\{Redshift\} = Z(\lambda)$

$t_1 = \{time\ across\ linear\ "d"\ at\ "c"\}$; $t_2 = \{time\ across\ curved\ "\chi"\ at\ "c"\}$; $\Delta v = \{apparent\ recession\ velocity\}$

$d/t_1 = c$; $\chi/t_2 = c$; $d/t_1 - \Delta v = d/t_2$; $\Delta v = d/t_1 - d/t_2$

$$\frac{d}{t_1} = c = \frac{9.715612e\text{-}15\ Mpc}{sec.}\ ;\ t_1 = \frac{d}{9.715612e\text{-}15\ Mpc}\ ;\ \frac{\chi}{t_2} = c = \frac{9.715612e\text{-}15\ Mpc}{sec.}\ ;\ t_2 = \frac{\chi}{9.715612e\text{-}15\ Mpc}$$

$$\Delta v = \frac{d}{d/9.715612e\text{-}15\ Mpc} - \frac{d}{\chi/9.715612e\text{-}15\ Mpc} = \left(1 - \frac{d}{\chi}\right)9.715612e\text{-}15\ Mpc = \left(1 - \frac{1}{Z}\right)c$$

That is, the apparent change in velocity " Δv " for the speed of light in a static, curved-space universe might be mistaken for Doppler recession velocity "v" in an expanding universe.

It is generally not recognized that Hubble's Constant —as used to convert redshift to the distance to the light source — is actually a time value (*distance/velocity=time*[130] ; as above). Hubble's Constant is measured in "velocity *per* unit of distance." Specifically, it is velocity in "kilometers *per* second (km/ sec)" *per* distance in "megaparsecs (Mpc[131])":

d = distance ; v = velocity ; t = time

$$H_o = \frac{v}{d}\ \ ;\ \ \frac{d}{v} = t\ \ ;\ \ \frac{v}{d} = \frac{1}{t}\ \ ;\ \ H_o = \frac{1}{t}$$

$$v = H_o d = d/t$$

Hubble's Constant is generally not thought of as a time value because its units of measure are different for "velocity (kilometers *per* second)" and "distance (megaparsecs)." Its application is thought restricted to "recession velocity" for an expanding universe as so many "kilometers/ second" *per* "megaparsec." Distance can be easily found by converting redshift to recession velocity (by Doppler formula) and dividing it by Hubble's Constant. The different units of measure are irrelevant to this conversion:

$$d = \frac{c}{H_0}\left(1 - \frac{1}{Z}\right)\ \ \ \ \{see\ formula\ on\ opposite\ page\}$$

The fact that Hubble's Constant is a time value becomes extremely significant because it supplies a time period during which the universe has been expanding and, therefore, an alleged time since the hypothesized "Big Bang."

---

[130] A common formula for velocity, distance and time.

[131] 1 Mpc=3 .261 63626 10^6 light years =3.08568025 10^{19} kilometers.

95

Convert Hubble's Constant from "kilometers *per* second" to "Mpc per second":

$$H_o = \frac{500\,km/\sec}{Mpc} = \frac{1.6204e\text{-}17\,Mpc/\sec}{Mpc} = \frac{1.6204e\text{-}17}{\sec} = \frac{1}{t} \quad \text{[132]}$$

$$t = (6.1713157245e16 \text{ seconds}) = (1.955572e9 \text{ years})$$

$$\{\text{Time of the universe's expansion is approximately 2 billion years}\}$$

$$\{recession\ velocity\} = H_o d = \frac{d\,(in\ Mpc)}{(6.1713157245e16 \text{ seconds})}$$

$$\begin{Bmatrix} \text{Velocity equals the distance of the stellar light source from the Earth} \\ \text{divided by the amount of time the universe has been expanding.} \end{Bmatrix}$$

The "Apparent" Recession Velocity of the Curved Space Model

Recession velocity of Hubble's "Big Bang" cosmology is the distance to the foreign stellar object divided by the time since the beginning of the universe's expansion. The "apparent recession velocity" of the static, curved space model is the change in velocity calculated by light traveling the path of "kinked curvature" rather than the direct path. That change in velocity is equal to the negation of subdivision for the speed of light with the negation factor being the "Z" factor. This is exactly the same formula as for Doppler redshifting velocity.

$$t_1 = \{time\ across\ linear\ "d"\ at\ "c"\}; \quad t_2 = \{time\ across\ curved\ "\chi"\ at\ "c"\}; \quad \Delta v = \{apparent\ recession\ velocity\}$$

$$d/t_1 = c; \quad \chi/t_2 = c; \quad d/t_1 - \Delta v = d/t_2; \quad \Delta v = d/t_1 - d/t_2 = \left(1 - \frac{d}{\chi}\right)9.715612e\text{-}15\ Mpc = \left(1 - \frac{1}{Z}\right)c$$

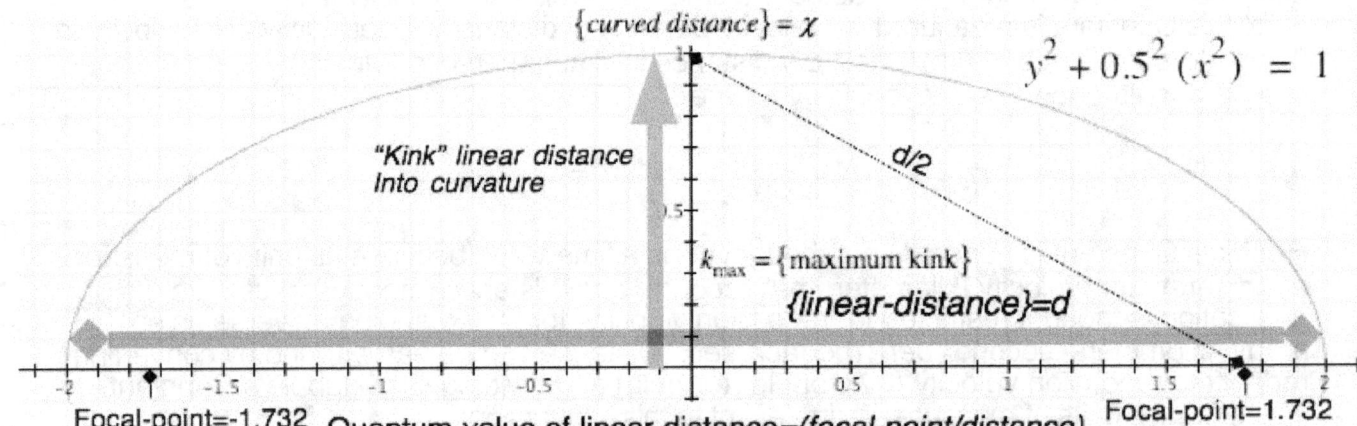

$$\{curved\ distance\} = \chi$$

$$y^2 + 0.5^2\,(x^2) = 1$$

"Kink" linear distance
Into curvature

$d/2$

$k_{max} = \{maximum\ kink\}$

$\{linear\text{-}distance\} = d$

Focal-point=-1.732   Quantum-value of linear distance=(focal-point/distance)   Focal-point=1.732

The Quantum Law of the Elliptical Curvature of Space

$$\varepsilon = \{elliptical\ eccentricity\} = \frac{2(\phi)}{d} = \{\text{Quantum-value of linear distance}\}; \quad \phi = \{elliptical\ focal\text{-}point\}$$

$$\varepsilon^2 = \frac{4(\phi^2)}{d^2}; \quad \phi^2 = (d/2)^2 - k_{max}^2; \quad \varepsilon^2 = \frac{4\left[(d/2)^2 - k_{max}^2\right]}{d^2} = 1 - \frac{4k_{max}^2}{d^2} = \frac{4(\phi^2)}{d^2}; \quad \frac{d^2}{4} - \phi^2 = k_{max}^2$$

---

[132] 1 km = 3.24077929e-20 Mpc

The Quantum Limit on Visibility across the Universe

The curvature of space restricts the distance an object may be seen from any single viewpoint. All cosmological distances are "kinked" into curvature and the maximum ratio of curvature to linear distance ("Z") is "$\pi/2=1.5708$." Greater distances cannot be kinked into greater curvature. This establishes the limit at which an object is visible and provides the maximum macro quantum value.

$$\varepsilon = \left(1 - \frac{d^2}{Q_{Max}^2}\right); \quad Q_{Max} = \{Maximum\ visible\ distance\ across\ universe\ from\ any\ viewpoint\};$$

$$\frac{d}{Q_{Max}} = \{light\ source\ distance\ as\ \%\ of\ "Q_{Max}"\}$$

$$\varepsilon = \{elliptical\ eccentricity\} = \{quantum\ value\ of\ distance\ "d"\} = \frac{2(elliptical\ focal\text{-}point)}{d}$$

$$\varepsilon^2 = 1 - \frac{d^2}{Q_{Max}^2} = \frac{4\phi}{d^2} = \frac{4\left[(d^2/4) - k_{Max}^2\right]}{d^2} = 1 - \frac{4k_{Max}^2}{d^2}$$

$$1 - \frac{d^2}{Q_{Max}^2} = 1 - \frac{4k_{Max}^2}{d^2}; \quad \frac{4k_{Max}^2}{d^2} = \frac{d^2}{Q_{Max}^2}; \quad \frac{k_{Max}^2}{d^2} = \frac{d^2}{4Q_{Max}^2} \quad [133]$$

$$2\chi = \{circumference\ of\ ellipse\}; \quad r_1 = \{minor\ axis\} = k_{Max}; \quad r_2 = \{major\ axis\} = d/2$$

$$\chi = \sqrt{3k_{Max}^2 + (d/2)^2}\left(\frac{2k_{Max}}{\sqrt{k_{Max}^2 + 3(d/2)^2}}\left(\frac{\pi-3}{3}\right) + 1\right) + \left[k_{Max}\left(\frac{\pi-3}{3}\right) + d/2\right] \quad [134]$$

With a Known Maximum Quantum, Redshift "Z" is a Function of Distance "d"

$$Z = \frac{\chi}{d} = \sqrt{\frac{3}{4}\left(\frac{d}{Q_{Max}}\right) + \frac{1}{4}}\left(\frac{2d}{\sqrt{d^2 + 3Q_{Max}}}\left(\frac{\pi-3}{3}\right) + 1\right) + \left(\frac{d}{2Q_{Max}}\left(\frac{\pi-3}{3}\right) + \frac{1}{2}\right)$$

Hubble's Constant is Revised, without Data, to give a Longer Age of the Universe

Since Hubble's death in 1953, the "Constant" has been continuously adjusted downward from Hubble's empirically determined "500 km/ s/ Mpc" to the current estimate of "65-50 km/ s/ Mpc.[135]"

In 1956, Allan Sandage, Hubble's successor at the Mt. Wilson and Palomar Observatories, began the revisions downward. Sandage revised Hubble's "500 kg/ s/ Mpc " to "180 km/ s/ Mpc." In 1958 Sandage published a value of "75 km/s/Mpc," and by the early 1970's estimates from Sandage and his longtime collaborator Gustav Tammann were hovering around "55 km/s/Mpc[136]," or very near the modern accepted range.

In the National Institute of Standards and Technology report "CODATA Recommended Values of

[133] The maximum "kink" is a direct function of source distance and visible maximum quantum across the universe.
[134] "The Quantum Formula for the Perifery of an Ellipse;" Dawson,L http://paradigmphysics.com/masters-thesis.pdf
[135] Harvard University web page "Hubble's Constant."
[136] ibid.

the Fundamental Physical Constants: 2006"[137] the authors give two reasons for the revision of scientific constants besides improvement in measurement and lab techniques. One of these is "time variation of the constant[138] " However, they admit that "there has been *no laboratory observation of time dependence* of any constant that might be relevant to the recommended values" (italics mine). The Hubble revisions are "time variation" revisions without "laboratory observation."

Light redshift due to quantum curvature produces an "apparent recession velocity" because of the variance in time across the curved pathway relative to time across the linear pathway. This "apparent recession velocity" is the equivalent of Hubble's recession velocity in that it resolves to the same formula as the negation of subdivision of the speed of light with the negation factor being red-shifting "Z." The "Z" factors are not equivalent due to the time variations for expanding vs curvature theory.

$$\{recession\ velocity\} = \left(1 - \frac{1}{Z_{Hubble}}\right)c \neq \{apparent\ recession\ vel.\} = \left(1 - \frac{1}{Z_{curvature}}\right)c$$

$$Z_{Hubble} \neq Z_{curvature}; \quad 1/t_{actual\ rec.} = H_o; \quad 1/t_{appear\ rec.} = \left(t_{curve} - t_{linear}\right)/\left(t_{curve}\right)\left(t_{linear}\right); \quad v_{rec.} = d/t$$

Both the static, quantum curvatured universe and the expanding, Doppler Effect universe provide the same redshift mathematics. The great divide between the two hypothesis —and the only effective test of them— is in their distinct treatments of time.

The hypothesis that the volume of vacuous space is determined by the quantum squared —with redshift being explained by the forced curvature of space— replaces Edwin Hubble's expanding universe concept. Although the quantum curvature model supplies an "apparent recession velocity," it can be seen that the curvature model will predict a different "Z" value than Hubble's model when applied to the Doppler effect formula. This is so because the two models will supply different time values for the same distance. The time value of Hubble's recession velocity is a constant while curvature time is a variable.

Redshift can be explained as Doppler Effect for both. Using the standard Doppler Effect formula[139] , the difference between the Hubble and curvature "Z" values is the following:

$$Z = \frac{c}{\left(c - v_{rec}\right)}; \quad Z_{Hubble} = \frac{c}{\left(c - \dfrac{d}{t_{expansion}}\right)}; \quad Z_{curvature} = \frac{c}{\left(c - \dfrac{d}{\left(t_{curve}\right)\left(t_{linear}\right)/\left(t_{curve} - t_{linear}\right)}\right)}$$

Hubble's "Z" is a direct function of distance "d" because recession velocity time is a constant. It is the time of the universe's expansion. For the curvature model, "apparent recession time" is a function of both "d" and the increase in curvature as "d" increases. At some point, quantum "Z" will increase faster than Hubble's at equivalent distances. Even so, Hubble's original distance-exaggerating time constant was later thought to be too low for the assumed age of the universe[140].

The Proposed Geometry of Hubble's "Big Bang" Cosmology

Hubble's "Big Bang universe" requires expansion over time in order for the universe to have reached its current size. As the distance to any galaxy "G" increased relative to us its velocity of recession also increases. Current velocity is determined by the time of expansion.

---

[137] National Institute of Standards and Technology, Gaithersburg, Maryland 20899-8420, USA; report authors: Peter J. Mohr , Barry N. Taylor , and David B. Newell:
http://physics.nist.gov/cuu/Constants/codata.pdf

[138] *Ibid.* Page 5

[139] $f_{rs} = f_o\left(c - H_o d\right)/c$; f_{rs} = redshifted frequency ; f_o = original frequency.

[140] Harvard web site. *op. cit.*

Hubble's "Big Bang" as Curvature Around an Expanding "4-D" Radius

$$r = \frac{\Delta r}{sec.}t \quad ; t = \text{time of expansion} \quad d = \frac{\theta}{2\pi}2\pi r = \theta(r) = \theta\left(\frac{\Delta r}{sec.}t\right); \quad \frac{d}{t} = \frac{\theta\Delta r}{sec.}$$

$$\frac{d}{t} = \{\text{recession velocity}\} = H_o(d) = \frac{\theta\Delta r}{sec.} \quad \langle\text{See pg. 96}\rangle$$

Space said to curve back upon itself.

Fourth dimensional radius, "r," around which universe is said to be expanded

Distances between stellar formations and our view represent expansions from the singularity point over the time which the universe has been expanding. Time is "1/(Hubble's Constant)."

Radial Expansion Velocity is Calculable from Model

$$\langle\text{Maximum expansion is at "}\theta = \pi."\rangle \quad \frac{d}{t} = c = \frac{\pi(\Delta r)}{sec.}; \quad \frac{\Delta r}{sec.} = \frac{c}{\pi} = 9.5426903e7 \ meters \ / \sec.$$

Distance "d" is the Integral of Expansion Velocity over the Time of Expansion

$$D(d) = \frac{\theta\Delta r}{sec.}; \quad d = \int_0^t \frac{\theta\Delta r}{sec.}d(t) \quad t = \frac{1}{(Hubble's \ Constant)} = \frac{1}{H_o}$$

Any two points in Hubble's universe have a fixed angle of curvature "θ." This angle remains constant as the distance between the two points expands. The velocity of recession between the two points also has a fixed rate because the expansion of the 4-D axis "r" is at a fixed velocity. This angle *times* fourth dimensional radial expansion velocity *times* the time since the beginning of the universe *equals* the distance factor for the recession velocity. That is, current distance "d" is the integral of radial velocity *times* the angle as integrated over the time since expansion commenced.

"Hubble's Constant" is not actually a *"constant."* It is a *"variable."* It is the time factor since the "Big Bang" and, as such, it is only a *"contemporary value."* That is, Hubble's "Constant" will change as time progresses.

As the time since the alleged "Big Bang" increases, Hubble's "Constant" will get smaller because the constant is actually equal to "1/ t." As time increases, the constant will get smaller. Hubble's "Constant" was actually an empirically determined estimate of the time since the alleged "Big Bang" as based upon measured stellar distances and the measured redshifting of their light.

Hubble had used a recently discovered method of determining stellar distances greater than those which could be determined by standard parallaxes. Parallaxes uses the visual angles to nearby stars from opposite sides of the earth's orbit to triangulate distances. However, the limit on parallaxes is "0.0001 Mpc" which is much too close to acquire valuable redshift data.

Hubble's Expanding Universe Hypothesis and his Empirical Estimation of the Time of Expansion using Stellar Distance and Redshift Data

In the early 1900s, the Harvard Observatory astronomer, Henrietta Leavitt, had discovered that the pulsation periods of variable Cepheid stars were directly related to their absolute luminosity. Using the Leavitt Cepheid brightness relationship, Hubble was able to measure distances up to 2 Mpc which were "20,000 *times*" greater than parallax maximum distances.

In a study of periodic variables in both the Large Magellanic Cloud as well as the Small Magellanic Cloud, Henrietta Leavitt had classified 47 Cepheids of measurable periodicity[140B]. From this data, she recognized that the brightness of the Cepheid was related to its periodicity. The longer the period of the Cepheid pulse, the greater the luminosity. From this discovery, Leavitt had developed her periodicity to luminosity table.

It was to others that the realization that Leavitt's discovery could measure stellar distance would fall. The Danish astronomer, Ejnar Hertzsprung, first used Leavitt's data to calculate the distance to Cepheids in the Small Magellanic Cloud (SMC) using the Law of the Inverse Square to compare Leavitt's absolute luminosity with the apparent luminosity from our view.

The American, Harlow Shapley statistically recalibrate the Cepheid absolute magnitude scale using 230 Pulsating Cepheids located in globular star clusters ("*extremely remote and highly concentrated stellar systems, arranged in a spherical form and consisting of tens of thousands of stars*").[140C] Shapley's large number of globular cluster Cepheids revised the absolute luminosity scale and modified Hertzsprung's distance calculations.

Although Shapley's globular cluster data is best known for correctly identifying the shape and magnitude of our Milky Way galaxy, his data also extended the stellar distances to which absolute magnitude could be applied. He used his more accurate Cepheid luminosity calculations to identify the absolute luminosity of the brightest stars of globular clusters which contained Cepheids of known luminosity. This extended the range of known luminosity to global clusters which were too distant to view detectable Cepheid periodicity. By using globular luminosities, Shapley was able to identify the actual scale of the Milky Way galaxy for the first time (currrently about 30 Kpc or 100,000 light years).

In 1924, Hubble was using the absolute luminosity scales to measure distances of "0.276 Mpc" or many times the distances measured by Shapley in determining the size of our galaxy. By 1929, Hubble was able to push "absolute luminosity" stellar distance measures to "2 Mpc" or 7.25 *times* his 1924 measurements. This allowed him to construct a 24 entry data table of Cepheid absolute luminosity distances which ranged from the SMC's "0.032 Mpc" to the "2 Mpc" of the galaxies "4382, 4472, 4486 and 4649." These distances gave a testable range against measured redshift of light which could then be compared with distance.

Hubble's data comparing redshift with distances as determined by absolute luminosity did not establish an absolute relationship due to variations in "peculiar velocities" (velocities which are determined by gravitation influence). However, relationship between measured distance and measured redshift still suggested the possibility of an expanding universe.

[140B] *"Cepheid Variable Stars & Distance Determination;"* The Australia Telescope National Facility...
http://www.atnf.csiro.au/outreach/education/senio/astrophysics/variable_cepheids.html
[140C] *"Shapley, Harlow "* Complete Dictionary of Scientific Biography, 2008, Charles Scribner's Sons.
http://www.encyclopedia.com/topic/Harlow_Shapley.aspx

A problem occurred when Hubble's time of expansion, as determined by his data, proved incompatible with the earth's age as determined by the radioactive decay of rocks[141].

Hubble's empirically-determined value for his "constant" was $1.95556e9$ years. His theory had determined a time "constant" for which the universe was expanding of approximately 2 billion years. However, the half-life of radioactive decay in rocks which were encrusted in the earth measured four billion years since the radioactive material had been deposited..

After Hubble's death, the constant was shifted downward 90% so that "time since Big Bang" would be increased by an equivalent 90% :

$$H_o = 1/t; \quad \{ Let \text{ "}H_o\text{" become "70 } km/sec/Mpc \} \quad (70 \ km/sec/Mpc) = 2.2687e - 18 = 1/t; \quad t = 13.97e9 \ years$$

To reduce Hubble's Constant by 90%, increases "distance" by 90% and "time" by 90%, since time and distance must vary proportionally according to the Expanding Universe model.

If the Hubble constant, as determined by actual measurement, is incompatible with other-source time measurements, science should have recognized this fact as a defect in the expanding universe theory. However, since they had no alternative to an "expanding universe," they responded by deserting Hubble's empirical foundations for his constant in favor of a better time "fit."

The time-scheme shifting of Edwin Hubble's "constant" downward began only after his death in 1953. Hubble remained faithful to his measurements —to the data by which he had established his constant and recession time frame.

The 1929 Hubble Data Table Presumes Doppler Redshift and Estimates "H_o"

Object Name	Dist. (Mpc)	Vd. (km/s)	Object Name	Dist. (Mpc)	Vd. (km/s)	Object Name	Dist. (Mpc)	Vd. (km/s)
SMC	0.032	+170	5194	0.5	+270	1055	1.1	+450
LMC	0.034	+290	4449	0.63	+200	7331	1.1	+500
6822	0.214	-130	4214	0.8	+300	4258	1.4	+500
598	0.263	-70	3031	0.9	-30	4151	1.7	+960
221	0.275	-185	3627	0.9	+650	4382	2.0	+500
224	0.275	-220	4826	0.9	+150	4472	2.0	+850
5357	0.45	+200	5236	0.9	+500	4486	2.0	+800
4736	0.5	+290	1068	1.0	+920	4649	2.0	+1090

SMC = Small Magellenic Cloud; LMC = Large Magellenic Cloud; All object numbers are preceded by "NGC." 1 parsec = 3.26 light years; 1 Mpc = megaparsec = 10^6 parsecs.

Edwin Hubble calculated his constant value by his 1929 data table[142] arraying redshift measurements from nearby galaxies to the distances which were measured by an

---

[141] Harvard web site. *op. cit.*
[142] *"A Relation between Distance and Radial Velocity among Extra-Galactic Nebulae"* by Edwin Hubble. 1929 PNAS Vol 15, Issue 3, pp. 168-173

independent method. Hubble primarily used individual Cepheid[143] stars he had detected in the foreign galaxies.

Characteristically, Hubble data tables present redshift measurements as converted to "recession velocities." The formula for this conversion is absolutely quantum. Specifically, apparent recession velocity is the negation of the subdivision of the speed of light using the redshift value "Z" as the subdivisional unit:

$$v = \left(1 - \frac{1}{Z}\right) c \quad \textit{as "Z" increases, "v" is a larger percentage of "c."}$$

The measurement of redshift is also characteristically quantum. The primary index is the Rydberg visible frequency (Balmer Series) absorption lines for hydrogen, hydrogen being the primary stellar material. The absorption lines for the Balmer Series are harmonically regular. In star light, the two highest frequencies in the Balmer Series (n'=8, n'=7) are output as light. The four lowest frequencies in the series (n'=6, n'=5, n'=4, n'=3) are "absorb light frequencies" (wavelengths missing in spectrum) :

Balmer Series Formula

$$\left(\frac{1}{2^2} - \frac{1}{n'^2}\right) \frac{1}{(\lambda_r = 91.14 \text{ nm})} = \frac{1}{\lambda}$$

Example

Calculated Wavelengths missing from spectrum	Measured Wavelengths (redshift=1.0036)
n'=6; 410.13 nm	n'=6; 411.61 nm
n'=5; 434.00 nm	n'=5; 435.56 nm
n'=4; 486.08 nm	n'=4; 487.83 nm
n'=3; 656.21 nm	n'=3; 658.57 nm

When Balmer Series missing wavelengths (absorption lines) are spectrographically measured at the wavelengths in the right column above, redshift "Z" is determined by dividing the measured wavelengths by the calculated wavelengths.

In his 1929 data table, Hubble had independent measures for distance to the galaxy (from Cepheid calculations) and redshift as measured above.

From these 24 independent determinations of redshift (given as velocity) and distance determined by Cepheids detected in the galaxies, Hubble selected approximately 14-17 cases to determine his constant value using the formula:

$$H_o = \frac{v.}{d} \quad v=Vd. \text{ and } d=Dist. \text{ in table}$$

Hubble rejected data points which were obvious aberrations to the expansion pattern. For example, both the Large and Small Magellenic Clouds give much to high Constant values (H_o =5312.5 for Small Magellenic Cloud ; H_o =8529.4 for Large Magellenic Cloud) indicating possible high-energy event redshifts.

It is crucial to recognize the difference between continuous light redshift due to distance and high energy event redshift. Scientists are currently reporting redshifts in the range of "Z=6" with gamma-ray bursts and high-energy quasar events. A redshift of "6" would shift the highest visible light wavelength in Balmer series (388.9 nm) to the mid-infrared Brackett Series (2333.4 nm). All visible frequencies would be shifted out of the visible range into mid and far infrared. Before distance interpretations are made of such high-energy redshift events, scientist should explain why hydrogen-fusion bombs are also known to cause characteristic redshifts in the light flashes released[144] .

In point of fact, redshift measurements of continuous-light sources are consistent with the

---

[143] Cepheids are pulsing, variable light-intensity stars. The period of pulse establishes a known and absolute brightness.

[144] Natural observation of the author.

quantum geometric maximum of "Z=1.571." NASA telescopes focused on continuos light sources measure such redshift ranges. The Sloan Digital Sky Survey (SDSS), is ongoing as of 2005 and aims to obtain measurements on around 100 million objects. The highest redshifts SDSS has recorded for galaxies (continuous light sources) is "Z=1.4.[145]" Further, the Two-Degree Field (2dF) Galaxy Redshift Survey of the Anglo-Australian Observatory measured redshift for 221 thousand Galaxies obtaining a maximum "Z" value of approximately 1.25. [146]

The Magellenic Cloud redshifts were much to high to explain by distance alone and were eliminated from the data.

Other data points were eliminated for a second reason. The galaxies NGC 6822, NGC 598, NGC 221, NGC 224, and NGC 3031 are not receding but closing towards us at velocities determined by blue shifting. For these galaxies, gravitational effect is producing an opposite motion to that of the proposed expansion. They also are rejected as aberrations to the hypothesis being tested.

The remaindered cases from the 1929 data table were used to estimate the constant. They are presented in the following table, with the redshift measures reinserted. . Nearly all redshift measurements were within the "threshold range" for redshift detection which the contemporary 2dF Galaxy Redshift Survey[147] had identified. That range was "lower case 'z[148]' <.003 and >.0000 (margin of error ±.0003)."

Presumed Recession Velocities

velocity (km/s) $= (1 - 1/Z)c$; *from Doppler equation*

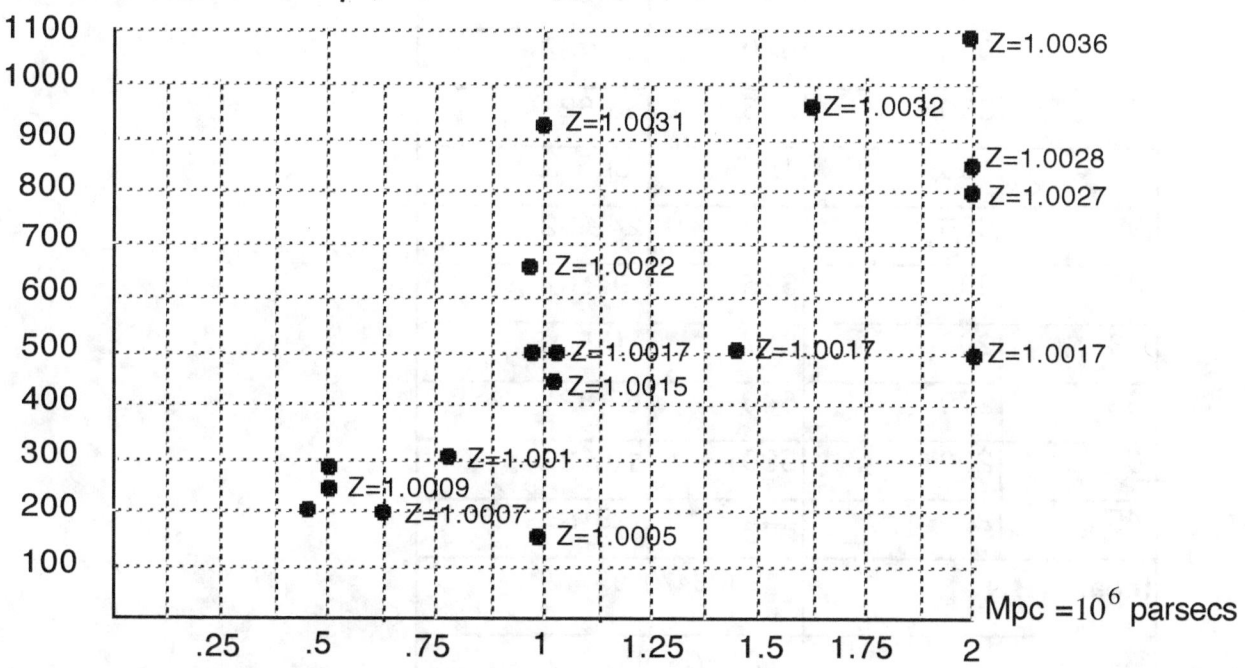

Source: *Hubble 1929 data table.*

[145] The Sloan Digital Sky Survey home page; *http://www.sdss.org/*
[146] The 2dF Galaxy Redshift Survey: Final Data Release, —June 30, 2003; *magnum.anu.edu.au/~TDFgg/*
[147] op. cit.
[148] lower case 'z'=Z-1 and represents the precentage of wavelength increase.

Hubble's conversion of redshift "Z" to apparent recession velocity is accurate, but not for the reason he supposed. It does not represent a real recession— as with his expanding universe concept— but is caused by light being forced to travel across the curvature of space. There is an "apparent recession velocity" because light travels over the arc appearing to slow down relative to the linear distance. From this data, an estimated value for the constant can be calculated for every data point. The value which Hubble settled upon "$H_o \approx 500$ km/ sec./ Mpc" is close to the the mean of the data points:

$\overline{x} = 472.2$ km/ sec./ Mpc

See data table below.

d (in Mpc)	v (in kg/s)	$H_o = \dfrac{v}{d}$	Var. $= H_o - \overline{H_o}$
2	1090	545	72.82
2	800	400	-72.18
2	850	425	-47.18
2	500	250	-222.18
1.7	960	565	92.82
1.4	500	357	-115.18
1.1	500	455	-17.18
1.1	450	409	-63.18
1	920	920	447.82
.9	500	556	83.82
.9	150	167	-305.18
.9	650	722	249.82
,8	300	375	-97.18
.63	200	317	-155.18
.5	270	540	67.82
.5	290	580	107.82
.45	200	444	-28.18
mean ($\overline{H_o}$)		472.17647059	
standard deviation $\sigma = \left(\dfrac{\sqrt{\sum \text{Var.}^2}}{\sqrt{n}} \right)$		177.11727587	

The current constant estimate is "65 km/sec/Mpc." It is 2.3σ (standard deviations) from Hubble's 1929 mean estimated constant of 472.2 km/sec./Mpc". Using the confidence interval for " 2.3σ (approx. .97)[149]," the current estimate of "H_o =65" has only a 0.015 probability of occurring by chance within the Hubble data set[150].

Further, the greatest redshift value for the Hubble data using the revisionist "65 km/sec/Mpc" constant value would only be Z=1.00043, almost outside the detectable range as determined by the 2dF Galaxy Redshift Survey. All other "Z" values from the Hubble data table, using "65 km/sec/Mpc" would fall outside the 2dF threshold.

After Hubble's death in 1953, assaults were made upon his measures, claiming that he had underestimated distances. Walter Baade argued that Hubble's Cepheids were part of "star clusters" and therefore had a much greater light intensity than originally estimated by Hubble. In the 1950's, Baade[151] had discovered the dimmer "Population II Cepheids." He had tried to recalibrate the "Classic Cepheids" which Hubble had used[152] — arguing Classic Cepheid's were brighter and Hubble's calculation of his constant too high. However, Baade's recalibration has not been universally accepted.

It wasn't until 1997 that an actual empirically founded challenge was made to the Cepheid brightness scale which Hubble had used. Feist and Catchpole took data from the parallaxes satellite, Hipparcos[153] and compared parallaxes triangulations to the nearest Classical Cepheids with distance measurements based upon the Cepheidic period. Their study proposed an upward revision of original Cepheid brightness by .2 magnitude[154]. The Feist and Catchpole revision would downshift Hubble's original calculations from a constant of "500 km/sec./Mpc" to nearer "400." However, the Feist and Catchpole revision does not fully meet the test of scientific reliability because the distances to nearby Cepheids (1000 light years or more) are on the verge of being outside parallaxes triangulation range for the Hipparcos satellite and therefore subject to error.

Curiously, it is only the Feist and Catchpole study which offers any experimental evidence to contradict Hubble's 1929 calculations. Yet that data was issued forty years *after* Hubble's Constant was revised downward by Sandage *et. al.* The Feist and Catchpole Cepheid brightness revision does not warrant anything like the 90% emendation which Hubble's Constant suffered after his death. Feist and Catchpole say their revision results in a 10% increase in distances. By extension, it would therefore result in a 10% decrease in Hubble's 1929 constant value.

---

[149] Weisstein, Eric W. "Standard Deviation." From MathWorld--A Wolfram Web Resource. http://mathworld.wolfram.com/StandardDeviation.html

[150] Below I will show that the revisionist constant of "65-70 km/s/Mpc" predicts distance measures for the 1929 Hubble data set which are so high as to be beyond the range of possibility.

[151] Wilhelm Heinrich Walter Baade (March 24, 1893–June 25, 1960) was a German astronomer who emigrated to the USA in 1931. He took advantage of wartime blackout conditions during World War II, which reduced light pollution at Mount Wilson Observatory, to resolve stars in the center of the Andromeda galaxy for the first time, which led him to define distinct "populations" for stars (Population I and Population II).

[152] Harvard University "The Hubble Constant" web page. Op. Cit.

[153] *"The Cepheid PL Zero-Point from Hipparcos Trigonometrical Parallaxes (1997);"* M. W. Feast, R. M. Catchpole; http://astro.estec.esa.nl/Hipparcos/pstex/feast_ceph.ps

[154] magnitude $= \left(100^{.2}\right)^n$ "standard candle"

The desertion of the rigorous empiricism of the 1929 data table for "age of the universe" schemes is misfortunate, if accuracy in measurement is the goal. Because quantum curvature measures redshift as apparent change in velocity, "age of the universe" is an irrelevant time factor, and Hubble's data can be evaluated independently.

Hubble's original data identifies a contradictory motion to apparent recession velocity. The "negative recessions" identified by blue shifting (NGC 6822, 598, 221, 224 and especially 3031) are motions of contraction due to gravitational influence. Most of the gravity "negative recessions" galaxies are the closest foreign galaxies, being in the ".2-.3 Mpc" distance range. There is one exception. NGC 3031 is at .9 Mpc, within the distance ranges used to estimate "H_0." Since gravitational force equals the multiple of the two masses divided by distance squared, "3031" is obviously of greater mass than the other closer "negative recession" galaxies.

Within the data set used to estimate the constant, a galaxy of sufficient mass can be affected by reverse gravitational motion and its "H_0" calculation will be low. Gravitational contraction provides a "negative variance bias" to the "H_0" estimations.

The greatest negative variation is NGC 4826 at a variance of -305.18 from the mean "$\overline{H_0}$." Significantly, "NGC 4826" is at the same distance (.9 Mpc) as "NGC 1068" which has a negative apparent recession velocity. This indicates that motion of galaxies at this distance can be influenced by gravitational interaction, depending upon their mass.

"Positive variation bias" to the "H_0" calculations is probably not explained by the Magellenic Cloud data which Hubble obviously included for that purpose. A better explanation is probably offered by the Baade observation that some of Hubble's Cepheid indicators may have been contained within star clusters and thus were brighter and gave an artificially low distance estimation. For example, NGC 1068 has a higher redshift measurement (higher apparent recession velocity) at distance of "1 Mpc" than all galaxies of greater apparent distance, except one.

This resulted in the highest positive variance from the mean for NGC 1068 at + 447.83 or two standard deviations from the experimental mean value of the constant. If the actual constant value were the mean, NGC 1068 would be "1.9484240688" times as far as the Cepheid measurement made it out to be. NGC 1068 would actually be four times as bright (indicating a cluster) than the single phase measured Cepheid it was assumed to be.

It is true that bias from Cepheid/ star-clusters misidentification will have greater influence on the mean than bias from gravitational negation of (apparent) recession velocity. The actual value of Hubble's Constant will be lower than the experimental mean.

However, the true constant could not possibly fall to the currently presumed value of "65 km/ sec./ Mpc." This is lower than the constant as calculated from gravitationally-biased "NGC 4826" (167 km/ sec./ Mpc). The true constant could not be lower than that calculated from a measured redshift — known to be biased by negative motion from gravity. Statistically, "NGC 4826" establishes the lowest possible limit of the true value of Hubble's Constant.

The post-Hubble revisions represent the desertion of empirical science in the defense of incorrect theory. The post-Hubble constant was not chosen by the empirical method by; by the measure of redshift determining (apparent) recession velocity and an independent measure of distance to the galaxy which originated that redshift:

$$H_o = \frac{(1 - 1/Z)\,c}{d}$$ *Accurate measures of "Z" and "d" give "H_o."*

Instead, the post-Hubble revision was chosen by an "age of the universe" time-factor required by the expanding universe model. That model is incorrect and the proof of this is that it has required abandonment of Hubble's 1929 data.

The quantum curvature model explains distance-proportional redshift and Hubble's mathematical description of it as an apparent difference in the velocity of the speed of light traveling a quantum-produced curvature over a linear distance.

Any distance, "d," between ourselves and stellar light source is a partial or subdivision of the quantum, "Q." "Q" establishes the diameter of the visible universe by providing maximum curvature which is the circumference of the semicircle with a "diameter=Q."

Any subdivision of "Q," designated as "d," Follows the quantum law of elliptical curvature. The distance "d" is kinked into an elliptical curvature with an eccentricity squared equal to the negation of the square of "d as subdivision of Q" [155] :

$$\frac{d}{Q} = \text{subdivision of Q} \; ; \; \varepsilon = \text{elliptical eccentricity}$$

$$\varepsilon^2 = 1 - \left(\frac{d}{Q}\right)^2 \quad \textit{The negation of subdivision}^2 \textit{ of Q}$$

This kinking of stellar distances into elliptical curvature is the actual source of the redshift as light follows the path of curvature. The redshift can be predicted for all values of "d" if the value of the quantum "Q" is known:

$$Z = \frac{\text{elliptical periphery}}{2d}$$

The peripheries of ellipses of known eccentricity and major axis can only be estimated using conventional three-dimensional Euclidean geometry[156] . However, an exact formula for the periphery of the ellipse has been developed using quantum geometry[157] :

$$\chi = \text{ circumference of ellipse} \; ; \; r_1 = \text{ minor axis} \; ; \; r_2 = \text{major axis}$$

$$\chi = 2\sqrt{3\,r_1^2 + r_2^2}\left(\frac{2\,r_1}{\sqrt{r_1^2 + 3\,r_2^2}}\left(\frac{\pi - 3}{3}\right) + 1\right) + 2\left(r_1\left(\frac{\pi - 3}{3}\right) + r_2\right)$$

This formula can be used to provide a predicted redshift "Z" for any stellar distance "d," as a subdivision of the diameter of the visible universe, "Q"[158] :

$$Z = \sqrt{\frac{3}{4}\left(\frac{d}{Q}\right)^2 + \frac{1}{4}\left(\frac{2d}{\sqrt{d^2 + 3Q^2}}\left(\frac{\pi - 3}{3}\right) + 1\right) + \left(\frac{d}{2Q}\left(\frac{\pi - 3}{3}\right) + \frac{1}{2}\right)}$$

Hubble's 1929 data table has proved inconsistent with the expanding universe model and was ignored by later generations of astronomers. They proposed a constant revision which, I will show, is completely incompatible with Hubble's 1929 data. That data confirms the

---

[155] *The Quantum Law of Ellipses and the Elliptical Kink* in Appendix

[156] The best estimation uses the MacLaurin Derivative Series which is based upon derivatives of eccentricity.

[157] *Quantum Determination of Elliptical Periphery and the Detection of Systemic Error in the Maclaurin Derivative Series;* master's thesis, The Virtual University; Dawson, Lawrence

[158] *The Quantum Formula for Redshift* in Appendix.

quantum curvature model and eliminates the post-Hubble revision.

 Quantum geometry seeks a scientifically accurate measurement of the physical world. The 1929 data table represents the serious attempt of very competent astronomer to compare measured redshift with measured distances in the relatively close distances such measurements are possible. There is however a variance between measured distance and mathematically predicted distance for the measured redshift This is true whether that mathematical prediction is made by Hubble's expanding universe model and Doppler effect or it is made by the quantum curvature model. Clearly other factors are affecting the redshift measurement, the primary one being gravitational influence between nearby galaxies. Galaxies rushing towards one another from gravitational influence produce blue shifting of the light, a phenomenon clearly identified by Hubble's data (objects 6822, 598, 221, 224 and 3031 have antirecession velocities).

Any redshift measurement can be affected by extraneous influences and will generate a variance between measured distance and the distance mathematically predicted for the measured redshift. In the competition between the expansion universe mathematical model and the quantum curvature mathematical model, the model which best eliminates the variance is the correct one. Since variance is caused by modification of the predicted redshift at the distance, the more correct the mathematical determination, the less the mean variance will be. The quantum curvature model wins this competition.

Hubble's "expanding universe" constant and its predicted redshift at distance can be compared with redshift at distance predicted by quantum curvature.

Quantum Curvature Graph of Distance at Redshift (d=10 at Z=1.571)

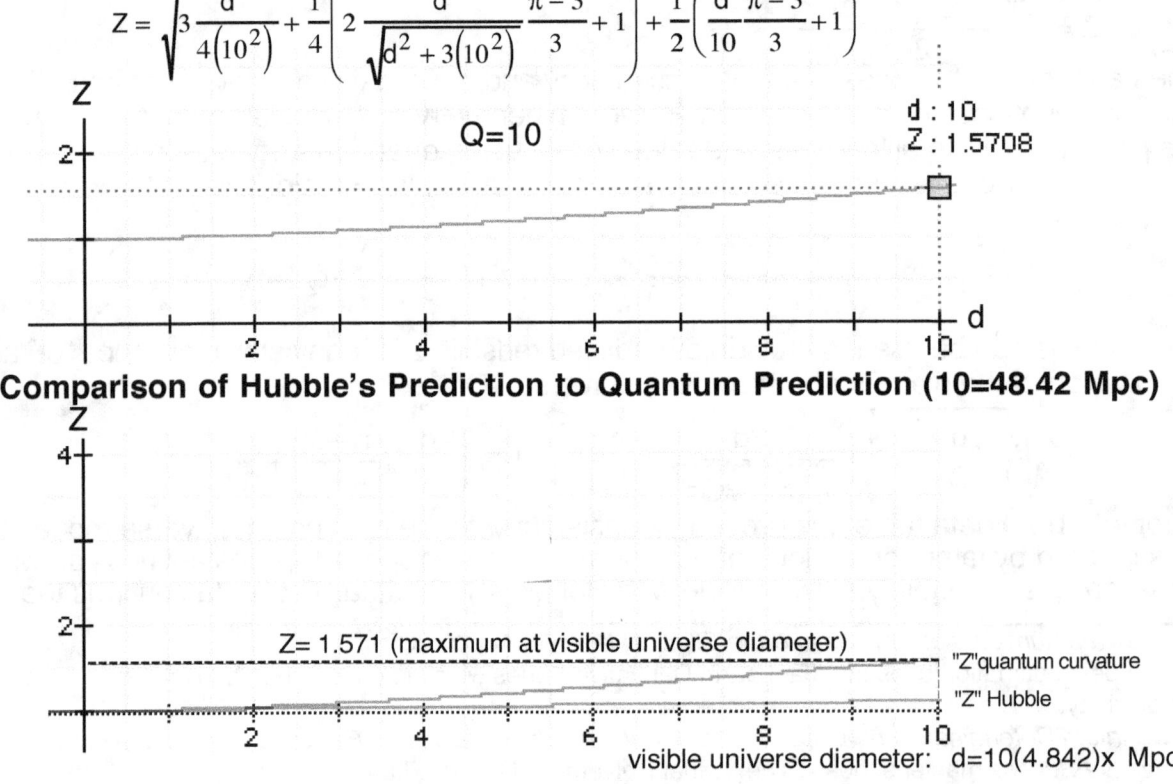

$$Z = \sqrt{3\frac{d^2}{4(10^2)} + \frac{1}{4}\left(2\frac{d}{\sqrt{d^2+3(10^2)}}\frac{\pi-3}{3}+1\right) + \frac{1}{2}\left(\frac{d}{10}\frac{\pi-3}{3}+1\right)}$$

Q=10

d : 10
Z : 1.5708

Comparison of Hubble's Prediction to Quantum Prediction (10=48.42 Mpc)

Z= 1.571 (maximum at visible universe diameter)

"Z"quantum curvature

"Z" Hubble

visible universe diameter: d=10(4.842)x Mpc

Hubble's 1929 data were used to find the most probable visible universe diameter for the quantum formula. Starting with Hubble's predicted distance at "Z=1.571" (217,8773 Mpc), trial-and-error was used to establish a quantum "Q" value which provided the best quantum curvature "fit" to the 1929 data. The best fit was found to be a partial of Hubble's predicted "Q" distance of 217.8773: Q=(217.8773)/ 4.5=48.42 Mpc

$$Q=\frac{1}{4.5}\text{Hubble at redshift Z=1.571.} \quad Q=\frac{1}{4.5}(217.8773)=48.42 \text{ Mpc}$$

Hubble's Formulation; v=recession velocity ; H_o= Hubble's Constant

$$v = \left(1-\frac{1}{Z}\right)c \qquad H_o = \frac{v}{d} \;\; ; d = \frac{v}{H_o}$$

measured Z	predic. "d" Post-Hubble revision H=70 km/s/Mpc (in Mpc)	predic. "d" Hubble (in Mpc) H=500 km/s/Mpc	predic. "d" elliptic. (in Mpc) Q=48.42 Mpc	Measured "d" (in Mpc) object name
1.0005	2.14	0.3	0.424	0.9 obj. *4826*
1.00057	2.42	0.34	0.4713	0.032 obj. SMC
1.00066	2.86	0.4	0.536	0.63 and 0.45 obj *4449/ 5357*
1.0009	3.86	0.54	0.707	0.5 obj *5194*
1.001	4.29	0.6	0.769	0.8 obj *4214*
1.0015	6.43	0.9	1.075	1.1 obj *1055*
1.0017	7.14	1	1.19	.9/ 1.1/ 1.4/ 2.0 obj *5236, 7331,4258, 4382*
1.0022	9.29	1.3	1.452	0.9 obj *3627*
1.0027	11.43	.1.6	1.69	2.0 obj. *4486*
1.0028	12.14	1.7	1.74	2.0 obj *4472*
1.0031	13.14	1.84	1.878	1.0 obj *1068*
1.0032	13.71	1.92	1.92	1.7 obj *4151*
1.0036	15.57	2.18	2.09	2.0 obj. *4649*

The above table identifies several important facts. In the first place, the modern revised Hubble's constant of "65-70 km/s/ Mpc" is completely rejected by the empirical data. Alll distance prediction for "Z" values by the modernist revision are between 2 and 10 times the actual measured distances. The predicted distance for the lowest "Z" value (Z=1.0005) is greater than the furthest distance measured in the table. All variances are whole number multiples of measured distances. The modernist revision simply cannot estimate the measured distances with any credibility.

There is too much variance within the 1929 Hubble data set to establish anything but a trend. The higher redshift "Z" values tend to be at greater distances. However, the inclusion of the quantum curvature model along with Hubble's distance predictions may have refined that trend. Notice that the quantum predicted distances are higher than Hubble predicted distances starting at "Z=1.0005" and increasingly so until the two graphed lines reach "Z=1.0017." At this point, the two begin approaching one another again and cross at "Z=1.0032." After this point the Hubble predicted distances will always be greater than the quantum predicted distances. Tantalizingly, Hubble's predicted distances begin to pull away from measured distance at the "Z=1.0036." At higher redshifts from sources outside the measurable range for distances, Hubble's Constant may be giving unrealistically high distance predictions. The visible universe may actually be much smaller than modernists currently believe. If the quantum curvature model is correct, these data indicate that the edge of the visible universe may only be 1.5793 (10^8) light years away from the earth. This is in contrast to the 4.65 (10^{10}) light years currently believed. Modernists may have overestimated the visible universe by a factor of 294 times.

Differences Between the Hubble and the Quantum Distance Predictions

measured Z	Variance squared (σ^2) between Hubble prediction and measured distance	Variance squared (σ^2) between quantum prediction and measured distance	Difference in predicted distance between Hubble and quantum diff $= d_Q - d_H$.
1.0005	0.36	0.227	0.124
1.00057	0.094864	0.1929	0.131
1.00066	0.0529 0.0025	0.008836 0.0074	0.136
1.0009	0.0016	0.042849	0.167
1.001	0.04	0.000961	0.169
1.0015	0.04	0.000625	0.175
1.0017	0.01 0.01 0.16 1	0.0841 0.0081 0.105625 0.6561	0.190
1.0022	0.16	0.305	0.152
1.0027	0.16	0.0961	0.09
1.0028	0.09	0.0961	0.04
1.0031	0.7056	0.770884	0.038
1.0032	0.0484	0.0484	0
1.0036	0.0324	0.0081	-0.09
$\overline{\sigma}^2$	**0.1746**	**0.1564**	

Expanding Universe May Be in Error

It is possibility that the expanding universe assumption, which has guided science for over 70 years, may be in error. This possibility is confirmed by the above table; by the comparison of the mean variance of Hubble's expansionist model and that of quantum curvature. The variance between Hubble's prediction and actual measured distance is greater than the variance between prediction and measurement for quantum curvature. The mean variance for Hubble's prediction is 12% greater, at 0.1746, than the mean variance for quantum curvature at 0.1564.

Quantum curvature is a better fit with Hubble's own data. This is not definitive proof in that the variance between the two means is not statistically significant using the two tailed t-test for matched pairs. However, the t-test may not be adequate under conditions which might be termed "biased variance." 60% of the variance between predicted and measured distance is on the low side. The measured distance is less than predicted and this bias favors the lower of the two tested models which, in this case, is the Hubble model. The t-test assumes that variance outside tested variables must be random and unbiased.

The possibility of a non-expanding universe is especially significant because the Hubble expansionist concept has proved detrimental to theoretical physics in general. Specifically, Einstein's "cosmological constant," which emerged from the field equations for General Relativity, had to be abandoned. The "cosmological constant" is a tension attached to space itself, sometimes described as "nonzero vacuum energy."

> *Einstein included the cosmological constant as a term in his field equations for general relativity because he was dissatisfied that otherwise his equations did not allow, apparently, for a static universe: gravity would cause a universe which was initially at dynamic equilibrium to contract. To counteract this possibility, Einstein added the cosmological constant. However, soon after Einstein developed his static theory, observations by Edwin Hubble indicated that the universe appears to be expanding......Since it no longer seemed to be needed, Einstein called it the '"biggest blunder" of his life, and abandoned the cosmological constant. However, the cosmological constant remained a subject of theoretical and empirical interest. Empirically, the onslaught of cosmological data in the past decades strongly suggests that our universe has a positive cosmological constant.*[159]

Quantum geometry has identified the cosmological constant as a time force of separation sustaining the spatial quantum. Heaviside's magnetic permeability/ electric permittivity formula for free space equals time force ($1/c^2$) and is given as a field value, in Newtons per geometric unit of space. The constant force of time sustaining spatial vacuoles is shown to be the equivalent of one Newton *per* meter squared of vacuum[160] . The "Newton" unit of force seamlessly integrates with the time force constant. This is not directly transferable to the constant as used by astronomers because their cosmological constant is not given in standard international units (SI units) for fields, in force per unit of area or volume.

---

[159] *Cosmological Constant;* Yale University Wiikipedia entry; en.wikipedia.org/wiki/Cosmological_constant#cite_note-Yale-0

[160] *The Quantum Dimensional Review of the Einstein vs Newton Gravitational Controversy;* Dawson, Lawrence, The Paradigm Company. ISBN 978-1516918096

The Theory of Time-Enforced Quantum Space

In the distant past, an inexplicable change in a steady Primary state of existence required the construction of two distinct geometries and the universe was born of two geometric elements, the elements of vacuous volume or void; and density volume or mass.

Mass is a strict three-dimensional Euclidean construction. It has density because it is composed of a continuum of points in each and every direction within its volume.

Vacuous space is composed by a strict Euclidean plane which is intersected by quantum distances from a point laying outside the plane's surface. The Euclidean plane and the quantum lines produce quantum volume. Quantum volume is defined as "quantum planes separated by 'some' distance." The volume is only the the space of separation between quantum planes; between a curved quantum plane and a linear quantum plane. Vacuous volume is only two dimensional Euclidean and therefore has no density. The volume is completely void. It is only the empty space between planes which are themselves made from lines of empty distance.

Time Origins of the Universe

The inexplicable change in the steady Primary state was an anomaly in time. It consisted of time acquiring a potential energy component. Time split into potential and kinetic energy. The Big Bang occurred as the result of time acquiring this potential energy component. The construction of vacuous volume and dense volume preserved the potential time-energy thus acquired.

Potential time energy is *conserved* time energy(retained or stored time energy). Potential time energy is the absent energy from an existing time flow or, alternatively, the reserved energy from an existing time flow.

For time energy to be reserved, there must be two separate time values existing simultaneously.

Potential time energy is created when a second "stalled" time value is introduced which initiates a time variance between itself and the original time value. The variance between the two offset time values reserves time energy. The "stalled" time value conserves the time-energy normally expressed by the flow of time. Time-energy builds up behind the stalled time value.

The introduction of potential time energy to a time flow is the introduction of this stalled component to that time flow. The stalled component reserves increasing amounts of time energy. This "stored" time energy increases in direct proportion to the increasing time variance between the stalled component and the moving or flowing component. This paper intends to supply the mathematics for this conception and to demonstrate that those mathematics are a superior model of geometric reality.

With the introduction of a potential energy component, two different time values try to occupy the same time position. This creates the aforementioned anomaly which requires resolution. The energy differential forces a rupture between the two components.

As the kinetic component moves forward in time, there is an increasing time differential between it and the "stalled" potential component. There is also an increasing energy differential between the two. The time differential and the energy differential are not the same thing, however. The energy differential between the two points is in direct proportion to the time differential (although the proportion is not one-to-one).

The energy differential is the difference between expressed kinetic energy (+ value) and unexpressed potential energy (– value): $[(+1) - (-1) = 2]$. Time differential is the difference between expressed time by the kinetic component and the expressed time of the potential component: $[(+1) - (0) = 1]$. The energy differential increases twice as fast as the time differential. time differential=(1/2) (energy differential) :

   Let energy differential = ΔT ; time differential = $\Delta T/2$
   Let the time-value at the moment time acquires a potential component = T_1;
   Let the time-value of the moving component = T_2;   Let the time differential= $\Delta T/2$.

$$T_2 = T_1 + \Delta T/2$$
$$-\Delta T/2 = T_1 - T_2\overset{.}{}$$

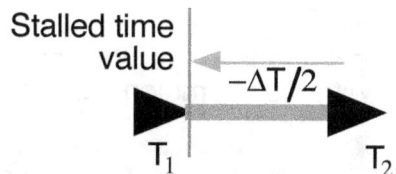

The energy differential, however, is the following:

   Energy expressed by $T_2 = \Delta T/2$

   Energy conserved by $T_1 = -\Delta T/2$

   $\Delta Energy = Energy(T_2) - Energy(T_1)$

   $\Delta Energy = \Delta T/2 - (-\Delta T/2) = 2\Delta T/2 = \Delta T$

Therefore, the time differential=$-\Delta T/2$, but the energy differential= ΔT. The energy differential increases twice as fast as the time time differential between the two values.

Let us assume that a single time position has a restricted capacity to sustain an energy differential caused by diverging time values. The increasing energy differential is a pressure building within the time position.

The energy differential is a force expressed against the two time positions. It is a "negative vacuum pressure," an absence which repels. The advancing time point cannot be "sucked back" to the stalled time position without reversing time. The stalled time point cannot be "sucked forward" towards the kinetic time point without accelerating the rate of time, thus creating an anomalous dual time rate.

The only possible solution is a "separation" such that the disparate time values are not trying to define time for the same position. The "time force" thus "invents" spacial distance to separate the time values. The definition of time force is the following:

F_t = time force ; D = distance of separation

$$\Delta\text{Energy} = \Delta T = F_t(D)$$

$$F_t = \frac{\Delta T}{D}$$

The postulated time force is non-Newtonian in character. It is not "mass *times* acceleration." It is a proposed force component of the Einsteinian "cosmological constant," that is, the force component of "nonzero vacuum energy" — or the energy value for quantum volume. Einstein proposed that space itself —devoid of matter— has an energy value which he deemed the "cosmological constant.[161] "

Very briefly, a few points about Einstein's cosmological constant need to be made before proceeding with the analysis of time force. Einstein postulated an energy with an associated force which was attached to space itself and which stabilized the universe. He abandoned the concept under pressure from the expanding universe theory and in the face of Edwin Hubble's redshift data from the far universe. The redshift, explained as Doppler Effect, seemed to confirm the expanding universe idea and Einstein renounced the cosmological constant.

However, quantum geometry provides an alternative explanation for the red shift and a stable universe model which is not only compatible with the cosmological constant, but actually requires it.

The quantum model identifies red shift as due to the stretching of light waves by dimensional distortion. If the dimension in the direction of wave travel is stretched relative to the other two dimensions, the frequency will be shifted downward. It is the principle used by the slide of a trombone.

The straight line of any Euclidean distance is a curved line in quantum space. The straight line on the Euclidean planar component of quantum volume is forced into a curved line upon the Euclidean plane by the construction of quantum volume. The straight Euclidean line is systematically curved by quantum volume.

Light from the far universe is shifted downward in frequency because the light which reaches us has had to travel the arc of the quantum to transverse the linear distance between us and the source of the light. The greater the distance, the more the curvature and the greater the frequencies shift downward towards the red. Amount of red shift is related to distance just as Hubble discovered it to be.

We don't live in an expanding universe, we live in a stable quantum universe. Einstein's cosmological constant is correct

The formula for time energy fits the universal equation which relate force and energy:

$$\text{Energy} = (\text{Force})(\text{distance}) \quad ; \quad E_t = F_t(D)$$

When the differential limit for time energy is reached, the associated time force establishes a distance of separation.

---

[161] "The cosmological constant has the same effect as an intrinsic energy density of the vacuum, ρ_{vac} (and an associated pressure)."; *Source: Wikipedia.* Time force (pressure) is a substitute for density in quantum volume. Einstein's concept anticipated this necessity.

This forcing of a second time position brings dimensional space into existence. Two differentiated time values become geometrically defined as points separated by the potential time-energy established by their differentiation($-\Delta T$).

Potential time-energy is conserved as the variance between differentiated time values located in quantum points. That potential energy sustains spacial separation between the quantum points.	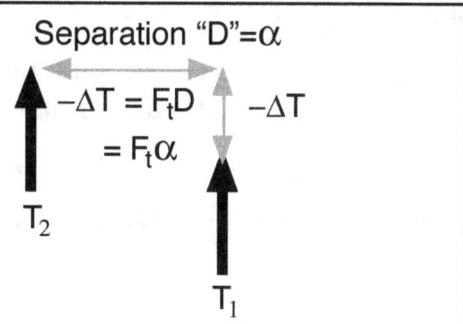

Time-Force *times* Alpha Space is the Primary Quantum

A quantum is defined as the separation of two points by unstructured space. It is contrasted with an "Euclidean length" the end points of which are separated by a continuum of points. For the quantum, there are no geometric points within the space of separation. Elsewhere I have given mathematical proof that our physical geometry cannot be measured without the existence of a fourth "quantum dimension" *(See my paper "Dawson's Theorem......")*.

The Alpha Space (α) is the primary quantum of this fourth dimension. The "unstructured space" separating differentiated time points is sustained by the potential time energy generated by this differentiation. A quantum dimension can only exist because force sustains separation between two points;..

The conversion of energy ($-\Delta T = F_t\alpha$) result in a potential energy differential between time points($PE = -\Delta T$).

The geometric space constructed by time-energy conversion is not actually linear distance.

Area is constructed by a field of energy ($-\Delta T^2/2$). The linear force separating actual time-points is the derivative of the field:

$$D(-\Delta T^2/2) = 2(-\Delta T) / 2 = -\Delta T = F_t\alpha$$

The total energy available for conversion to geometric separation of the components is the energy differential (ΔT) *times* the potential energy possessed by the stalled time component ($-\Delta T/2$):

$$E_t = (\Delta T)\left(-\frac{\Delta T}{2}\right) = -\frac{\Delta T^2}{2}$$

$$= \frac{F_t^2\alpha^2}{2} \; ; \; \alpha = \text{separation required by time force}$$

$$Q^2 = F_t^2\alpha^2$$

$$\frac{Q^2}{2} = \frac{F_t^2\alpha^2}{2}$$

Time force *times* the space of separation (Alpha) equals the quantum. The two quantum points are time values and the space separating the quantum points is produced and sustained by the force of time. Mathematically, the space separating any two quantum points is unstructured space and must be sustained by a force. Time force is that force sustaining the quantum.

The time force *times* "α" establishes the quantum squared. Therefore, the initial space generated by the time differential is the quantum squared divided by "2."

A unit of area defined from a single point and equal to a strict Euclidean value (rather than a quantum value) does exist for "$Q^2/2 = x^2/2$; x=Euclidean value of Q."

Euclidean Defined Area for $x^2/2$

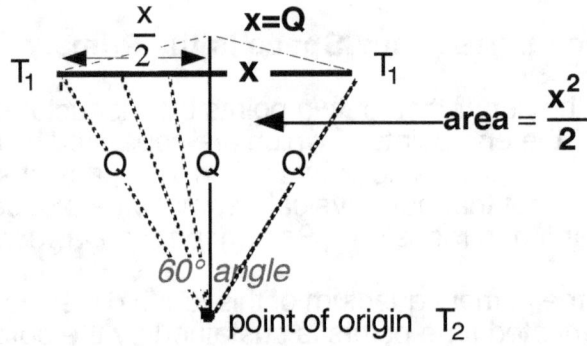

The Euclidean area for one of the back-to-back triangles in the figure above is the following:

$$\text{area of one triangle} = \frac{1}{2}\text{base} \times \text{height} \; ; \quad \text{height} = \frac{1}{2}x \; ; \quad \text{base} = x$$

$$= \left(\frac{1}{2}\right)\left(\frac{1}{2}x\right)x = \frac{x^2}{4}$$

The area for the whole figure is twice the area of one of the back-to-back triangles:

$$2\left(x^2/4\right) = x^2/2$$

However, time force cannot construct an Euclidean unit of area. It must be a quantum unit of area. The quantum unit of area does not equal the Euclidean unit of area:

$$\frac{Q^2}{2} \neq \frac{x^2}{2}$$

The Euclidean unit of area is composed of a lines which are themselves composed of a continuum of points —that is, the plane is composed of Euclidean lines in any and all directions.

In contrast to the Euclidean line, the quantum line is composed of only two points separated by a geometrically undefined distance. The force of time structures the quantum distance and the two points separated by this force are alien time values.

The surface area of any plane composed by quantums contains no points within that surface area. The only lines are quantums composed of the point of origin and points along an Euclidean line which does not intersect the point of origin. The summation or "integration" of these quantum lines determines the plane surface:

A Proposed Quantum Plane
Plane is summation of quantums

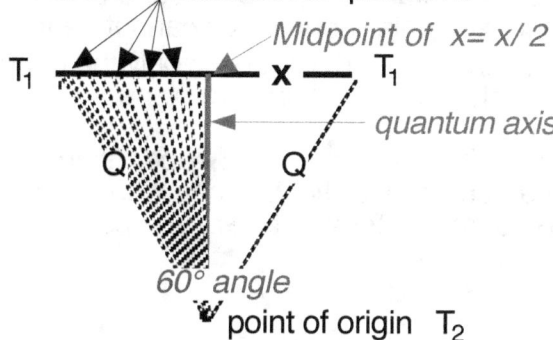

Such a plane could be constructed by integration of the shortest quantum which is indicated as the "quantum axis" in the above illustration. If this quantum axis is the derivative of a function of the midpoint subdivision "x/2," the integral produces 1/2 the area:

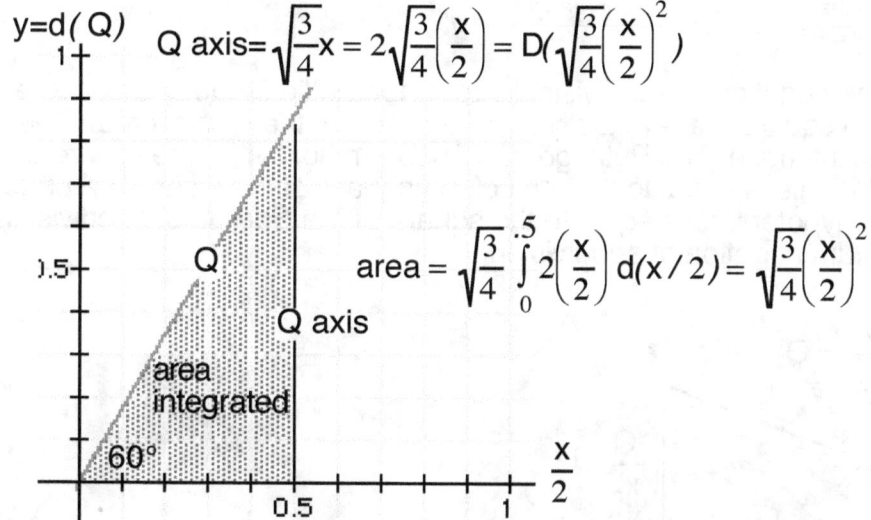

$$y = d(Q)$$

$$Q \text{ axis} = \sqrt{\frac{3}{4}}x = 2\sqrt{\frac{3}{4}}\left(\frac{x}{2}\right) = D(\sqrt{\frac{3}{4}}\left(\frac{x}{2}\right)^2)$$

$$\text{area} = \sqrt{\frac{3}{4}} \int_0^{.5} 2\left(\frac{x}{2}\right) d(x/2) = \sqrt{\frac{3}{4}}\left(\frac{x}{2}\right)^2$$

The area so determined is 1/2 the area of the whole quantum plane. To determine the area for the whole of the quantum plane, we multiply the above area resolution by "2":

$$2\sqrt{\frac{3}{4}}\left(\frac{x}{2}\right)^2 = \sqrt{\frac{3}{4}}\frac{x^2}{2} = \sqrt{\frac{3}{4}}\left(\frac{Q^2}{2}\right)$$

Euclidean determined area ≠ quantum determined area.

$$\sqrt{\frac{3}{4}}\left(\frac{Q^2}{2}\right) \neq \frac{(x = Q)^2}{2}$$

Even though this proposed quantum plane is achieved by the integration of a quantum across the x axis and thus is an area composed of "quantum space," it still cannot be the

117

quantum plane. The quantum distances between the point of origin and the points along the x axis do not resolve the time variance correctly.

With the exception of end points, all quantums intersecting the x axis are partials of the "α space." It is only the full quantum distance "α" which can resolve the time variance. There needs to be a full "α space" of separation to locate the time point "T_1" along the x axis.

The continuum of points along the x axis do not define an "α space" of separation with the point of origin and cannot locate "T_1" along the x axis. This linear x axis cannot be the Euclidean line which establishes the actual quantum plane. The x axis is only the quantum differential of the true Euclidean line which establishes the real quantum plane.

A quantum unit of measure must be differentiated by the *negation* of subdivision rather than by subdivision. This is the fundamental mathematical distinction between quantum and Euclidean units of measure. For the quantum squared, differentiation by the *negation* of subdivision is the following:

$$d(Q_{x_n}^2) = \left(1 - \frac{1}{x_n^2}\right)Q^2 \quad ; \quad "\, x_n = n + \text{all possible partials of 1 }"$$

as "x_n" $\xrightarrow{\text{approaches}} \infty$, $Q^2 \xrightarrow{\text{approaches}} 1$

as "x_n" $\xrightarrow{\text{approaches}} 1$, $Q^2 \xrightarrow{\text{approaches}} 0$

While the principle of the negation of subdivision is clear for the linear quantum, it is less obvious for the quantum squared. It is actually an application of the Pythagorean Theorem to a quantum/Euclidean interface. The Pythagorean Theorem holds that the square of the hypotenuse of a right triangle is equal to the sum of the square of the sides. The perfect square of the quantum hypotenuse is equal to the square of the Euclidean subdivision *plus* the square of the quantum negation of subdivision:

$$\left(1 - \frac{1}{x_n^2}\right)Q^2 + \frac{Q^2}{x_n^2} = Q^2$$

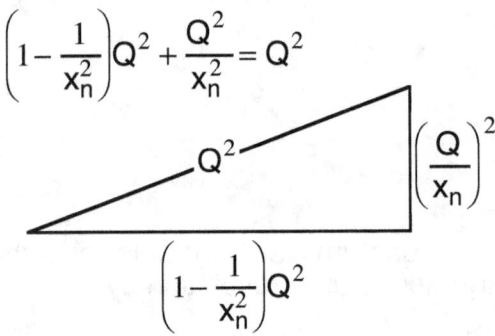

The Pythagorean Theorem produces a right triangle in which the square of the linear quantum is equal to an Euclidean subdivisional value (the subdivision squared) plus an equivalent quantum value (the negation of that subdivision squared).

If all the quantums intersecting the "x axis" were *negations* of subdivision for the quantum squared[162], then they would be missing an equivalent Euclidean "subdivision squared" in

---

[162] The linear quantum would be the square root of the negation of subdivision: $\sqrt{\left(1 - 1/x_n^2\right)Q^2}$

118

order to equal the quantum squared. The quantum force must construct a "y axis" and a resultant Euclidean plane. It must do this in order to provide the missing Euclidean subdivision squared to add to the *negation* of subdivision. The geometric form so constructed would be the following:

The Construction of the Actual Quantum Plane

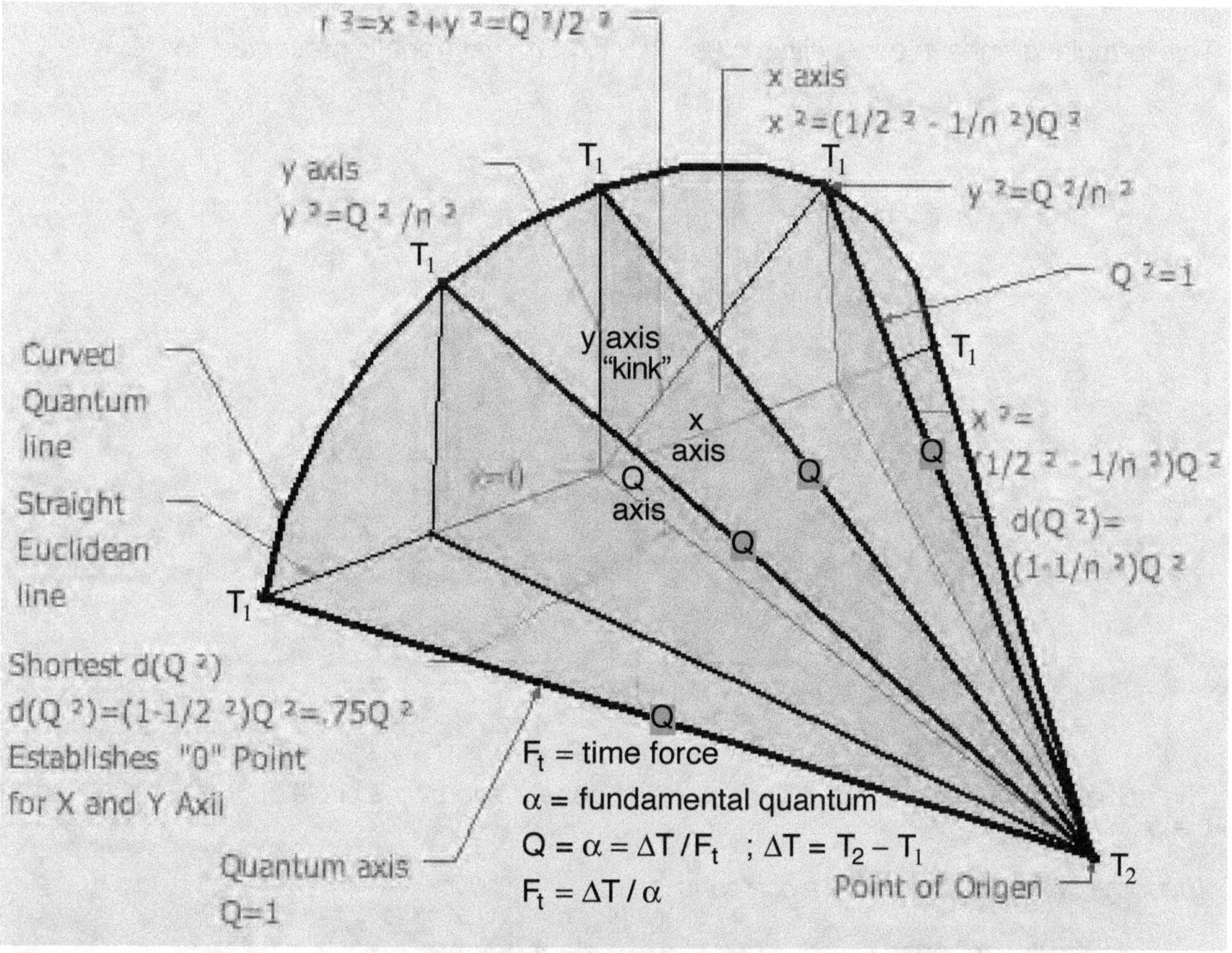

The quantum plane is the conical surface curved above the linear plane made by the x axis and the point of origin. The Euclidean line which the quantum intersects is the circumference of a circle with a radius "r=x/2" and a center at "x/2."

This is proven in the following manner. Any quantum intersecting the x axis is a *negation* of subdivision for the quantum squared. The value on the y axis is the subdivision squared. This also provides a value for the x axis because a second right triangle is produced by the quantum axis and the *negation* of subdivision upon the x axis plane:

$$(Q\ \text{axis})^2 = \left(1 - 1/2^2\right)Q^2$$
$$x^2 + y^2 = (Q/2)^2$$

119

$$\left(1-1/n^2\right)Q^2 = (Q\ axis)^2 + x^2 = \left(1-1/2^2\right)Q^2 + x^2$$

$$x^2 = \left(1-1/n^2\right)Q^2 - \left(1-1/2^2\right)Q^2 = \left(Q^2/2^2 - Q^2/n^2\right)$$

The Euclidean line which the quantums intersect is now the circumference of the circle made by "$x^2 + y^2 = (Q/2)^2$."

This formula graphs a circle with radius= x/ 2 ; x=Q. The graph is presented for "Q=1"

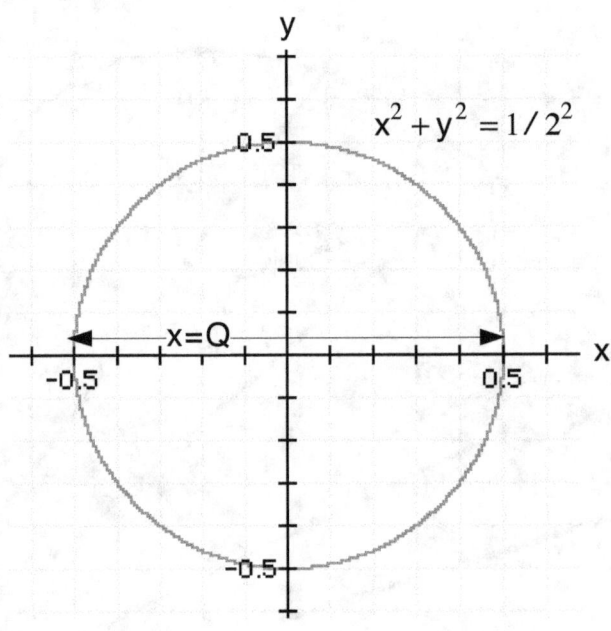

The actual quantum plane is the surface of the cone made by the circle on the "x,y" axis and the point of origin.

The formula for the surface area of a cone is:

r = radius of base = x/2 = Q/2 ; l= slant height = Q

$$\text{surface area} = (\pi\ r\ l) + \left(\pi\ r^2\right) = \frac{\pi Q}{2}Q + \pi\left(\frac{Q}{2}\right)^2 = \frac{\pi Q^2}{2} + \frac{\pi Q^2}{2^2}$$

$$\frac{\pi Q}{2}Q + \pi\left(\frac{Q}{2}\right) = \left(1+\frac{1}{2}\right)\frac{\pi Q^2}{2} = \frac{3}{2}\pi\left(\frac{Q^2}{2}\right) = 1.5\pi\left(\frac{Q^2}{2}\right)$$

The Euclidean dimensions expand or "distort" the quantum plane "$Q^2/2$" by a factor of "1.5π." This is 3 *times* the area of the Euclidean circle:

$$\text{quantum area} = 3\pi r^2 = 3\pi\left(\frac{x}{2}\right)^2 \ ; \qquad\qquad r = \frac{x}{2}$$

$$D\left(3\pi\left(\frac{x}{2}\right)^2\right) = (3)\left(2\frac{x}{2}\right)\pi \quad ; \qquad \left(2\frac{x}{2}\right)\pi = \textit{cirumference of circle}$$

The derivative of quantum area is three *times* the circumference of the "x,y" circle. Therefore, quantum area equals 3 *times* the integral of circumference by the radius "x/ 2":

$$3\pi\left(\frac{x}{2}\right)^2 = 3\int (\text{circumference})\ d(\ r) = 3\int_0^{.5} 2\pi\left(\frac{x}{2}\right) d(\frac{x}{2}) = 1.5\pi\left(\frac{Q^2}{2}\right)$$

The integration of circumference by the radius is the expansion of circular area from the point at which the quantum axis intersects the x axis at "x/ 2." The circular area is expanded as the radius changes from "x/ 2=0; area=$0^2\pi$" to "x/ 2=x/ 2; area= $(x/2)^2\pi$."

In contrast, the integration of the "x/ 2" subdivision for *the quantum axis length* produced the linear quantum area subscribed by the quantum axis and the quantum.

The integration of the "x/ 2" subdivision from the *point of quantum axis intersection* to the *point of quantum intersection* produces the area of the "x,y" circle.

The quantum points established by quantum intersections *are the only quantum information which the Euclidean x axis possesses*. The Euclidean x axis has no informational access to the lengths of quantum intersections. Those lengths are non-Euclidean and only composed of two points.

It is only quantum "points of entry" and "distance between quantum points of entry" to which a Euclidean dimensional axis has access. Area is expanded by integration of circumferences. Area is established by integration of the circumferences for all circular values between "r=0" (quantum axis entry point) and "r=x/ 2" (end quantum entry point) . The quantum/Euclidean point for the quantum axis is expanded to a circle using quantum information.

The area of the curved quantum plane is three times the circular area determined by the integration of circumferences from the quantum axis point:

$$3\int_0^{.5} 2\pi\left(\frac{x}{2}\right) d(\frac{x}{2}) = 3\left(x^2/4\right)\pi = 1.5\pi\left(x^2/2\right)$$

$$\int_0^{.5} 2\pi\left(\frac{x}{2}\right) d(\frac{x}{2}) = \text{ area determined by integration of circumferences}$$

The area of the curved quantum plane (surface of cone) is exactly 3 times the area of the Euclidean plane which constructs it.

The surface area of the cone is also a function of the quantum axis squared. Specifically, it is π *times* the quantum axis squared:

$$(Q\ \text{axis})^2 = d(Q^2_{n=2}) = \left(1 - \frac{1}{2^2}\right)Q^2 = \frac{3}{4}Q^2$$

121

$$\text{surface area} = \pi(\text{Q axis})^2 = d(Q_{n=2}^2)\pi = \frac{3}{4}Q^2\pi = 1.5\pi\frac{Q^2}{2}$$

$d(Q_{n=2}^2)$ " *is the negation of subdivision for the quantum squared at the quantum axis*

The area of the quantum plane is exactly equal to the area of a circle with the quantum axis as radius. Such a circle, however, could not be constructed by the initial time force:

$$1.5\pi\left(Q^2 / 2\right) \neq Q^2 / 2 \quad \text{\textit{Construction of quantum area requires Euclidean intervention.}}$$

Quantum Space Must Be Expanded by Euclidean Space

Quantum space is expanded by Euclidean intervention upon that quantum space. This intervention takes the form of the addition of another dimension. Thus the linear quantum which begins as the equivalent of an Euclidean length (x=Q) becomes a curved line:

$Q = \pi x / 2$

In the conical construction of quantum area, the quantum cannot acquire Euclidean definition along the x axis. Even though "linear x=linear Q," all the quantum values defined by the continuum of points along the x axis are differentials without sufficient length to position the time point along the x axis. They must be pushed upward along a second Euclidean "y axis" to gain sufficient length. The x axis must become curved circumference. The one dimensional x axis expands to two dimensions to position the quantums.

The situation is similar for the quantum plane. The linear quantum plane composed by the x axis and the point of origin is pushed upward to become the curved conical surface. The two-dimensional plane is expanded by distortion into three-dimensional volume.

A quantum plane cannot exist without the intersection of an Euclidean line. That Euclidean line of intersection is distorted along the Euclidean plane. The distortion expands the quantum plane into a form of volume.

The quantum squared is the *forced* separation of two time points into a multidimensional space of separation:

$$Q^2 = F_t^2 \alpha^2$$

On the surface, it appears that the square of time-force acting upon the square of the alpha space— has produced a conventional three dimensional cone. This is deceptive, however, and a dangerous assumption to make.

The quantum cone cannot define volume in the sense we are use to thinking about volume. The volume of the quantum cone is constructed by vectors of force and these vectors are two-dimensional, not three dimensional. The time force is vectored along three planes: the linear plane made by the x axis and the point of origin; the curved quantum plane of the conical surface; and the base "xy" plane. The force vectors would become three dimensional if the linear x plane were spun around the Q axis to form the cone, but this is not how the quantum cone (vacuole) is constructed.

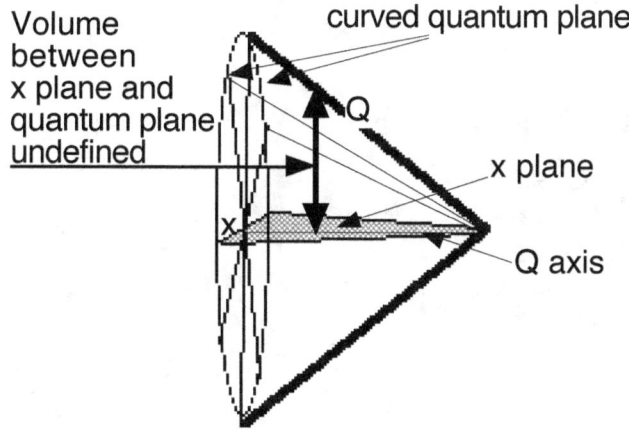

The magnetic -force vectors which are illustrated below, demonstrate that three dimensional force is produced by the rotation of a plane around an axis contained upon that plane.

.

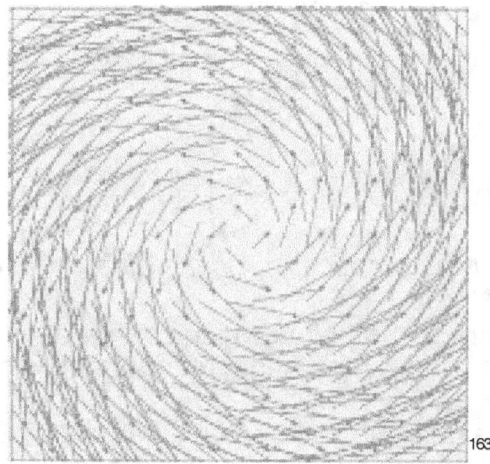

Magnetic Force Vectored Around a Polarity Axis

In contrast to the magnetic field which vectors force into volume by rotation around the polar axis of a plane containing that force, the vectors of time force are restricted to immovable planes. Those planes are: the "x plane;" the "x,y plane;" and the "curved quantum plane." The planes are locked into position relative to one another and there is no external geometric referent within which the whole structure could rotate.

Vectors of force are restricted to the sides of the triangle illustrated below. There are no vectors of force within the area subscribed by that triangle. Therefore, there are no vectors of force within the volume of the cone.

---

[163] Source: Wikipedia

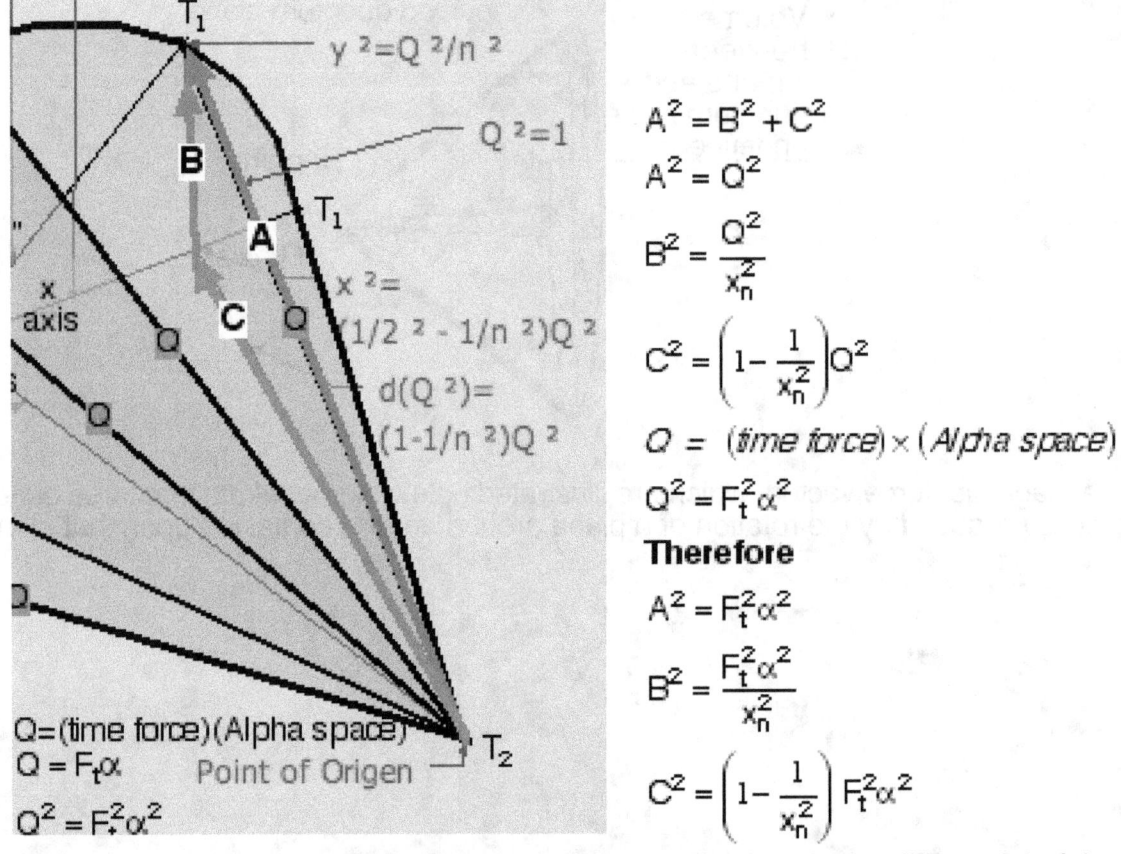

- T_1
- $y^2 = Q^2/n^2$
- $Q^2 = 1$
- B
- T_1
- A
- x axis
- $x^2 = (1/2^2 - 1/n^2)Q^2$
- C
- Q
- Q
- $d(Q^2) = (1-1/n^2)Q^2$
- Q = (time force) × (Alpha space)
- Q = (time force)(Alpha space)
- $Q = F_t\alpha$
- $Q^2 = F_t^2\alpha^2$
- T_2
- Point of Origen

$$A^2 = B^2 + C^2$$

$$A^2 = Q^2$$

$$B^2 = \frac{Q^2}{x_n^2}$$

$$C^2 = \left(1 - \frac{1}{x_n^2}\right)Q^2$$

$$Q = (time\ force) \times (Alpha\ space)$$

$$Q^2 = F_t^2\alpha^2$$

Therefore

$$A^2 = F_t^2\alpha^2$$

$$B^2 = \frac{F_t^2\alpha^2}{x_n^2}$$

$$C^2 = \left(1 - \frac{1}{x_n^2}\right)F_t^2\alpha^2$$

The line labeled "C" lies on the linear "x" plane. Mathematically, it is the differential of the quantum squared for "$n = x_n$."

The quantum is defined as a length composed only of two points which cannot be differentiated by subdivision but must be differentiated by the negation of subdivision.

"C" is the negation of subdivision for "$n = x_n$." As a quantum differential, the line "C" is shorter than the quantum space of separation required to locate a time point at its intersection with the "x" axis. "C" is always a function of the quantum *minus* a fraction.

Therefore, between the quantum end points for the x axis, all intersecting quantum differentials are shorter than the required space of separation. The linear quantum plane as defined by the x axis is composed of quantum differentials which do not provide enough space of separation to locate time points along the x axis.

The force of separation which composes the "C" quantum differential can only be a vectored partial of the total squared force:

$$C^2 = \left(1 - \frac{1}{x_n^2}\right)Q^2 = \left(1 - \frac{1}{x_n^2}\right)F_t^2\alpha^2 \quad ; \quad Q^2 = F_t^2\alpha^2$$

The line labeled "B" lies on the Euclidean "x,y" plane or base of the cone. As an Euclidean length it is subject to differentiation by subdivision. "B^2" is the subdivisional value which has been subtracted from the quantum squared differential "C^2." It composes the vectored partial of the total squared force which must be added to "C^2."

$$B^2 = \frac{Q^2}{x_n^2} = \frac{F_t^2 \alpha^2}{x_n^2} \qquad ; Q^2 = F_t^2 \alpha^2$$

"B" represents the amount of squared time force which must be vectored off the x axis to located the time point. This is proven by the Pythagorean Theorem "$A^2 = B^2 + C^2$. "
Therefor:

$$F_t^2 \alpha^2 = \left(1 - \frac{1}{x_n^2}\right) F_t^2 \alpha^2 + \frac{F_t^2 \alpha^2}{x_n^2} \qquad ; \quad Q^2 = F_t^2 \alpha^2$$

These vectors of squared time force can only lie upon two planes; the linear x axis plane and the "x,y plane" (the base of the cone). The required equality would not hold for any other vector lines located within the "A,B,C" triangle. The area subscribed by that triangle is not defined by time force. By extension, neither is the whole of the volume between the x plane and the curved quantum plane.

The volume is undefined geometrically and is devoid of all time value. By definition, it is a complete "vacuum" or void. It is devoid not only of matter but of geometric definition and even of time. it is "nothingness" since "existence" requires time.

This complete vacuum or "nothingness" is only a mathematical variance between the curved quantum plane as defined by the surface of the cone and the linear quantum plane as defined by the x axis and the time point of origin.

The variance explains that, for any Euclidean line intersected by a quantum, the force of time must curve the Euclidean line relative to the quantum dimension. The points along any Euclidean line do not provide adequate spacial separation needed by the different time values. Any point identified by quantum intersection of the x axis must be forcibly vectored at 90° and acquire a position off the x axis which does provide adequate separation.

The Force of Time Explains Light as Waveform Energy

If we assume that any linear distance in space is composed of back-to-back multiples of the quantum-squared cone, then we are provided with a model which can identify light as wave form energy. Multiples of these quantum squared units also multiply the quantum time force.

Back-to-Back Quantum-Squared Soliton

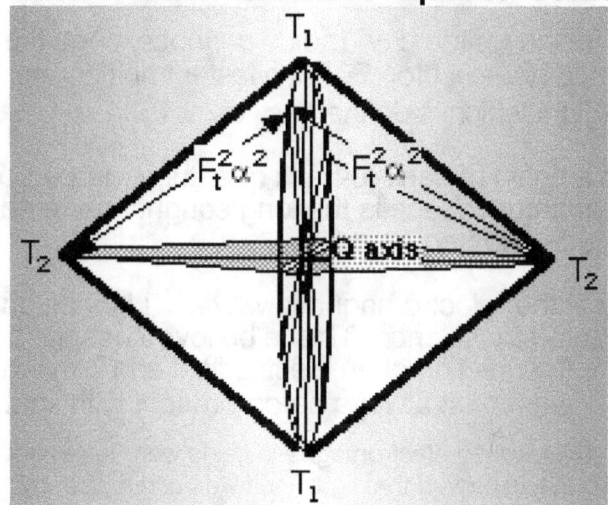

A "pulse" can be sent down a line of back-to-back vacuoles by pressing upon "T_2."

125

If an external force should press upon " T_2 " and thereby shorten the Q axis, the radius of the " T_1 " circle must expand. Time force is vectored along the Q axis and the radius. The square of the Q axis plus the square of the circle's radius equals the the quantum squared (time force squared). If the Q axis is shortened, the circle's radius must be lengthened by an equivalent amount to keep the equality.

In physics, a wave which exchanges length for height in a motion 90° to the motion of the wave is known as a "transverse wave." The most common example of the transverse wave is the water wave.

Geometrically, the length containing water from the trough is "shortened" increasing the height of the crest. It is a dimensional exchange which translated into an up and down motion of the water at 90° to the motion of the wave. Light is also a transverse wave. The Q axis is shortened which increases the height of the circular radius. The motion of this exchange is 90° to the motion of the wave.

In this manner, an energy wave may be sent down the line of back-to-back cones. If " T_2 " is compressed, the radius of " T_1 " is forcibly expanded. The expanded radius can only forcibly require its original position by forcibly expanding the next Q axis. The forcible expansion of the second Q axis forcibly contracts the Q axis in the next set of back-to-back cones— and so forth.

Both the frequency and amplitude of the wave thus generated is controlled by the originating external force. Amplitude is the amount the Q axis is compressed. Frequency is amplitude over time or the rate at which the amplitude is acquired:

$$f = \text{amplitude} / t = (\text{compression Q axis}) / t$$

If the Q axis is compressed by 1/4 and it takes 1/10 of second to reach this compression, the frequency is 10 hertz. This frequency, however, is communicated at the speed of light. The changes in amplitude will pass any point down the chain at the speed of light. In the same manner, changes in amplitude for a constant-frequency sound wave will pass any point in its path at the speed of sound.

The transference of force across the Q axis and up the "x,y" radius cannot exceed the time variance between " T_2 " and " T_1 ." Time would not exist for " T_1 " at a faster transfer rate. This restricts force transference to the speed of light which is defined as " $\alpha / \Delta T$."

This waveform transference of force down a chain of back-to-back quantum-squared cones is obviously light energy. Time structured quantum space is the long sought after medium which explains light as a wave with amplitude and frequency.

Science may now desert the near insanity of the "electromagnetic wave." "Light as an electromagnetic wave" is incessantly repeated by science. This is believed despite the fact that— if light were a wave conducted by the intersection of an electric field and a magnetic field as Maxwell[164] proposed— then such fields must always be co-present with light. Yet,

---

[164] Maxwell, James Clerk proposed the theory in 1864. When electromagnetic fields were later used to *induce* radiation such as radio, it was supposed that this proved that light itself was *conducted* by electromagnetism. The distinction between "induction" and "conduction" became permanently confused.

they never are.

The "light as pulsed time-force" thesis, however, requires something in the material world which is capable of "pushing" the quantum axis. The electron is this link.

Elsewhere I have shown that the electron is a quantum particle which acquires mass by attachment to a Euclidean particle. The electron/proton bond is a string which exchanges energy the nucleus. The string itself is a replication — in the material world — of quantum-squared space. As such, it can interact with quantum space readily.

The Proof of Quantum Curvature; The Shape of Galaxies

A three-dimensional vacuum or "void" is created by the enforced curvature of the quantum plane by its intersection with Euclidean space. That vacuum defines the two dimensional quantum or the "quantum squared."

The linear quantum is two points separated by "some" distance. The two dimensional quantum is two plains separated by "some" distance. This space of separation is the variance between the curved quantum plane (cone surface) and the linear quantum plane (x axis plane). The variance or "space of separation" is generated by a quantum-squared expansion of a single quantum point.

Vacuum, or more accurately, the volume of vacuous space, is the space of separation between two dimensional quantums. What happens, however, when this space of separation is imposed upon by Euclidean volume?

Usually there is a clear separation between the material solids which compose Euclidean volume and the "vacuum" composed by quantum planes. After all, volume which is completely devoid of matter is the definition of "vacuum." Vacuum can't exist unless it is absent of all Euclidean matter.

There is one exception to this however. Clusters of stars attempt to impose Euclidean definition upon quantum space.

The line of opposition between any two stellar bodies is the gravitational line. The gravitational line is and must be completely Euclidean. Gravity is the compelling force attempting to unify matter into one geometric solid. In quantum geometry, gravity is the force of Euclidean unification and is an element of the Euclidean geometric definition of both bodies. In essence, they are already geometrically "unified" by the force of gravity.

The gravitational lines between any three stellar bodies define an Euclidean plane:

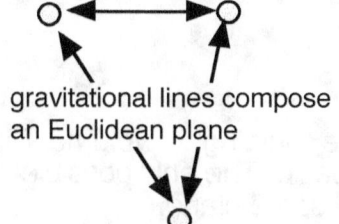

gravitational lines compose
an Euclidean plane

Such an Euclidean plane establishes vacuous space as the separation between quantum planes. The Euclidean plane produces the curvature variance between the linear quantum plane (x axis plane) and the curved quantum plane (surface of cone). The volume of space is this curvature variance as established by our Euclidean plane.

However, when you introduce a fourth stellar body located outside the plane, the new Euclidean formation establishes competitive spacial definition. The gravitational lines established by the fourth stellar body define a three dimensional Euclidean form competitive with quantum curvature definition of that same space.

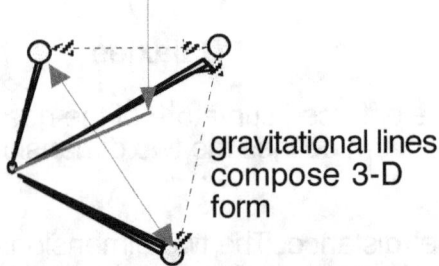

Third Euclidean axis
now intersects
original plane.

gravitational lines
compose 3-D
form

Quantum volume and Euclidean volume now define the same space. Volume is defined along the third Euclidean axis by both quantum and Euclidean calculation. The original stellar plane defines quantum volume as curvature above the new Euclidean axis. Euclidean volume is defined by that same axis.

The fourth star establishes four intersecting Euclidean planes where there had been one. Each one of these intersecting planes now locate a competitive quantum curvature variance as the definition of volume. Intersecting Euclidean planes produce quantum curvatures which try to expand the same volume.

Multiple Intersecting Euclidean Planes
Seriously Distort Quantum Space

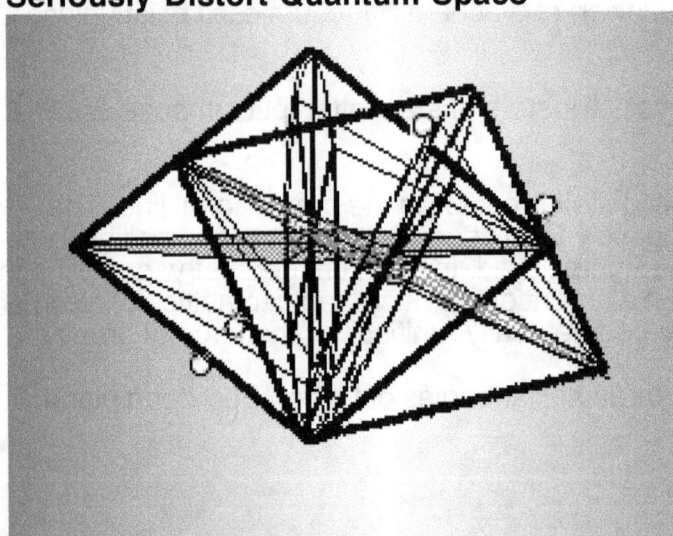

The end result of these competitive quantum curvatures is the "bulging" of volume. Since curvature is a function of time force, such "bulging" contracts back. The only possible resolution is to force the intersecting Euclidean planes into the same plane.

In the above illustration, the circles represent intersecting Euclidean planes. If the angle of aperture between them were eliminated, they would become the same plane and the distortion of quantum volume would be eliminated. Quantum space contracts upon stellar masses pressuring the lines of gravitation between stellar objects into a single plane. Vacuous space can only be two-dimensional Euclidean. It cannot be three. The shapes of the stellar masses in a common gravitational field known as galaxies prove that this is true.

With the exception of gaseous masses surrounding a few heavy "blue stars," all galaxies fall into two categories. They are either "elliptical" or "spiral."

A Spiral Galaxy

What astronomers have failed to recognize, primarily because they lack four-dimensional quantum geometry, is that the elliptical galaxies are in the process of becoming spiral galaxies.

The force of quantum space is contracting back upon these stellar masses pressuring them into a single plane. This tendency is seen in its completed form as the "spiral galaxy."

The spiral galaxy is flat. The motion of the stars contained within the plane to which it has flattened have a new vector of motion defined by increased gravity. That is, they orbit around a denser center in a plane extending from the center's equator at 90° to the axis of rotation.

Spiral galaxies were spherical formations of stars which were compressed by quantum space into a plane around a denser center of gravity. The stars in the plane were forced into orbital motion in order to compensate for the gravitation pull created by the compression of the center.

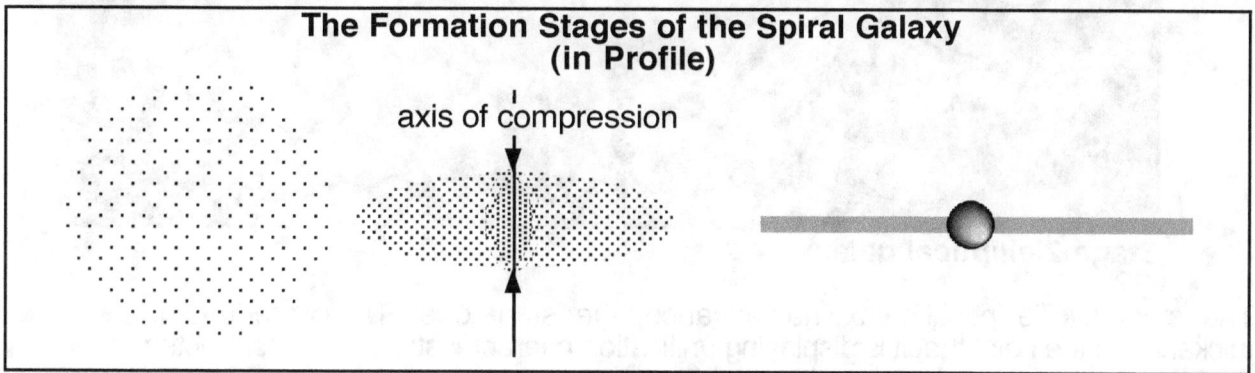

The Formation Stages of the Spiral Galaxy (in Profile)

axis of compression

In the above illustration, it can be seen that the "flattening" of the galaxy increases density of stars in the center over the density of stars in the periphery. There are more stars along the axis of compression than along peripheral axii.

The new star system constitutes a more refined Euclidean formation. The "flattening" of the system has made it more coherent with the two dimensional Euclidean definition of vacuum. Becoming "more" Euclidean has unified separate and distinct lines of gravitation into a common field. As the result of flattening, stars are brought into a stronger gravitational field relative to the center and this stronger field requires orbital motion along the Euclidean plane in order to compensate.

Stage 1 elliptical galaxy

The stage one elliptical galaxy shows the thickening of the center relative to the periphery. While there is evidence that the plane is beginning to form, the center is not yet advanced enough to constitute a unified gravitational force. There is not yet evidence of orbital motion.

Stage 2 elliptical galaxy

The "stage two" elliptical galaxy has advanced the "stage one." The center has now thickened to the point that it is displaying unification characteristics. It has contracted to form clearer edges and has increased in brightness in comparison to the poorly formed "stage one" center. While the two-dimensional plane is not completely formed, it is closer than the "stage one." Further, evidence of orbital characteristics have begun to appear in ring-like lines in the most advanced areas of the plane.

For comparison's sake, let us review the well formed spiral galaxy. The center has completely contracted. Both plane and orbital motion are all well formed.

Final Stage

In nearly spherical clusters of stars with a low elliptical eccentricity, the motion of individual stars is random and disassociated. As the galaxy approaches the spiral stage, however, with a well defined center of gravity, that motion becomes orbital. When a spherical collection of stars begins to acquire a center of gravity, the gravity becomes a system which begins to influence the motion of all stars contained within the system.

While astronomers have catalogued galaxies as "spiral and elliptical", few would accept the assertion that the elliptical galaxy is a stage in the formation of the spiral galaxy. It simply wouldn't fit what they believe they know about gravity. Gravity must pull toward the center of the formation, not towards a two dimensional plane.

What scientists are missing is the fact that the oppositional lines of gravitation between stars are also geometric definitions. They do not recognize what vacuous space actually is; that it is composed by the force of time; that volume is only the variance between linear and curved quantum planes. They do not understand that the geometry of gravitational lines can disturb and distort this forced curvature; that space is composed by force, a force which can be applied to shape the very space in which the stars themselves swim.

Appendix

Appendix Table of Contents

The Discovery of Negative Radiation
and the Four-Dimensional Quantum Atom

We have looked at the three most important quantum measurements made in the late nineteenth and early twentieth centuries. I have demonstrated that those empirical discoveries were either misunderstood (Rydberg), misapplied (Planck) or not comprehended (Millikan) by twentieth century science. Further, I have suggested that the corrections require four dimensional quantum geometry.

Understanding Rydberg's formula as identifying wave harmonics based upon a root frequency requires a new quantum "string model" of the electron bond.

Correctly applying the time value for Planck's Constant requires substitution of the "canonized" photon theory of light with a functioning wave model. This in turn, requires finding a credible medium for the replacement of the discredited "ether" as a medium for conducting light waves (electromagnetic wave theory not withstanding)[165].

Quantum geometry identifies vacuous space as the time-enforced curvature of Euclidean lines and planes. The capacity of such a space to exchange linear distance for volume makes a transverse wave possible. I have shown that the water wave "soliton" is explained by volume pulses through quantum defined vacuous space. This identifies space itself as the *succedaneum* for "ether."

Comprehending how Millikan's oil drops could ionize by whole number multiples of a single particle charge —regardless of the mass of the drop— requires knowledge of negative radiation and the identification of x-ray as a negative radiation band with respect to hydrocarbons. The possibility of negative radiation pressure was revealed by both four-dimensional quantum mathematics and "1+1 dimensional" soliton mathematics. It was confirmed experimentally by studies of black light.

The facts are these: the Rydberg frequencies are an harmonic distribution of an unrecognized root frequency; Planck's Constant must be multiplied by time in seconds to give energy —the "static" value of Planck's Constant can only be used to measure the movement or change in energy states between quantum electron orbits; Millikan's ionization by quantized elementary charges cannot be explained by Einstein's photoelectric equation— x-ray is negative radiation and only negative radiation pressure can explain the quantized changes in elementary charge which Millikan observed.

We now must turn to the four-dimensional atom to reveal in detail and by experimentation what was missed by twentieth century science.

The Electron Bond is a Quantum String which Outputs Light

1) A particle charge is an attachment through the fourth dimension relative to the particle's volume. The electron/proton bond is a tensioned string defined by the quantum squared (Q^2). The frequency of vibration of this string determines the radiation frequency which the bond length can absorb and output. Frequency is a function of string

---

[165] The alleged " electromagnetic wave" as the explanation of radiant energy across space is sheer fantasy. Electromagnetic fields are never detected in the presence of light. Electromagnetic field fluxes can *generate* low frequency radiation, but they cannot *carry* radiation. No amount of mental gyrations will make the fact that such fields are never found in the presence of light go away.

tension which itself is a function of string length. Higher frequency of light output indicates longer electron radials (string lengths).

The Quantum String Produces Quantum Light Harmonics

2) String lengths are restricted to quantum mathematical differentiations of the "root frequency." The root frequency is produced by the shortest wavelength (and highest frequency) in the spectrographic Lyman Series of hydrogen ultraviolet (91.14 nm).

Five bands of hydrogen spectrographic lines have been identified by the Rydberg formula. I have demonstrated that the Rydberg Constant is only the inverse of the the root wavelength and that Rydberg's identification of hydrogen spectrographic lines actually identifies a harmonic series of light based upon the quantum differentiations of the root wavelength (91.14 nanometers).

Light harmonics are built upon quantum differentiation of string lengths (by the negation of subdivision), not subdivision of string lengths as with sound harmonics. Rydberg unknowingly identified the formula for the quantum harmonic series.

The Electron String Exchanges Energy with the Nucleus

3) The electron string is a bond which produces an energy exchange between the nucleus and the electron of a molecule. The formula for this exchange was identified as Planck's Constant by experimentation with negative radiation. The Electron/ nuclear bond (the string) absorbs negative radiation by negative impedance of the frequency. Five months before the existence of N-radiation was confirmed experimentally, mathematical theorists predicted the existence of "negative radiation pressure."[166]

Negative radiation pressure forces the nucleus to neutralize string tension with the nucleus's own energy. The radiation is negatively impeded by the electron string. Negative impedance forces absorbing electron strings to radiate off the heat of the nucleus.

Research has shown this process is most efficient with hydrocarbons. Temperature reduction was discovered to be a function of the number of hydrogen bonds in the hydrocarbon molecule. Negative radiation is concatenated by Rydberg frequencies and the Rydberg frequencies are indexed on hydrogen.

Planck's Constant identifies the drop in temperature per hydrogen bond times the mass of the proton attached to the electron string. Planck's Constant determines the change in temperature which can be provided the nucleus per electron string. When multiplied by the frequency natural to the string length, Planck's Constant adjudicates the total potential energy which can be exchanged between an electron— at that radius— and the nucleus.

Planck's Constant determines only the potential energy value of the string and is equivalent to the energy in the light which can be output by the string. It does not represent either the energy of the light being absorbed by the string nor the total energy expressed by the vibration of the string.

All of the above are confirmed either by mathematics or hard empirical data. Taken as a

---

[166] *Negative radiation pressure exerted on kinks;"* Péter Forgács, Árpád Lukács, Tomes Roma ń czukiewicz; Phys.Rev.D77:125012,2008

whole, they represent accurate scientific models made available by recognition of a fourth quantum dimension and applying it to existing science.

Negative Radiation Pressure

In the article "*Negative radiation pressure exerted on kinks*[167] " the authors illustrate what they mean by "negative radiation pressure" with the figure below (modified for clarity purposes).

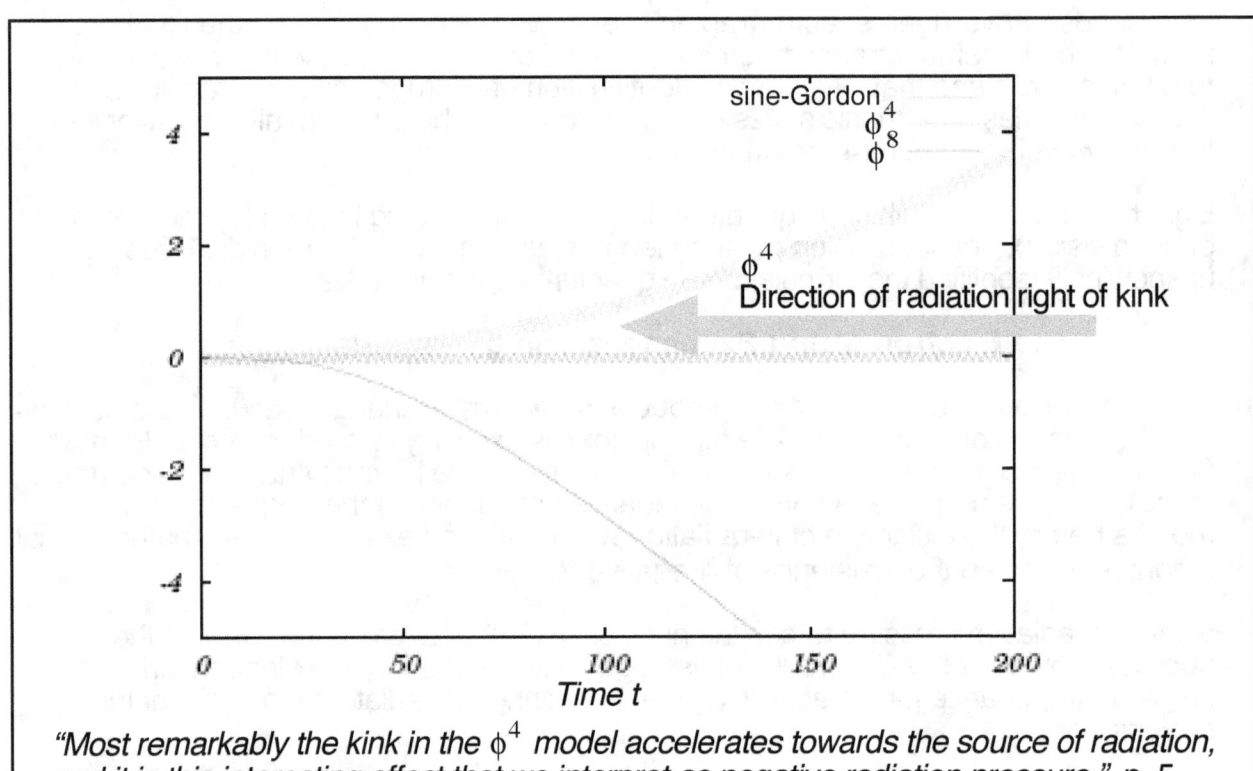

"Most remarkably the kink in the ϕ^4 model accelerates towards the source of radiation, and it is this interesting effect that we interpret as negative radiation pressure." p. 5

Before we can discuss the significance of the findings of Roma ń czukiewicz *et. al.* we must first turn to the geometric description of the soliton "kink" to which the findings apply. Essentially, a "kink" is a strict Euclidean attempt to identify the quantum squared.

Vongehr likens "kinks in 1+1 dimensions" to clothespins attached to a twisted rope[168] . The "kinks" are projections into vacuum along a single dimension but a dimension which is under tension (the twists on the rope to which the kinks are attached). Essentially, the amount of twist on the rope is differentiating kink-influenced vacuum by tension. An untensioned vacuum is likened to the rope with no twisting torque and all clothespin "kinks" pointing downward. The greater the twist on the rope the more tension the "kinks" are applying to vacuum. The 1+1 dimensional kink is a conceptual attempt to supply a field value to vacuum.

The kink so described is the quantum squared turned on its head. Quantum geometry identifies quantum squared vacuum as applying tension to the Euclidean dimension

---

[167] op. cit.
[168] *"Solitons"* p.5 op. cit.

(Vongehr's rope). It is not the rope which is applying tension to vacuum. In fact, it is the tension being applied to the "rope" which is causing it to "kink."

Time-force tension applied to the x axis "rope" by the quantum squared is forcing the x axis to "kink" upward to relieve the tension. Roma ń czukiewicz *et. al.* mathematically modeled these "kinks" as "accelerating towards the source of radiation...that we interpret as negative radiation pressure."[169]

In the above illustration, "the kink in the ϕ^4 model" (ϕ^4 is a specialized mathematics based upon a "Schrodinger-type operator"[170] which will not be consider) is undergoing acceleration in the direction of radiation travel. The radiation is causing the kink to rush towards it as if the radiation were a vacuum, a "negative radiation pressure."

In keeping with pressure analogy, high pressure matter rushing towards vacuum is releasing pressure potential energy as motion. By doing so, it acquires a less energetic lower pressure state. The kink is giving up its own energy to the radiation. In quantum geometry this is called "negative impedance."

To illustrate this we can use the pressure differential sine wave used previously (realizing that the model is not adequate to light radiation).

The Pressure Differentiated Scaled Sine Wave
The number of density peaks and density troughs passing a fixed point over time is frequency

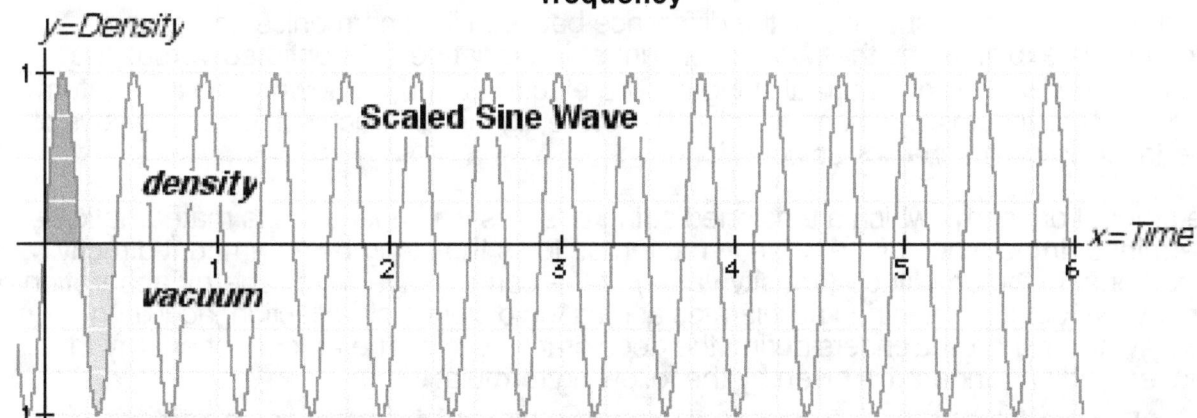

Every density peak is mass in motion and represents a positive energy state. It is followed by a negative density trough or vacuum energy state of equivalent value. Density has a mass value but vacuum is the relative absence of mass.

Roma ń czukiewicz *et. al.* are arguing that the vacuum trough is not just the absence of energy, that is the removal of "mass" from the motion of the wave. They are arguing that the vacuum trough is negative energy in that it can force an impeding "kink" to surrender its own energy. Research by the Snake River N-Radiation Lab has proven the theoretical mathematicians to be right. However, this is possible only because of the special characteristics of quantum-squared "kinks."

A fixed mass —something standing in the way of the motion of the density peaks — can

[169] Roma ń czukiewicz *et. al.* p. 5 op. cit.
[170] ibid. p. 3

"impede" or absorb the energy of high density peaks in motion.

The best model for an impedance device may be a tensioned, vibrating string. However, the tensioned Euclidean string, such as that on a violin, is restricted to impedance high density "peaks" and cannot really respond to "negative radiation pressure."

A tensioned, vibrating string is an impedance device which can convert a compression carrier wave into a standing wave of the same or related frequencies. This is called "the impedance wave." That is, the string is made to vibrate by the carrier wave, the cycle of vibration identifying the impedance wave.

The standing impedance wave can also be graphed by the scaled sine wave with an x axis defining time. The y axis, however, no longer indicates density as with the carrier wave, but energy. The positive energy or "peak" impeded from the carrier wave is balanced by the negative energy given up through the acceleration/deceleration cycle of the string.

Tensioned string impedance is based upon frequency harmonics. A string of fixed tension and length vibrates at a set frequency —its "root" frequency. The string can absorb or impede frequencies which are harmonically related to this root frequency. In sound, the harmonic series consists of whole number multiples of the root frequency.

The frequency harmonics of light, however, are not the same as the frequency harmonics of sound.

The current writer has shown that the difference between light harmonics and sound harmonics is explained by the way a quantum string must be differentiated versus the harmonic subdivisions of a normal string.[171] The electron bond is shown to be a quantum string which distributes light harmonically. Therefore, light is as much subject to the laws of impedance wave mechanics as sound.

The amount of energy which a tensioned string absorbs via impedance is mathematically determined. Impedance of the wave in motion deflects the string by an amount directly proportional to the amplitude (density value) of the carrier wave. After energy absorption of carrier wave density, the string accelerates across the distance of deflection, acquiring energy which it then surrenders during the deceleration cycle. The amount of energy thus acquired and surrendered is given by the following formula:

$$E = 8(m)\left(d^2\right)\left(f^2\right) \; ; \; m = \text{mass of string} \; ; \; d = \text{deflection} \; ; \; f = \text{frequency}$$

It will be objected that, tensioned string impedance may be appropriate for measurement of sound waves, but certainly not appropriate for light. The objection is spurious, founded upon ignorance of the difference between a sound wave and a light wave and the difference between the Euclidean tensioned string and the quantum tensioned string.

Sound vs. Light; Compression Waves vs. Pulsed Transverse Waves

Sound waves are compression waves moving through matter, usually a fluid. Light waves are transverse waves pulsing through linked quantum-squared "solitons" which compose vacuous space. Sound waves may be impeded by a tensioned Euclidean string. Light waves are impeded by "kinks" in the quantum squared electron string which can be "stretched" in unison with the rise and fall of "kinks" in the solitons composing vacuous space.

---

[171] See Appendix, "Quantum Harmonics by Negation of Subdivision vs. Euclidean Harmonics by Subdivision"

The exact mechanism by which quantum strings absorb energy pulses from quantum space presents conceptual difficulties , even if the mathematics are clear. There may be some hint, however, in "electromagnetic wave theory" of light.

The invention of radio does not "prove" that light wave are the intersection of electric and magnetic fields being projected through space. It proves only that the intersection of such fields can interface with space and send pulses of regular frequency through it.

The electromagnetic wave theory of light is a confusion of cause and effect. The proof of this is obvious. If light were pulses conducted through space as fluxes in electromagnetic fields rather than pulses in vacuous space generated by fluxes in electromagnetic fields, then the fields themselves should be measurable in the presence of light. They never are.

The standard argument that the "fields are moving to fast to detect" is spurious. If light is a high-energy electromagnetic field flux passing any point in space at a moment in time, the fields should become more visible to any point in space as frequency increases. The field would be "present" for a greater number of times per second. An electromagnetic field detection meter doesn't move in the presence of light no matter how high the frequency of light.

The electromagnetic wave theory is perpetually and continuously disproved by the absence of electromagnetic fields in the presence of light.

Light waves are pulses of energy conducted by two intersecting fields at 90°, as Maxwell surmised. Those fields, however, are geometric "fields" or units of area in a quantum squared soliton. They are the intersection of a geometrically "dense" Euclidean line "kinked" into area and a geometrically "vacuum" quantum field. They are fields of strict geometric definition, and the energy pulses are the exchange of vacuum for density —producing a volume pulse — and density for vacuum —producing a volume subtraction. That exchange is a force vector exchange governed by the principle of the negation of subdivision for the quantum squared.

Maxwell's electromagnetic field pulses are similar enough to the spacial field pulses that electromagnetic pulses can generate sympathetic resonances with space, just as a tuning fork will begin vibrating in sympathetic resonance with a sound frequency to which it is tuned.

Maxwell argues that if a magnetic field is rotated with its poles set at a 90° angle to the direction of an electric field, the force of the electric field will begin to "flux" or pulse at a frequency determined by the velocity of rotation. Since a field force is, by definition, a force projected into space, these fluctuations in field force will cause space itself to begin vibrating in sympathetic resonance. Maxwell's fluctuating field force has become a "string" vibrating at frequency.

Electromagnetic fields fluxing at frequency explain lower bands of radiation and the explanation is so elegantly simple that one is tempted to abandon electron string theory for a Maxwellian application to the atom itself as the explanation of all light. This, however, is not possible because the data won't support it. The Rydberg harmonic frequencies and the clear geometric principles involved precludes returning to the conceptual vagueness and the dubious mathematics of conventional physics.

It can be said, however, that the electron string establishes a capacitance field which is tuned to the frequency of the string. Frequency of string is equal to Planck capacitance divided by electron capacitance "*(38.74 / eC)μF*." It is possible that light of the same frequency which invades this field can cause the string to sympathetically vibrate much like a radio receiver which composes fields tuned to a certain frequency impedes radiation of the same frequency.

Empty space, that is volume without matter (vacuum), is composed by the quantum dimension[172] . This quantum dimension has characteristics which simply cannot be anticipated by strict three dimensional Euclidean geometry (with the exception of soliton "kink" geometry). Empty space can be "contracted" and "expanded" by very small quantum units called "alpha spaces."[173] The expansion and contraction of these small quantum units allows vacuum to conduct light radiation as waveform energy[174] .

Similarly, three dimensional physics has no knowledge of the mechanics of light propagation. Light radiation is propagated by the electron bond which is string-like, a string which can be partially addressed by conventional tensioned-string mathematics.

If these things are true— light is conducted as a wave form by empty space and is propagated and absorbed by string-like characteristics of bonded electrons —then the Planck mystery is solved (light is a frequency only, no amplitude energy value).

If light absorption is the deflection of an electron string by wave impedance, then the energy so absorbed will be a positive and negative amplitude cycling around a potential energy value. Whereas energy averages out to "0" for the carrier wave, this is not true of impedance wave. Impedance wave energy averages out to a potential energy value, not "0". Planck's "frequency only" energy is a function of the potential energy of a tensioned string.

This potential energy value of a tensioned string is the derivative of Energy for "$E = f(d^2)$." The derivative differentiates "d^2" to "0."

> The derivative for "$E=f(d^2)$" is the following:
>
> $E = 8(m)(d^2)(f^2)$; m = mass of string ; d = impedance amplitude ; f = frequency
>
> $E = f(d^2)$;   Potential Energy $= PE = D\left(\left[E = f(d^2)\right]\right) = 8(m)(f^2)$

Potential Energy is the energy value of a stilled (non vibrating) string. It must be multiplied by some string deflection (i.e. d^2) to become kinetic energy.

The vibrating string acquires energy by accelerating over the impedance amplitude until it reaches the "potential energy point" ($d^2=0$). Then it surrenders the same amount of energy by decelerating as the string is restretched. This is a positive-negative energy cycle across the potential energy point.

It can be empirically proven[175] that *Total Energy* is greater than *positive-negative vibrational energy*. That is, there is more total energy than is acquired and surrendered by

---

[172] See *"The Theory of Time Enforced Space"*

[173] $\alpha = .50214\left(10^{-15}\right)$ quantum meters .

[174] *The Theory of Time Enforced Space"* op. cit.

[175] By the fact that an impeding musical string vibrates (vibration energy) and gives off sound.

acceleration/deceleration of the vibrating string. During the impedance of a sound wave by a tensioned string tuned to the frequency of the sound, the string begins to broadcast the same pitch. The way that a tuned string absorbs and rebroadcasts a sound wave proves that total energy equals vibrational energy plus impedance energy(Potential Energy).

An energy value in excess of that required to restretch the string is emitted as a sound wave. Therefore, the formula for *Total Energy* and *Rebroadcast Energy* is the following:

Total Energy = (Vibrational Energy) + (Potential Energy = Impedance Energy)

$$\text{Total Energy} = E + PE = \left(d^2 + 1\right)PE$$

also

Total Energy = E + (*Re* broadcast Energy)

E + (*Re* broadcast Energy) = E + PE

Re broadcast Energy = PE

The proof for this is given by the standard formula for string impedance:

Z = impedance ;   F = force ;   V = velocity ;   A = acceleration;   2 t = time of vibration

$$Z = \frac{F}{V} \; ; \; F = m(A) \; ; \; V = A(t)$$

$$Z = \frac{m(A)}{A(t)} = \frac{m}{t} \; ; \; \frac{1}{2t} = \text{frequency} \; ; \; \frac{1}{t} = 2f$$

$$Z = 2\,m\,f$$

Impedance is the frequency factor for the Potential Energy derivative:

$$D(m\,(2f)^2) = 2(2\ m\ f) = 4(m\ f)$$

Momentum is always the derivative of energy[176] . Impedance is 1/2 the derivative of Potential Energy. Therefore, twice impedance may be considered "Potential Momentum." Multiplied by distance of deflection "d," Potential Momentum gives string momentum as the vibrating string crosses the "0" acceleration point (no deflection on the string):

d=distance of string deflection

$$V = \frac{2d}{t} = (2d)(2f) = 4\ d\ f$$

$$\text{momentum} = m\,v = d\left[4(m\ f\ )\right] \; ;$$

(Potential Momentum)(d)=string momentum

Impedance is the frequency factor for the Potential Momentum of a string which vibrates at a certain frequency(determined by tension and length). The unit of tensioned-string impedance is kilogram-hertz (kg-hz). The weight of the string is measured in kilograms and the frequency of the string is measured in hertz.

Any tensioned string has a predetermined kilogram-hertz impedance of a carrier wave of the same frequency. The impedance is the same regardless of the "loudness" of the carrier wave.

---

[176] A math fact largely unrecognized in physics: $E = m\ v^2 / 2 \; ; \; D(E) = 2m(v / 2) = m\ v$

The amount of energy absorbed by the string is not the same as the energy retained by the string. Carrier wave amplitude is completely absorbed by string deflection. Only the impedance energy value of the carrier wave is retained by the string. Impedance determines the energy which is left as the vibrating string crosses and recrosses the "0" acceleration point. This retained impedance energy is the energy which the string must surrender or "rebroadcast" as a sound wave.

If a string is tuned to a certain frequency carrier wave, the string begins to rebroadcast the carrier wave frequency. The loudness of the rebroadcast emission, however, is never a function of the carrier wave amplitude (Total Energy). It is always is a function of the frequency and the frequency alone(Impedance or Potential Energy).

Carrier wave amplitude is invested in string deflection and string deflection alone. It does not effect the rebroadcast of the wave. If the carrier wave is "loud" the rebroadcast wave will have the same energy amplitude as if the carrier wave were "soft." The amplitude of the original carrier wave is irrelevant to the rebroadcast emission which is determined by the string's impedance value.

This is exactly what Planck found, but was unable to recognize as the Impedance Energy of an tensioned string. By his own testimony, Planck's Constant is determined by subtracting light-frequency replication (rebroadcast emissions). A black body is one which subtracts light frequency replication (rebroadcast emissions) and invests the energy as heat in the atom. The body is "black" because it does not rebroadcast light frequencies.

If the electron string theory is correct, Planck energy can only be the impedance energy ("frequency only") portion of total energy. The rest of Total Energy, which incorporates carrier wave amplitude, is invested in stretching and restretching the electron string to sustain vibration. The intensity of the received light wave is irrelevant to Planck energy.

Planck's "no amplitude, frequency only" energy formulation was mathematically determined by multiple experimental measurements and subsequently experimentally confirmed many times. Still, its operational principles were unknown, its underlying impedance mathematics were a complete mystery. Ignorance allows for misapplication and Planck's Constant soon suffered this fate.

Planck's energy value was assumed to be the energy contained in light. That is, it was treated as if it were Total Energy. But this was not true. It can only be the impedance energy or Potential Energy portion of Total Energy. It is a measure of frequency and frequency alone. It is a measure of the impedance energy of an electron string.

The Quantum Geometric Determination of Planck's Constant
This determination was made by a quantum study which was conducted through the Snake River N-Radiation Lab. In a previous four-year study of a radiant cooling[177] the current author had co-discovered several bands of "negative impedance radiation." Negative impedance radiation (N-radiation) operates as the inverse of positive impedance radiation.[178]
It is the equivalent of the string's "kink" responding to negative radiation pressure.
Black bodies operate by positive impedance. They add Impedance Energy to the nucleus in the form of heat. Instead of being rebroadcast, impedance energy adds nuclear heat.

---

[177] Proprietorial study of a radiant cooling system invented by thermodynamics engineer David Rule.
[178] The full explanation of negative radiation requires quantum mathematics.

"White bodies" absorb negative radiation and operate by negative impedance. They subtract impedance energy (heat) from the nucleus and invest it as broadcast radiation output. White bodies "glow" while their nuclei cool down.

Black bodies absorb P-radiation by heating and emitting no light. White bodies absorb N-radiation by glowing and subtracting heat. Black bodies and white bodies are the logical inverses of one another.

The most familiar form of "N-radiation" is "black light," the soft band of ultraviolet at approximately 370 nanometers wavelength.

Negative Impedance

The negative impedance frequencies are quantum phenomenon. Their existence and measurement is definitive proof that the electron-proton bond is a string and that particle charges are projections into a forth dimension. There simply is no more room left for debate.

Negative impedance operates on the principle that the radiation frequencies which can be absorbed and output by the electron string are restricted by the quantum. Frequencies are functions of string lengths which are restricted mathematically to quantum-squared negation of subdivisions of the root frequency.

Negative impedance occurs when the negative radiation pressure peaks[179] of a non quantum frequency overlap the positive radiation pressure peaks of a quantum frequency.

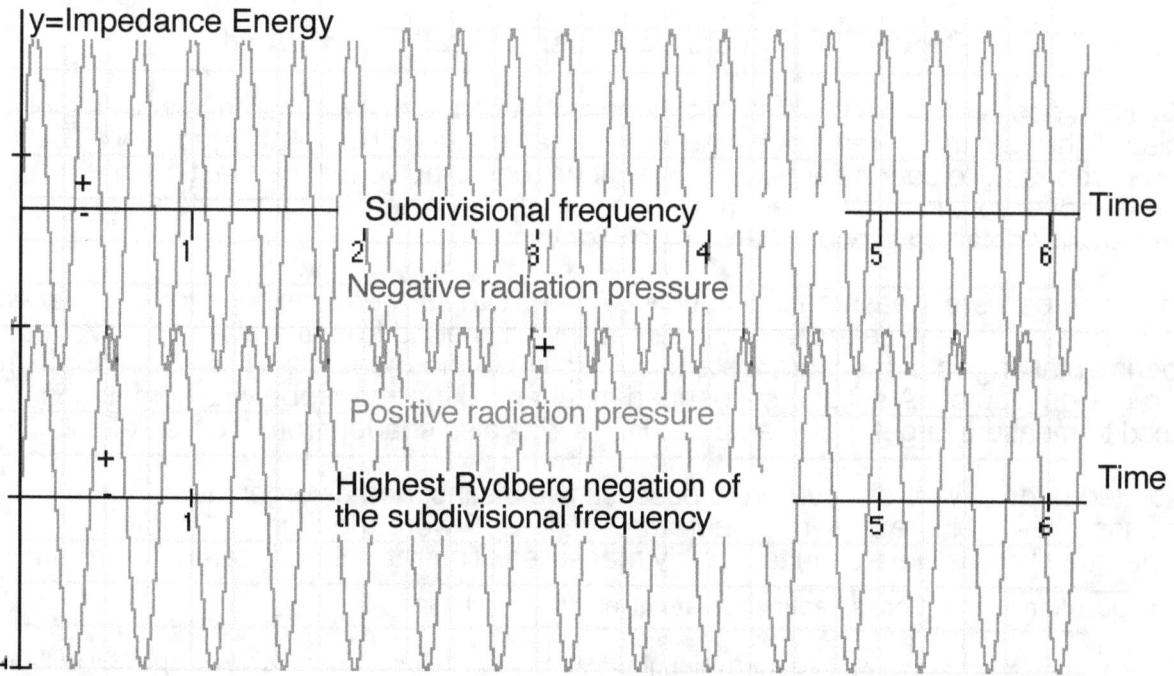

Black light is an excellent example of wave overlap. At wavelength 370 nanometers, it gives a frequency at just above the last quantum frequency in the Balmer Series of visible

---

[179] "Negative radiation pressure" is the subtraction of volume from vacuum or the contraction of space. It literally "accelerates" the kink towards the contraction.

frequencies which are available to the hydrogen atom.[180]

The highest Balmer quantum frequency is at 388.9 nm. The 388.9 nm wavelength represents the longest string length the electron can acquire in the visible light frequencies and is shorter than needed by the black-light frequency:

$$\frac{1}{\lambda} = \left(\frac{1}{2^2} - \frac{1}{8^2}\right)\frac{1}{\lambda_r} = \frac{1}{388.9 \text{ nm}}$$

Using the formula for Impedance (Potential) Energy, the Impedance Energy for the quantum frequency is 90.5% of the Impedance Energy for the black light . Amplitude for the quantum frequency would need be less than 5% of the frequency's Impedance Energy $[d^2=.05(PE)]$. This would allow the waves to overlap. 5% is a very easily obtained wave amplitude.

The next longest string length available to the electron is much too long. It impedes the ultraviolet Lyman Series wavelength of 121.5 nm[181] . (Shorter wavelengths produce higher frequencies and longer string lengths). The electron cannot acquire the exact string length needed for positive impedance of the black-light frequency.

Restricted to its quantum string length, the electron can only impede the black-light carrier wave in its negative radiation pressure peaks. The positive impedance peaks for the string are negative peaks for the carrier wave. This has the effect of inverting the impedance formula; of outputting impedance amplitude as light and subtracting Potential or Impedance Energy from the atom.

Negative Impedance Experimentally Confirmed

This effect has been experimentally confirmed.[182] A series of hydrocarbons which "glowed" in black light were measured for change in temperature during black-light irradiation. The hydrocarbons were cotton fibers, polyester fibers and a polyethylene plastic. The hydrocarbons were irradiated from a distance of approximately 30 centimeters using a commercially obtained Woods-filtered black-light source.

Temperatures were measured in two ways. A sensitive infrared thermometer was placed at approximately 20 centimeters from the target. It was shaded from the black-light. Various experimental angles of measure were used relative to the most direct line of exposure from the radiation source. A second suspended mechanical thermo probe was sometimes placed to measure target "glow temperature." It was also shaded from the black-light.

Irradiated targets were discovered to possess two distinct and divergent temperature readings. "Glow" increases in temperature" of 1°-2 °F were measured. "Body temperature" drops were simultaneously measured. The rate of body temperature drop

---

[180] The quantum Balmer Series frequencies are given by the formula:

$$\frac{1}{\lambda} = \left(\frac{1}{2^2} - \frac{1}{n^2}\right)\frac{1}{\lambda_r}; \ \lambda_r = \text{root wavelength} = 91.14 \text{ nm}; \quad 3 \leq n \leq 8; \text{ Formula dev eloped by author.}$$

[181] The Dawson formula for Lyman Series 121.5 nm wavelength is:

$$\frac{1}{\lambda} = \left(1 - \frac{1}{2^2}\right)\frac{1}{\lambda_r}; \ \lambda_r = \text{root wavelength} = 91.14 \text{ nm}$$

[182] *"The Determination of Planck's Constant by the Negative Irradiation of Cotton Fibers,"* See Appendix.

per second was used as a measure of energy loss.

A thermo probe suspended in the emitted "glow" (but shaded from the black light) determined that the "glow" was radiating off energy at a fixed temperature. Simultaneously, the IR thermometer determined that the body temperature of the target was cooling. The theory of negatively impeded radiation was confirmed. The "white body" target was indeed radiating off its own energy under black light stimulation.

Measuring the underlying body temperature drop was not simple. Methods had to be developed to prevent the "glow temperature" from masking infrared measurement of "body temperature." Two methods of unmasking were discovered.

If the black light were pulsed, the IR thermometer would often "lock" onto body temperature during the off phase of the pulse. With glow temperature eliminated during the off phase, the IR could "see" the body temperature. This method, however, was not experimentally feasible. The drop in body temperature would also be "pulsed," eliminating accurate energy-loss measurements.

A second method of unmasking body temperature was discovered which would prove viable for energy-loss measurement. If the angle of the IR temperature reading were made oblique enough to direct glow, the thermometer would ignore glow temperature and lock onto body temperature.

The major problem associated with using an oblique angle to get only indirect "glow" radiation was that greater angles increased the sample size of the target surface area. Care had to be taken that the sampled area did not exceed the target surface area. In the illustration the sample surface area for the oblique measure is much greater than the sample area for the direct measurement.

Angles of Measurement
with Infrared Thermometer

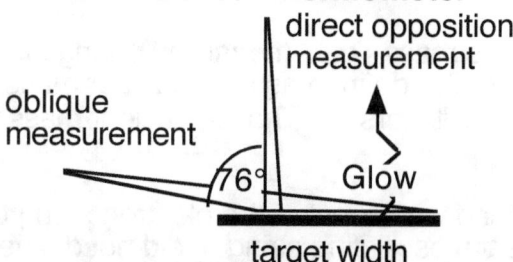

Care was taken that the measured surface area would not exceed the target surface area.

As much as possible the fibrous targets were flattened into planes with approximately 100 square centimeter surface areas. This allowed an oblique measurement angle of 76° from a distance of 20 centimeters.

Of the materials tested, cotton was chosen as the index material. There were several reasons for this. First, cotton glowed brilliantly in black light. Secondly, and most importantly, the molecular structure of cotton was established and known. The molecular structure of the artificial hydrocarbons was problematic. The number of esters and ethylenes in the molecule can change, depending upon manufacture and were unknown.

Cotton did present methodological problems of its own. The fibers are highly absorbent and prone to acquire coatings which are non reflexive with respect to black light. Even careful handling could leave dark spots in glowing cotton from skin oils. Commercial cotton which had been chemically treated proved useless as a sample.

Finally, precautions were taken to assure that dynamic ambient heat exchange approached "0." A room of static temperature which is being simultaneously heated from the walls and air conditioned does not meet this criteria.

Measurement of exact temperature changes in a known molecular structure was required to test the electron string theory. The drop in body temperature under black light was hypothetically due to negative impedance.

Just as black bodies under positive irradiation add the potential energy from the impedance formula to the nucleus, so "white bodies" under negative irradiation hypothetically subtract impedance potential energy from the nucleus.

Quantum harmonic frequencies were originally calibrated for the hydrogen atom. The hydrogen bonds of the hydrocarbon were thought to be "oscillators" which exchange energy with the nucleus. Bonding hydrogen electrons are not "locked" into orbital positions as are the electrons for the heavier elements and are free to acquire the 388.9 nm orbit .

The presence of covalent hydrogen bonds was considered the reason that hydrocarbons "glowed" in black light and non-organic molecules did not. Bonded hydrogen electrons were assumed to be free to acquire the orbital levels needed to emit light frequencies (as opposed to merely reflecting light frequencies.) The glow was presumed to be light emission since it illuminated nearby darkened objects more than could be accounted for by simple light reflection. The results proved this presumption to be correct.

Quantum geometry predicts the sum of hydrogen bonds can influence the whole of the nucleus. Each hydrogen "oscillator" had to theoretically move a fixed amount of nuclear weight.

The energy equation used was mass *times* measured change in temperature per second and converted to joules. Mass was defined as total nuclear mass of the molecule divided by the number of hydrogen "oscillators.[183] " That is, nuclear mass was defined as "per string oscillator."

The energy equation so defined was a test of the electron string theory. It defined mass as "per hydrogen bond" or for each oscillating string. It did not define mass as "per atom." Mass was not defined as that which is contained in an individual atom but that which must be moved by an oscillating electron string. The unit of measure became the proposed molecular string and not the atomic makeup of the molecule.

If the definition of mass as "per hydrogen bond" instead of "per atom" were artificial, then the energy equation used should have proven futile. Instead, it derived Planck's Constant with a high degree of accuracy.

If one divides the measured change in temperature for the cotton per second by the number of hydrogen bonds per molecule, one arrives at the change in temperature per hydrogen bond (oscillator) per molecule. If one multiplies this figure by the mass of a single proton(in kilograms) theoretically attached to the electron string, one arrives at the negative

---

[183] *"The Determination of Planck's Constant by the Negative Irradiation of Cotton Fibers,"*   op.cit..

of Planck's Constant with an accuracy of six tenths of a percentage point[184] :

Empirical Equation Derived from Cotton Data

ΔT = measured change in temperature per second;

n = number of hydrogen bonds per molecule;

(239) = conversion operator: *calories to joules*

m = mass of single proton

$$\frac{\Delta T}{n(239)}(m) = \text{Planck's Constant} = h$$

Planck's Constant is the change in molecular temperature per oscillating string.

Calculating Plank from Experimental Data

Experimental result:

$$-h = \frac{(-0.0057189542° \text{ centigrade} / \text{sec.})\left(1.6726\left(10^{-27}\right) \text{ kg } \triangleright \textit{mass of proton}\right)}{(60 \triangleright \textit{H bonds per molcule})(239 \triangleright \textit{joules conversion operator})}$$

$$= -6.6705177548\left(10^{-34}\right)$$

Actual value of $-h = -6.6260755\left(10^{-34}\right)$

experimental as percentage of actual Planck's Constant=1.006707176. There is less than a 1% variance (.006) (Millikan variance .005)

The negative of Planck's Constant was derived because negative impedance subtracts frequency energy from the nucleus. Planck's Constant is determined to be the change in temperature provided a proton in one vibration of an electron string. Positive Planck energy (frequency only) is this change in temperature provided the proton *times* the number of vibrations per second (frequency).

Change in Nuclear Temperature Determined by Number of Electron Oscillators Only

The change in temperature provided the nucleus is determined by the number of electron strings vibrating within the negative amplitude of the carrier wave and only by these vibrating strings. Said another way, number of oscillators and only the number of oscillators determine temperature change:

Change in Temperature Formula

let "n" = number of hydrogen bonds;

let "h" = Planck's Constant;

let "m" = mass of proton.   Independent variable="n"

$$\Delta Temp. = \frac{-(239)(n)\,h}{m}$$

The independent variable in this equation is "n." All other values on the right side of the equation are constants. Therefore, the dependent variable "change in temperature" is a function of the number of electron string operators(n) and only a function of the number of string oscillators.

---

[184] Robert Millikan's 1916 derivation of Planck's Constant while testing Einstein's photoelectric equation was .5% of actual. This is still considered the standard for derivations from "h dependent" formulas.

Change in Temperature is a Function of Number of String Oscillators; Energy is a Function of Molecular Size

Energy *per* string is the change in temperature *times* the mass being moved *per* vibrating electron string *times* the Calories to joules conversion factor:

Let "Wt."= atomic weight or number of protons and neutrons in the molecular nucleus

$$E = \frac{(57.34)(m)(Wt.)}{n(239)} \Delta Temp.$$

$$-h = \frac{(m)(\Delta Temp.)}{n(239)}$$

$$E = (-h)(Wt.)(57.34)$$

The loss of nuclear energy for each vibrating string is (−) Planck's Constant *times* the total Atomic Weight of the molecule *times* 57.34 (Caloric component of the conversion factor). While the number of string oscillator's absolutely determines the change in temperature, the amount of energy lost for each string is determined by the size of the molecule. The amount of energy lost per string equals (−) Planck's Constant *times* Atomic Weight of the molecule.

Two molecules with the same number of negative impedance oscillators — but of differing atomic weights — will drop in temperature at the same rate. However, the amount of energy lost will be greater for the heavier molecule than for the lighter. There will be a variance in energy at the same rate of temperature drop.

This is consistent with the original determination of the Constant by Planck. A brief review of that methodology will reveal why the current research results could have been expected[185] :

> *Planck hypothesized that the equations of motion for light are a set of harmonic oscillators, one for each possible frequency. He examined how the entropy of the oscillators varied with the temperature of the body, trying to match Wein's law, and was able to derive an approximate mathematical function for blackbody spectrum.[*
>
> *However, Planck soon realized that his solution was not unique. There were several different solutions, each of which gave a different value for the entropy of the oscillators. To save his theory, Planck had to resort to using the then controversial theory of statistical mechanics, which he described as "an act of despair ... I was ready to sacrifice any of my previous convictions about physics." One of his new boundary conditions was:*

> *"to interpret U_N [the vibrational energy of N oscillators] not as a continuous, infinitely divisible quantity, but as a discrete quantity composed of an integral number of finite equal parts. Let us call each such part the energy element ε ;"*

Planck's original methodology assumed a vibrating string. This fact has simply been ignored

---

[185] Source *"Blackbody radiation"* Main article: *"Planck's law"* ; Wikipedia

by science for over a hundred years. He assumed that, for the purposes of the energy equation, light was stimulating an "harmonic oscillator" with an "entropy" value (i.e. a measure of energy exchange efficiency) which varied with the "temperature" (i.e. higher temperatures greater energy exchange efficiency). He was trying to match Wein's Law which shows temperature as a direct function of light wavelength:

(wavelength in meters)(Kelvin temperature)=.0029

He could find this relationship (i.e. temperature is a function of wavelength) only with "black bodies" (i.e. for bodies that absorbed all radiation as increases in temperature). Even then the "entropy" or the availability of the thermal energy to transfer as work was inconsistent.

Why would this be so? Because the bodies varied in mass and mass was not part of his calculations. Planck then resorts to the statistical method, what we now would call manipulations of empirical data sets. He resorts to the Gaussian statistical method, using variables as exponential factors of the natural number.

Using these, Planck finds a statistical regularity, an energy constant which eliminates the problem of mass variations in the black bodies. He describes this as an "act of despair" which sacrificed all of his "previous convictions about physics."

In some ways, the negative radiation study of cotton reversed Planck's original journey, identifying the missing factor which had initially caused him "despair." It began with the assumption that negatively impeded energy was being exchanged through a discrete number of oscillators tuned to the frequency. In this case, however, tuned only to the "negative radiation pressure" amplitudes of the frequency.

The amount of energy lost— change in temperature times the mass of the molecule— was differentiated by the number of oscillators. This derived Planck's Constant as the change in temperature per oscillator multiplied by the mass of the attached proton. That is, the Constant was proven an actual energy value which incorporated both mass as well as change in temperature.

When Planck's Constant is multiplied by the total number of oscillators it gives the change in temperature for the whole molecule. This change in temperature must be multiplied by the "quantum size" of the molecule (number of nuclear elements devoid of their mass value) to give the actual energy value. Planck's Constant is the actual energy value. Size of the molecule is only a quantum value which must be used to determine total energy. No more reason for despair, Professor Planck.

Negative radiation has reversed the direction of energy transfer along the electron/proton bond string and negative radiation has "pierced the statistical veil," revealing Planck's Constant for what it actually is. Frequency has been factored out by subtraction[186] — a mathematical process available only to quantum geometry— revealing "h"as the fundamental energy value of the four-dimensional bond which is increased proportionally by tensioning of that bond.

This is not the first and only occasion that quantum mathematics has pierced the veil of statistical mathematics to reveal underlying physical reality. It also happened with the water soliton wave. These waves without leading or trailing pressure differential troughs could only be described as Gauss-like distributions of probability densities using the nonlinear Korteweg de Vries equation.

---

[186] See "The Planck Electron/Nucleus Energy Exchange " i n Appendix

Quantum geometry was able to replace these probability densities with quantum squared units of vacuous space which have actual physical dimensionality in that an exact number can be fitted into a meter cubed of volume. Wave motion through these quantum squared vacuum units could be modeled using the precedent of a Fermi nonlinear wave through loosely arrayed oscillators.

A non-linearity velocity scale for the quantum squared oscillators was built on the Fermi precedent, a scale which located actual soliton wave velocities as credible probability densities. Measured wave velocities were converted to credible probability densities, but the wave itself now had physical location.

With the Korteweg de Vries equation, the wave location was a probability density and wave velocity was an unrestricted independent variable. Quantum geometry pierced the solitons statistical veil, just as it has now done for Planck's Constant.

The Importance of the Negative Radiation Discovery
Losses to Science if the Quantum Dimension is Ignored

The prediction and confirmation of negative radiation opens three new pathways to science, all of which are important. Negative radiation holds promise as a powerful new technological tool. We will demonstrate one such practical application in the field of thermodynamics. Negative radiation is also a lens into the structure of the atom and offers new possibilities for unlocking the energy stored in the nucleus.

Perhaps most important of all, however, is the boost it gives to four dimensional quantum geometry. Negative radiation cannot be explained outside of quantum geometry. Negative radiation will either advance quantum mathematics or negative radiation will be swept into oblivion.

Without quantum mathematics, the phenomenon threatens to duplicate Russell's soliton wave experience. Despite strong experimental evidence and natural observation, science denied the possibility of solitons for over 50 years because solitons did not fit contemporary mathematical models. Similarly, if negative radiation cannot be explained by off-the-shelf three dimensional mathematics then the phenomenon may be resisted as "impossible."

Quantum electron-string theory was required to identify the soft-ultraviolet, "black light" frequencies as potential negative radiation. Those frequencies were not chosen out of some "intuitive feeling" about black light. They were chosen because quantum mathematics pinpointed the gap between the highest Rydberg visible frequency and the lowest Rydberg ultraviolet frequency as too great a span for available electron-string energy. The gap would necessitate negative radiation frequencies and black light was found to exist right where the equations said n-radiation must exist.

It is true that "1+1 dimensional kink" geometry predicted the existence of negative radiation. The location of the frequencies, however, was not predicted by the Roma ń czukiewicz *et. al* mathematical model. Frequency was modeled as a function of an exponential series based upon amplitude[187]. The higher the wave amplitude, the greater the frequency. Since negative radiation pressure resulted from lower amplitudes, frequency was "falling" toward the impedance kink when negative radiation pressure occurred. However, the authors did not identify specific locations of n-radiation frequencies.

The actual experimental results from n-irradiated cotton fibers —the derivation of Planck's Constant — cannot be explained outside of four dimensional quantum mathematics. The energy lost to the nucleus during n-irradiation is not simply the negative of the energy which would be gained by positive impedance of the frequency. The energy gained per string would be "$E=h(f)$" or Planck's Constant *times* frequency. The energy lost to n-irradiation was not "$-h(f)$" it was simply "$-h$"

Frequency was removed from Planck's Constant, indicating that change in temperature was not determined by the frequency of the negative radiation but only the number of string oscillators which the molecule had which were impeding the frequency. Other experimental

---

[187] Roma ń czukiewicz *et. al.* op.cit.. p.6

data proved this to be absolutely true.

Negative radiation had first been detected in what proved to be the Paschen (infrared) Series from the Rydberg distribution. The Paschen infrared frequencies (Rydberg "n=3") are between 954.48 nm and 1874.88 nm with an N-radiation frequency range of 820.26 nm $\geq f_N >$ 954.48 nm. The Paschen Series N-radiation was detected as a radiant heat exchange between polyethylene and water.[188] The water was being uniformly cooled as a recipient of Paschen infrared negative radiation generated by direct contact heat exchange between the polyethylene and the water.

The cooling of the water by Paschen Series "n=3" infrared was the exact equivalent of the cooling of the cotton by Balmer Series "n=2" black light. Despite the fact that the wavelength of Paschen Series N-radiation ($\leq$ 820.26 nm) is over 2.25 times longer than black light ($\leq$ 364.56 nm) both temperature drops were strictly determined by the number of hydrogen bonds. The "per bond" formula derived from the cotton study correctly predicted the drop in temperature for the water as measured over a great number of experiments. Frequency of the negative radiation was irrelevant. Only the number of bond oscillators determined temperature change.

The loss of temperature due to n-irradiation cannot be explained by conventional three-dimensional mathematics. It requires four-dimensional quantum mathematics to explain how frequency can simply be subtracted from the potential energy of a quantum string length, remaindering Planck's Constant.

Understanding the experimental results will require quantum calculus. Even though this calculus may be unfamiliar and difficult, full comprehension is not possible without it. Absenting quantum math, science will simply "scratch its head" and walk away like it did with Russell's soliton water wave for 50 years because off-the-shelf mathematics and physics could make no sense of it.

The Required Quantum Equations

Subtracting frequency from Planck energy requires that "frequency" be negatively integrated from "*frequency =frequency*" to "*frequency=1*." Negative integration is only available to quantum mathematics. Such negative integration also requires that time must be a quantum, another unfamiliar component of quantum geometry.

Frequency Defined by Time Quantum

$$PE = h(f) = \frac{h}{t} \; ; \; t = \frac{1}{f} = \left(1 - \frac{1}{x_n}\right) \; Quantum \; time \; definition$$

Potential string energy defined as Planck's Constant *times* frequency can also be defined by the time it takes for one vibrational cycle of the string: $t = 1/f$. If time is a quantum, then "t" must also be defined as a whole time value minus some "quantum tick." Since time is a quantum, it can only proceed by negation of subdivision. It cannot proceed by subdivisional "ticks" as with a standard "three dimensional" clock.

The derivative of a negation of subdivision produces an integral which can proceed negatively. That is, the subdivisional "tick" is equal to the whole of the area under the graph of the derivative *minus* the negative integration of the indeterminate area from "x_n" to "0":

[188] Proprietorial studies of David Rule, Pasco Poly, inc. Wieser, Idaho.

$$D(1 - \frac{1}{x_n}) = \frac{1}{x_n^2} \; ; \qquad \frac{1}{x_n} = 1 - \int_{x_n}^{0} \frac{1}{x_n^2} d(x_n) \qquad \textit{Quantum negative integration of time}$$

The subdivisional "tick" can also be defined by frequency:

$$\frac{1}{f} = 1 - \frac{1}{x_n} \; ; \qquad \left(\frac{f-1}{f}\right) = \frac{1}{x_n} \; ; \qquad x_n \left(\frac{f-1}{f}\right) = 1$$

$$x_n = \frac{f}{f-1}$$

We now have a frequency value for the quantum "tick" denominator which we can impart to the integral. What we want is a frequency value for "$x_n = \infty$; $d(1/x_n) = 0$"

$$\text{desired result} = \left\langle 1 - d(\frac{1}{x_n}) = 1 - 0 = \frac{1}{f} \; ; \; f = 1 \right\rangle \; ; d(\frac{1}{x_n}) = 0$$

as "$x_n \to \infty$", "$f \to 1$"

$$1 - \int_{x_n}^{0} \frac{1}{x_n^2} d(x_n) = \int_{0}^{\infty} \frac{1}{x_n^2} d(x_n) - \int_{x_n}^{0} \frac{1}{x_n^2} d(x_n) = \int_{x_n}^{\infty} \frac{1}{x_n^2} d(x_n) = \int_{f/(f-1)}^{f=1} \frac{1}{x_n^2} d(x_n)$$

In quantum geometry, the negative integral can stand for the positive differential since "differentiation is anti-integration" and "anti (anti-integration) is differentiation." In quantum calculus, the negative integral can stand for the positive differential since two negatives make a positive:

$$d(\frac{1}{x_n}) = 0 = - \int_{f/(f-1)}^{f=1} \frac{1}{x_n^2} d(x_n) \; ;$$

Negative integration to "frequency $= 1$" differentiates "$1/x_n$" to "0" and "$f = 1$"

Frequency can be negatively integrated from any frequency value to "1" which differentiates the subdivision by which the quantum is negated to "0." This negative integration also reduces Planck's Constant from "$h(f)$" to "$h(1)$."

$$t = \frac{1}{f = 1} \text{ (frequency subdivision)} = (1 - 0) \text{ quantum differentiation of subdivision to "0"}$$

$$PE = \frac{h}{1} \textit{ The negative radiation experimental result}$$

If time is a quantum and therefore subjected to differentiation by the negation of subdivision (not by subdivision as if with a continuum), then and only then can frequency of vibration be differentiated to "1" and all possible frequencies of vibrations can be differentiated to "1" irrespective of their original time value. Time is not restricted.

If time were a continuum, the only way that frequency could be made equal to "1" is by multiplying frequency by the time value of the vibration "$(f)(t) = 1$." Time cannot be restricted to "t" because it is also a constant for change in temperature in Planck's energy equation. Therefore, time as a continuum cannot produce the experimental results.

In four-dimensional quantum geometry, time is a quantum. Space maintains separation of a time differential which is the fundamental time quantum "ΔT." Time must proceed by "ticks" of 1.675×10^{-24} seconds.

153

David Rule's N-Radiation Dance with the Second Law of Thermodynamics

When the thermodynamics engineer David Rule first encountered Paschen Series infrared negative radiation[189] it perplexed him. He had designed a system to cool wine through the walls of polyethylene tanks and the system, unexpectedly, began working in ways he had never encountered before in his twenty years of experience with tank cooling systems.

Rule had invented a new system which could provide much greater and more uniform temperature gradients across the walls of his tanks. When he took this system to these new and higher temperature gradients, relative to existing systems, he made a completely unexpected discovery. The normal convection-current heat stratification of the tank suddenly disappeared. When the temperature variation between coolant and water (temperature gradient across the wall) hit a certain point, the tank ceased cooling as a conduction/convection tank.

Conduction/convection tank-cooling systems always lead to the temperature stratification of the stored liquids. Contact cooling by conduction of heat through the walls never completely penetrate the stored liquid, cooling only liquids proximate to the wall. These proximate cooled liquids typically drop to the floor of the tank due to the increases in specific gravity associated with lower temperatures. This generates convection currents which always temperature stratify the tank.

With lower temperature gradients across the tank wall, Rule's polyethylene tanks temperature stratified as expected. At higher temperature gradients, however, the stratification simply disappeared, leaving a nearly uniform temperature with a small variance of 1° F between the top and bottom of the tank.

The new performance produced previously unknown results. Rule was able to uniformly "super cool" 500 gallons of water to 28° F which immediately turned to ice when the water surface was slightly disturbed. The water had retained a liquid state below the freezing point because it had cooled uniformly, in a nearly perfect "still" state or one absent of convection currents. The theoretical reversal in the specific gravity of water at 39° F was proven by measuring a reversal in the 1° variance between top and bottom of the tank as temperature approached the freezing point of 32°. A tank of wine was turned into a complete block of ice, rather than forming the ice "jacket" typically produced on inner-tank walls by conventional low-temperature cooling.

David Rule's patented system applies a nearly uniform temperature to the outside walls of any storage tank. When the system was applied to stainless steel tanks, no escape from conduction/convection stratification occurred, regardless of how great a temperature gradient across the tank wall was used. The phenomenon only occurred when the tank was constructed of hydrocarbons— of hydrogen-bonded, high-density polyethylene.

A variety of sizes of hydrocarbon-based tanks were cooled by the Rule system and it was discovered that the temperature gradient required to initiate non-stratified uniformity differed with the size of the tanks. Larger tanks with thicker walls and lower ratios of wall area to volume mandated greater temperature gradients. However, once the initiating gradient was achieved, all tanks operated essentially the same, regardless of size. The whole of the volume contained within any sized tank was uniformly cooled at between 1° and 1.25° F per hour.

---

[189] See Chapt. 2

The uniform cooling required a temperature gradient between coolant and liquid which was over six times that which should have been necessary for thermo conduction across the tank walls. The Second Law of Thermodynamics[190] requires that energy must be exchanged to transfer cold to heat, but the energy being absorbed was much greater than could be explained by simple thermo conductivity.

The thermo conductivity coefficient "k" for high density polyethylene is .41-.51 W/ m °C (watts calculated by area/thickness in meters and temperature gradient in centigrade). By Fourier's Law[191] heat transfer is the following equation:

$$\text{Heat Transfer} = k(\text{area}) \frac{\Delta \text{Temp.}}{\text{thickness}} \; ; \; \Delta\text{Temp.} = \text{temperature difference across wall}$$

When the thermo conductivity coefficient is converted from "W/ m °C " to "BTU/(hr·ft·°F)" or "BTU's *per* hour," the heat transfer *per* hour across the tank walls, at gradient, can be compared to the actual loss of BTU's by the liquid in an hour. At the temperature gradient required to initiate uniform cooling, heat transfer was always multiples of temperature loss. The energy being absorbed across the tank walls was much greater than required for liquid cooling.

The current writer had known David Rule for over twenty years through a mutual interest in the wine industry. David brought the curious cooling performance to this writers attention. The best, and possibly only, explanation was a form of negative radiation.

It was assumed the best model for contact heat exchange was the Newtonian model. Using a conventional Newtonian "high mass, high velocity/ low mass, low velocity" momentum exchange model, it was concluded that the cooler, lighter polyethylene was surrendering all its heat energy before acquiring a new energy vector when heat was exchanged with the water. The momentum exchange between the hydrogen-bonded polyethylene's nuclear vibrations (its heat energy) and that of the higher energy state of the water forced the polyethylene nucleus to surrender all initial energy before acquiring a new energy state dictated by the water. It was assumed that the momentary fall to "0" energy could easily output negative radiation, via the hydrogen bonds, before reacquisition of a new energy state. The polyethylene hydrogen bonds could be pulsing Paschen negative radiation frequencies (at and just below 820 nm)— the highest infrared Rydberg negative radiation frequency range.

Subsequent research showed the radiological surmise to be essentially correct. David Rule designed a test of his system to test the radiological heat exchange hypothesis offered by the author's theory of negative radiation. The radiological character of the heat exchange was proven by restricting Rule's system to a uniform cooling of the planar floor of a specially constructed tank.

Temperature exchanges with the water in the tank penetrated much further above the planar floor than could be explained by direct contact. Further, all changes in temperature in the upper zones ceased when rapidly falling water temperatures on the floor of the tank— due to specific gravity convection— brought the gradient below the required variance. Radiological heat exchange shut down.[192] .

The n-radiation hypothesis was completely confirmed when the mathematics from the

[190] "Second form" or Clausius statement of the Second Law of Thermodynamics.

[191] Joseph Fourier , *The Analytical Theory of Heat* (1822).

[192] Proprietorial research, Pasco Poly, Inc.

black-light studies[193] correctly predicted the temperature changes measured in Rule's cooling system. That formula for change in temperature undergone by hydrogen covalent bonded molecules[194] absorbing n-radiation is the following:

n_s = number of hydrogen bonds; h = Planck's Constant

$$\frac{\Delta T}{sec.} = \frac{239\, n_s(-h)}{(mass\ proton)\ sec.} = \frac{239\,(2)(-6.6260755e-34)}{(1.6726e-27\ kilograms)\ sec.} = -\frac{\mathbf{1.\,2270639302°\ F}}{\mathbf{hour}}[195]$$

This figure is a good approximation of the temperature drops measured in the water when the Rule system achieves the required temperature gradient across the tank walls to initiate uniformity.

The experimentally deduced N-radiation formula for energy lost during radiation temperature exchange reveals why the Rule system requires thermo conductivity across tank walls which are multiples of energy lost by the liquid.

Thermo Conductivity of High Density Polyethylene.

k=0.7269087-0.88267485 Btu/(hr·ft·°F)
k=0.42 - 0.51W/(m·K) or Joules/ (sec.•m•K) *the SI unit*

1 Btu/(hr·ft·°F) = 1.730735 W/(m·K).
1 W/(m·K)=0.5777892052 Btu/(hr·ft·°F)
[Perry's Chemical Engineers' Handbook, 7th Edition, Table 1-4]

Heat Transfer=k(area) $\dfrac{\Delta Temp.}{thickness}$; $\Delta Temp.$ = temperature gradient across wall

---

Thermo conductive energy transfer is a function of an unmodified temperature variation across the tank walls. This unmodified temperature difference is not true of the drop in temperature in the water. Temperature drop by n-radiation is a constant *per molecule.* The energy directly transferring heat out of the water by thermo conduction is 4.167 *times* the amount of energy dropping temperature for the molecule. This difference is explained by a conversion factor. The factor which converts measured temperature difference to joules of energy is irrelevant to thermo conductivity but it is not to negative radiation conversion.

The "joule" energy unit is a direct conversion from Newtonian energy ($E = mv^2 / 2$) to the energy defined as "$E = m\,(\Delta temp.)$." That is, it converts "energy as motion" to "energy as temperature change." There must be a conversion factor from Calories which James Joule experimentally determined in 1847. In terms of watts that factor is the following:

$$239\,\frac{joule}{sec} = 57.35\,\frac{Calorie}{sec}$$

$$watt = \frac{1\ joule}{sec} = \left(\frac{57.35}{239}\right)\frac{Calorie}{sec} = \left(\frac{57.35}{239}\right)\frac{1\ kg(\Delta 1°\ C)}{sec}$$

---

[193] See *"The Drop in Temperature by N-Irradiation of Cotton"* in Appendix
[194] Water is assumed to be covalently bonded . An oxygen proton and the hydrogen nucleus share an electron. Citation: *"Water is polar covalently bonded within the molecule. This unequal sharing of the electrons results in a slightly positive and a slightly negative side of the molecule."*
http://www.emc.maricopa.edu/faculty/farabee/BIOBK/BioBookCHEM2.html
[195] See *"The Drop in Temperature by N-Irradiation of Cotton"* in Appendix

For thermo conductivity, the conversion factor "57.35/ 239" is irrelevant since transfer is a function only of the the temperature gradient across the tank wall:

Heat transfer= $f(\Delta Temp.)$; "tank surface area," "thickness" and "k" all constants

For the drop in temperature of the molecule due to negative radiation that is not the case:

W_m = watts per molecule *per* drop in temperature ;

N_n = number nuclear units in molecule (atomic weight)[196]

$$\frac{joules}{sec} = W_m = \frac{57.34 \; (mass \; proton)N_n}{239}(\Delta T / sec)$$

NOTE: This disregards the important equality discovered by N-radiation research:[197]

$$-(Planck's \; Constant) = \frac{\Delta Temperature(mass \; of \; proton)}{239(number \; of \; hydrogen \; bonds)}$$

The "joules to Calories" conversion factor determines drop in temperature of the molecule whereas it is irrelevant to thermo conductivity. The difference between thermo conductivity and negative radiation temperature drop is the following:

$$\frac{57.35}{239} = 0.239958159 \; ; \quad \frac{1}{0.239958159} = 4.1673931997$$

The n-radiation conversion to joules from a linear drop in temperature is only 24% of linear. Thermo conductivity transfer of heat is a function of the linear temperature gradient alone (temperature difference) across the tank walls. It takes an energy rate crossing the walls of the tank which is 4.167 *times* molecular temperature-drop energy in order to initiate negative radiant cooling. The energy transfer across the walls of the tank must be 4.167 times the energy represented by the drop in temperature due to N-radiaton absorption.

The Rule negative radiation cooling device is applied to various size tanks. Because wall surface area and thickness vary, the required temperature variance across the walls also will vary. We will illustrate the requirements for the Rule 500 gallon tank, using the commercial energy standard which Rule employs of "BTU's per hour."

BTU/ hr of Rule's 500 gallon tank.=500(8.345404 lbs per gallon)($-1.23°$ F)/ hr $\qquad\qquad$ =(-5132.42346 BTU)/ hr. Heat transfer required across wall=(4.167)(5132.42346 BTU)/ hr.=(21386.809 BTU)/ hr $\Delta Temp.= \dfrac{(Heat \; transfer)(thickness)}{k(area)}$; $\Delta Temp.$ = temperature difference across wall $\Delta Temp.= \dfrac{[(21386.809 \; BTU) / \; hr](.0278 \; ft.)}{[0.7269087 \; BTU / (hr \bullet ft \bullet °F)](68.722 \; sq. \; ft.)} = 11.892 \; °F$ gradient

It appears that it requires a 12° gradient across the tank wall in order to stimulate negative radiation temperature drop as *per* the Clausius statement of the Second Law of Thermodynamics. This, however, is not whole of the required variance. The calculated 12° gradient does not account for return "glow" radiation.

Negative radiation frequencies force absorbing hydrogen-bonded molecules to radiate off

---

[196] Ibid.

[197] Ibid.

heat energy from the molecular nucleus. The amount of return energy in the "glow" is equal to the energy lost as heat by the nucleus. That is, the hydrogen bonds in the tank wall are absorbing Paschen Series 954 nm radiation at an energy rate of 5132.42 BTU/hr. This is the rate by which the wall is being heated from return radiation "glow." Actually, the gradient must be adjusted for at least twice this amount. The gradient must not only neutralize the heating, but reduce it a second time to equalize with the temperature of the water. "Twice glow energy" is a minimum gradient increase as it does not account for the "Clausius Statement" energy required to move temperature from hot to cold.

The actual gradient across the walls of Rule's 500 gallon polyethylene tank which is required to initiate negative radiation cooling is calculated as a 17.61° F variance between coolant temperature on the outside wall of the tank versus the temperature of the water inside the tank.

Calculations of potentially required temperature gradients for other size tanks were made using the author's negative radiation mathematics, as developed and confirmed by the black-light cotton study. When David Rule was shown these figures, he asked if the author had forced the math to fit the experimentally determined model of tank performances. According to Rule, the fit between calculation and performance was nearly exact.

The author could not have "fudged" the math since, prior to the calculations, he had not been privy to the exact gradients which initiated uniform tank cooling. He had only been given Rule's generic recommendations that all tanks "worked better" with a gradient between 20° and 30° F.

The fact that both Rule and Dawson arrived at similar tank performance characteristics autonomously— Rule by careful experimental measurement and Dawson by calculation from a completely independent N-radiation study— validates the accuracy and interpretation of those tank performance characteristics.

David Rule's discovery of and the current author's mathematical confirmation of uniform negative-radiation induced liquid cooling by Paschen Series infrared will be lost to science if science chooses to ignore the negative radiation phenomenon because it was discovered outside a university context.

The possibility of such a desertion by establishment science is very real. During the early part of the Rule collaboration, the current writer offered the physics department of a major university in the area the possibility of sharing the research, since Rule's commercially based facilities produced an extremely restricted lab environment. The offer was rejected with a cavalier incredulity about the existence of N-radiation and the suggestion that the uniform cooling capacity could be explained by "heat escaping from the top of the tank." It will be left to the reader to detect the likely error in this "hypothesis" since there is no temperature gradient across the top.

As a result, the Snake River N-Radiation Lab[198] was founded and research conducted with extremely limited personal resources. Luckily, experimentation with black-light negative radiation required no expensive equipment. These initial experiments produced major results by experimentally deriving the negative radiation equations and applying them to the Rule discovery. The test for "amateur science" is lack of public funding. The test for science in general is significance of finding. We would all be poorer if the "amateur test' had been exclusively applied to J.S. Russell, Oliver Heaviside and the young Albert Einstein.

---

[198] Founded as the quantum physics department of the Virtual University.

Losses to Nuclear Physics if Science Ignores Negative Radiation Data[199]

The negative radiation studies revealed unknown aspects of the atomic structure.

In the first place, all of the nuclear elements in a hydrocarbon molecule must somehow be bonded together in such a manner that they are always attached to a single proton. This is the only way that "per string" energy loss can be determined as a multiple of the mass of the proton *times* the number of nuclear units in the whole molecule. In the case of cotton, all 972 of the protons and neutrons in the combined nuclei of the atoms incorporated in the molecule must be multiplied by proton mass to give energy per string:

E_s = energy per valenced string : N_m = number of nuclear units

n_s = number of valenced strings

$$E_s = \frac{(57.34)(\text{mass proton})N_m}{239}\left(\frac{\Delta T}{n_s}\right) = 57.34N_m(-h)$$

$$-h = \frac{(\text{mass proton})}{239}\left(\frac{\Delta T}{n_s}\right)$$

The factor "(mass proton) N_m" must be multiplied by change in temperature per string to give energy. The mass value of the equation is mass of proton *times* nuclear elements (*times* the Caloric factor from the joules conversion factor). This means that the full weight of all nuclear elements is given to a single string for the purposes of radiation impedance. The heavier the molecule, the more energy the molecule will absorb per string.

For negative radiation , change in nuclear temperature per string is a constant:
ΔTemp. per string= $-.00009468°$ Celsius / sec.
When multiplied by the total number of hydrogen bonds, this gives the change in temperature per second for the whole molecule. This is confirmed by both the cotton study and the Rule thermodynamic study.

Change in temperature is a function of the string and the string alone. Change in energy is a function of the nuclear mass and nuclear mass alone.

How all 972 nuclear elements in the cotton molecule —including neutrons — could be "chained" together and attached to the proton of a single hydrogen bond is a mystery. However, that is exactly what the data indicates.

Quantum geometry provides an emerging nuclear model which may explain how such nuclear "chaining" is possible. The electron's charge is composed of a missing Euclidean dimension. The neutron could supply this missing dimension as well as the proton since, as mass, the neutron's volume is also composed of three Euclidean dimensions. The electron could also attach to the neutron. In such an eventuality, the ratio of neutrons to protons in the nucleus would explain nuclear bonding. It could also explain the "strong force" as the total of all atomic forces binding the nucleus.

Such an attachment would be interesting since it would supply a negative charge to the nucleus in the following manner: The electron would be repulsed into orbit by the neutron"s

---
[199] see *"The Effects of N-Radiation on U-235 Emissions"*   p. 184

Euclidean definition of space, just as the electron attached to a proton is. The capacitance and electron voltage of the electron-to-neutron bond, however, would direct the negative charge towards the nucleus. Only the positive charge of the proton could be directed outward from the nucleus. Ionization could be measured but, "Antionization" probably could not be. Ionization would be the force binding atoms into molecules.

The chaining of such positive-negative nuclear elements would account for the data.

A-9: The Drop in Temperature by the N-Irradiation of Cotton

and the derivation of Planck's Constant

Several bands of negative frequency light radiation had been predicted by a system of four dimensional quantum mathematics which the author had developed over the past five years. The theory of negative radiation emerged to explain the radiological heat exchange which engineer David Rule had discovered and applied to the cooling of wine tanks.

The Rule N-radiation frequencies were in the high infrared band (Paschen Series of the Rydberg hydrogen distribution) —just below the Rydberg visible light Balmer Series.

Paschen Series (infrared spectrum). Formula:
$$\frac{1}{\lambda} = \left(\frac{1}{(3)^2} - \frac{1}{(n')^2} \right) \frac{1}{\lambda_r} \; ; \; 4 \le n' \ge 8$$

The water penetration and cooling uniformity of Rule's wine tank heat exchanges had proven that the exchanges were radiological in nature and composed of high infrared. The new system of quantum mathematics— mathematics which had been influenced in part by the Rule discovery — had suggested a theoretical explanation for the radiological heat exchange.

Electron bonds were identified by quantum mathematics as two dimensional strings vibrating across a third dimension. Light was absorbed and output by these vibrating quantum-squared strings. The strings could only acquire mathematically determined orbital levels based upon the Rydberg frequency distribution for hydrogen light emissions (the negation of quantum-squared subdivision of the root frequency). The Rydberg distribution is an array of frequency bands determined by the subdivision "n" of the root frequency (91.14 nm). All Rydberg frequencies within these bands were negations of those subdivisions:

Rydberg Distribution of Hydrogen Emissions

n = subdivisional frequency band, $1 \le n \le 7$; n' = negation of subdivision, $n < n' \le 8$

$$\frac{1}{\lambda} = \left(\frac{1}{n^2} - \frac{1}{n'^2} \right) \frac{1}{\lambda_r} \; ; \; \lambda_r = \text{root wavelength} = 91.14 \text{ nm};$$

The quantum squared string lengths can acquire only negations of subdivisional distances, not the subdivisional distance itself:

subdivisional wavelength distribution $= \left(\frac{1}{n^2} \right) \frac{1}{\lambda_r}$ and is always $> \left(\frac{1}{n^2} - \frac{1}{n'^2} \right) \frac{1}{\lambda_r}$

All orbital levels and associated light frequencies were negations of subdivision of the root frequency string length. It was mathematically possible that the negation of subdivision producing the highest frequency in any one subdivisional band could absorb the negative amplitude from the subdivisional wave frequencies. The theory of negative radiation was born:

negative radiation is found between $\left\langle \left(\frac{1}{n^2} \right) \frac{1}{\lambda_r} \right\rangle$ and $\left\langle \left(\frac{1}{n^2} - \frac{1}{8^2} \right) \frac{1}{\lambda_r} \right\rangle$;   *or*

between the subdivisional frequency and the highest frequency negating that subdivision which is always "n'=8."

Concurrently, theoretical physics had been developing a new system of geometry based upon solitons. Solitons are single waves moving along a dimensional axis. They are not two dimensional but described as "1+1" dimensional, in that they are only "kinks" in a single dimensional axis which are protruding into a second dimension. The soliton "kinks" as thus described are the closest that three dimensional physics have ever come to describing the actual interface between the Euclidean and quantum dimensions.

In February of 2008, four months before the Snake River N-Radiation Lab measured negative radiation, a theoretical article on soliton "kinks" also predicted the existence of such negative radiation[200].

Negative radiation would theoretically operate in near perfect alignment with the cooling behavior of David Rule's wine tank. Negative radiation would operate as a radiological heat exchange. Negative radiation frequencies would force the hydrogen bond to emit "glow" or radiate off the heat energy of the molecule. Negative radiation absorption would require that the nucleus supply its own energy to electron bond vibrations to output this "glow." Negative radiation pressure would extract or "suck" the energy from the nucleus along the electron string, emitting it as "glow.". The atom would be forced to radiate off its own heat.

In 2007, an experiment designed by David Rule had confirmed that his cooling design operated by such radiological heat exchange.[201] Rule applied his hypothetical negative radiation generator to the floor of a tank. If cooling exchange between the liquid and the floor were only by contact, there would be no penetration to liquids above. The measurement of cooling penetration confirmed its radiological character. The changes in temperature above the floor were roughly aligned with the inverse square law governing radiation.

Quantum mathematics identified David Rule's experimentally confirmed N-radiation band as between 820 nm and 955 nm (negative frequencies for Rydberg's Paschen Series "high infrared"). The hypothesis held that negative frequencies forced nuclei to radiate off their own heat energy. Theoretically water should be semitransparent to the frequencies being radiated out as "glow" as well as to the impinging negative radiation frequencies. Only one band met that condition for the Rydberg distribution, the Paschen Series (n=3).

Paschen Series N-radiation can be clearly recognized in the infrared absorption characteristics of water. Transparency is related to Rydberg quantum mathematics.

Paschen Series (Infrared) Negative Radiation

Second "glow" spike

$$1/\lambda = \left(1/3^2 - 1/5^2\right)1/\lambda_r \; ; \quad \lambda = 1.282 \; \mu m$$

First "glow" spike

$$1/\lambda = \left(1/3^2 - 1/8^2\right)1/\lambda_r \; ; \quad \lambda = 954.48 \; nm$$

Hale & Querry,
Appl Opt, 12, 555 (1973)

N-Radiation

Penetration of water
10mm

100mm

1m

10m

100nm   1µm   10µm   100µm

subdivision
820.26 nm

Wavelength (nm)

---

[200] *Negative radiation pressure exerted on kinks;*" op.cit
[201] Proprietorial research, Pasco Poly, Inc., Wieser, Idaho.

Negative radiation for any hydrogen bond exists in the variance between the predicted quantum-squared subdivision of the Rydberg root frequency and the highest actual frequency output as quantum-squared negation of subdivision for that subdivisional series. For the Paschen Series (n=3) of the Rydberg distribution, negative radiation exists between the 820.26 nm "subdivisional" frequency and the frequency which is the highest actual negation of the subdivision at 954.48 nanometers. *(see graph above)*

The 1973 Hale and Querry graph of water absorption of wavelengths shows an absorption pattern which reflects Paschen Series negative radiation. The red "absorption coefficient" line decreases through the visible spectrum between 400 and 700 nanometers wavelengths and begins rising again as it approaches the infrared wavelengths. However, the rise in absorption coefficient inexplicably "plateaus" when it reaches the Paschen Series subdivisional frequency of 820 nm wavelength[202] (the beginning of Paschen N-radiation). The absorption coefficients of water are relatively "flat" (at approximately 6 meters of penetration) through the probable N-radiation frequencies for the Paschen Series.

After the "N-radiation plateau" the red absorption coefficient line rises rapidly again until it reaches the highest actual frequency in the Paschen Series (n'=8 producing 954.48 nm wavelength) where it "spikes" downward[203]. That is, the absorption coefficient inexplicably falls or "spikes downward" at the first frequency which Paschen N-radiation could be expected to force "glow."

The water penetration for this spike in the first "glow" heat-exchange frequency approaches 1 meter which is very close the tank radius of the Rule radiant cooling device. Since N-radiation is applied from the tank walls, the Rule device could be expected to conduct exchange "glow" outside the tank.

"Spikes" in water penetrability are also found for two other "glow" frequencies within the Paschen Series. The "n'=5" frequency at 1.28 micrometers is "spiked" in water penetrability as well as a small spike at the "n'=8" frequency 0f 1.87 micrometers wavelength.

These data confirm that Paschen Series N-radiation exists for water and operates in a manner consistent with the Rule cooling device. Paschen Series N-radiation must be considered as authenticated by David Rule's radiant cooling system.

Negative radiation is formulated as the frequencies between the predicted quantum-squared subdivision of the Rydberg root frequency and the highest actual series frequency produced by the negation of that subdivision (always "n'=8"). This formulation has identified an N-radiation band responsive to observation and experimentation and no longer subjected to the mathematical inferences required for the Paschen Series. That N-radiation band is between 365 nm and 389 nm and is established by the subdivision/highest-series-frequency of the Balmer Series (visible light frequencies).

These N-radiation frequencies are popularly known as "black light." "Black light" is composed of 365 to 371 nm peaks produced by Wood's filtered fluorescent tubes using

---

[202] Careful measurement of the Hale and Querry graph identifies the "plateau" at exactly 820 nm.
[203] The Hale and Querry 1973 data are no longer available and all calculations must be made from the graph. The center of this first "glow spike" actually measures 940 nm and was obviously made by a singe data point. In 1973, however, there was no reason that the authors would choose the Paschen Series "954.48 nm" as their single point of assignment. Further, the width of the "spike" is measured at 56.25 nm and the Paschen Series "954.48 nm" hydrogen output would fall within this width.

either europium-doped strontium fluoroborate or europium-doped strontium borate[204]. These wavelengths reside in the N-radiation range for the Balmer Series. Black light is known to cause some hydrocarbons to "glow." It is a phenomenon which seems to be restricted to hydrogen-bonded organic molecules. Thus, black light is a known but unrecognized form of N-radiation. Specifically, it is N-radiation for the Balmer Series of the Rydberg hydrogen distribution.

N-radiation was predicted to exist in large frequency gaps in the Rydberg frequency distribution. By comparing string potential energy to possible kinetic energy, N-radiation frequencies should exist in the large gap between the Ultraviolet Lyman Series and the visible Balmer Series of frequencies. It is the largest gap in the Rydberg distribution and one which cannot be spanned by kinetic vibrational energy additions to the the potential energy possessed by the high-end Balmer-Series string.

Black light had never been tested as "N-radiation," although it has all the surface characteristics of a negative frequency. In the first place, hydrocarbon "glow" cannot be reflected radiation. Glowing hydrocarbons reproduce none of the color characteristics of the ultraviolet black light. Further, the "glow" partially illuminates. Hydrocarbon "glow" illuminates details of objects within its immediate vicinity. It is an actual light emission, not a reflection, although the "glow" is much softer and less intense than would be the case if the material were heated in a vacuum until it emitted light. This "ghost light" is obviously not a conventional light emission.

Finally, brilliance of color is irrelevant to the absorption of black light. Brilliant white non-organic pigments of paint do not glow, yet less brilliant whites of cotton and polyethylene do. Normal color reflexivity has been replaced by material composition in determining light output.

Black light induced "glow" is clearly the visible equivalent of Rule's Paschen Series negative-radiation-induced infrared "glow" at 954.48 nm.

A series of test of hydrocarbons which "glowed" under black light illumination soon confirmed this conclusion. Using a sensitive infrared thermometer, all absorbing materials were discovered to possess two temperatures; a "glow" temperature and a dropping body temperature underneath the glow. The "glow" temperature was a consistent 1-2 degrees or so above pre-irradiation body temperature of the cotton. The body temperature fell from pre-irradiation temperature at a measurable rate per second. Obviously, the "glow" was radiating off body heat at a consistent rate. Black light as Balmer Series N-radiation was fully confirmed.

It was significant that hydrocarbons, not inorganic molecules, were the primary absorbers of N-radiation. The theory of N-radiation was built upon the Rydberg frequency distribution which itself was built upon the hydrogen atom. The Rydberg formula describes the distribution of frequencies for the single electron of hydrogen. Rydberg frequencies are the absorption spectrographics for hydrogen.

Black light is a hydrogen sensitive form of N-radiation. It was no accident that hydrogen bonded organic molecules were absorbing it.

Since the glowing hydrocarbons were radiating off temperature at a fixed rate, it was important to discover what that rate might be. Cotton was chosen to conducted tests of the

---

[204] en.wikipedia.org/wiki/Black_light

rate of energy exchange(by measured temperature drop) between the nucleus and the hydrogen electron bonds which were presumed to be emitting the glow.

Natural cotton was chosen because it has an easily identified molecular structure, unlike the man-made hydrocarbons which were also being tested.

Under carefully controlled ambient temperature conditions[205], drops in irradiated cotton body temperature was simultaneously monitored by the current writer and a lab technician[206]. These temperature drops were time stamped, giving a "per second" mean rate of temperature drop. It was this mean rate temperature drop per second which was used in analyzing the data.

The results were completely unanticipated. When the measured temperature drop was calculated for the energy lost per hydrogen bond, the calculation resolved to Planck's Constant with an accuracy which disallowed any reasonable probability of coincidence. The resolution of correlation between Planck's Constant and measured results was ten times as great as Niels Bohr had accepted in his calculation for electron voltage. Bohr had allowed 7% variance between calculations and Planck's Constant. Our variation between calculations and Planck's Constant was six tenths of 1%. It is about the same variation obtained by Robert Millikan in 1916, the most accurate derivation of the time.[207] Millikan's highly acclaimed accuracy was five tenths of 1%, an accuracy which the n-radiation study of cotton duplicated.

The energy being exchanged between the "glowing" electrons and the nucleus was a function of Planck's Constant. All room for doubt is eliminated by the accuracy of the correlation between the Constant and the measurements.

That correlation showed that the nucleus was funding the glow of each bond at the rate of frequency energy minus the frequency— which resolves itself exactly to Planck's Constant. The "glow" of N-irradiated cotton is not as great as would be the radiation emitted from the same cotton heated "white hot." The difference explains the pale, "ghost light" appearance of black-light induced radiation.

That difference is the frequency factor. The nuclei of heated cotton contribute their energy to electron light emissions positively. The energy contributed to positive light emissions is Planck's Constant *times* frequency. The nuclei of N-irradiated cotton contribute their energy to electron light emissions negatively. The energy lost to light emissions is Planck's formula for light energy *minus* the frequency of the light. Heated cotton light is brighter because the emission energy is Planck's Constant *times* frequency. N-irradiated cotton light is duller because the emissions energy is Planck's Constant *times* a frequency of "1."

The significance of this should not be underestimated. Negative radiation stimulates the subtraction of energy from the nucleus to fund a "ghost light." That "ghost light" is a measure of a changing energy balance between the electron field and the nucleus.

---

[205] Tests were conducted only under condition that both external heat sources and air conditioning were not contributing to ambient temperature change. The twilight at sunset provided the best control.
[206] Jonathan Dawson assisted in all tests.
[207] *A Direct Photoelectric Determination of Planck's "h"*; R. A. Millikan; Phys. Rev. 7, 355 (issue of March 1916)

The Results

Cotton Irradiated by Black Light at 30 Centimeters, Temperature Measured by IR thermometer from 15 Centimeters

Black light wave length $\approx$370 nm.

Rydberg highest visible frequency (Balmer Series):

$$\frac{1}{\lambda} = \left(\frac{1}{2^2} - \frac{1}{8^2} \right) \frac{1}{\lambda_r}$$

$\lambda = 388.86$ nm

Temperature	Time (min.second)
73.5	7.01
73.1	7.11
72.8	7.37
72.7	8.13
72.5	8.2
72.3	8.36
72	8.51
71.8	9.22
71.6	9.32
71.4	10.20
71.1	10.21
70.9	11.31
70.7	11.33

Temperature dropped 2.8° F in 4 min. 32 second

Average change in temperature 0.0102941176 degrees F per second=0.0057189542 Celsius per second.   Celsius = F/1.8

239 joules per second = 57.34 Calories per second; Calorie= (1 kg)(1°c/sec)

$$x = \frac{57.34}{.001x} \ ; \ x^2 = 57340 \ ; \ x = 239.4577206941 \ conversion \ proof$$

cotton is polymer of $\left(C_6 H_{10} O_5 \right)_n$ n=6

60 hydrogen atoms=60 nuclear units (nuclear unit=proton or neutron)

30 oxygen=480 nuclear units

36 carbon atoms=432 nuclear units

Quantum molecular number=N_m=972 nuclear units

Number of hydrogen bonds(valence electron strings)=n_s=60

Each hydrogen string is "carrying" the weight of 16.2 nuclear units (N_m/n_s) in cotton molecule for a total mass of 16.2 *times* mass of proton (Quantum molecular number./Number of hydrogen bonds).

Mass of proton=1.6726e–27 kilograms

$$\text{watt=joules/sec.} = \frac{57.34}{239}\text{kg}\frac{\Delta 1°}{sec}$$

W_s = watts per valence string ; ΔT = change centigrade

$$W_s = \frac{57.34\ (\text{mass proton})N_n}{239}\left(\frac{\Delta T / sec}{n_s}\right)$$

$$= \frac{57.34\left(1.6726\ 10^{-27}\ \text{kilograms}\right)(972)}{239}\left(\frac{\left(-0.0057189542°\ c/sec\right)}{60}\right)$$

$$W_s = -\ (57.34)\left(6.4837434844\ 10^{-31}\right)\text{watts}$$

$$\frac{W_s}{(57.35)N_m} = 6.6705179881e\text{-}34 \cong \text{Planck's Wattage} = h/sec$$

$$\frac{W_s\ /\ (57.35)N_m}{h/sec} = 1.0067072112$$

$$(h/sec)(N_m)(57.35) = W_s$$

$$h/sec = \frac{(\text{mass proton})}{239}\left(\frac{\Delta T / sec}{n_s}\right)$$

$$\Delta T/sec = \frac{239\ n_s(h)}{(\text{mass proton})sec}$$

$$\text{molecular watts} = W_m = \frac{57.34\ (\text{mass proton})N_n}{239}(\Delta T / sec)$$

$$= \frac{57.34\ (\text{mass proton})N_n}{239}\left(\frac{239\ n_s(h/sec)}{(\text{mass proton})}\right)$$

$$= \frac{57.34(N_n)(n_s)h}{sec} = \frac{\text{joules}}{sec}\ ;\ E = \frac{57.34(N_n)(n_s)h}{sec}(\text{time})$$

6.6260755e-34= Plank's Constant (h) ; h /sec= Planck's Wattage

Wattage per hydrogen string = -*h/sec* (57.34) (N_m)

Energy per molecule= (-*h/sec*) (57.34) (N_m)(N_s)(time)

Total power (wattage) per hydrogen bond (valence string) equals (Planck's Constant)/ sec. *times* quantum molecular number times 57.34. Planck's Constant *per second* is a factor of the change in temperature per hydrogen bond (valence string) *times* the mass of the string's attached proton. Change in temperature (for the whole molecule is the constant

"239/(mass proton)" *times* number hydrogen bonds *times* Planck's Constant. Change in temperature is completely a function of the number of hydrogen bonds.

Converstion Watts to BTU/sec

$$\text{molecular watts} = W_m = \frac{57.34 \ (\text{mass proton kg})N_n}{239}(\Delta°\text{Centigrade}/sec)$$

$$BTU/sec = \text{lb.}(\Delta°F/sec)$$

$$2.205 \ \text{lb} = 1 \ \text{kg.} \quad ; \ 1.8° \ \text{Fahrenheit} = 1° \ \text{Celsius}$$

$$\frac{BTU}{sec(2.205)(1.8)}\frac{57.34}{239} = \frac{BTU}{sec}(0.060447548) = W$$

$$\frac{BTU}{sec} = W_m(16.5432682246)$$

Predicted Change in Temperature for Pasco Poly Tanks
(Theoretical Terahertz N-Radiation for H_2O)

$$\frac{\Delta T(H-O-H)}{sec} = \frac{239 \ n_s(-h)}{(\text{mass proton})sec} = \frac{239 \ (2)(-6.6260755e\text{-}34)}{(1.6726e\text{-}27 \ \text{kilograms})sec} = -\frac{\textbf{1. 2270639302° F}}{\textbf{hour}}$$

The above figure "-1.23° F per hour" has been empirically confirmed as the approximate cooling rate of the Pasco Poly wine tanks over many observations.

N-Radiation Determination of Planck's Constant

For N-radiation which is absorbed by a molecule, the molecule "glows" radiating off the energy (heat) from its own nucleus. N-radiation is determined by the Rydberg formulas for hydrogen electron string lengths. N-radiation is calibrated by black light (wavelength 370 nm) which is just above the last visible Rydberg frequency for hydrogen (Rydberg 388.9 nm).

N-radiation has been experimentally shown not to be absorbed by conventional black bodies. This is true of both organic and inorganic black bodies. Irradiation by black light does not raise the temperature of either blackened metal nor blackened hydrocarbons.

Black light absorption (causing glowing) is primarily restricted to hydrocarbons or organic molecules. Since the black light frequency is related to the last Rydberg visible frequency, it is assumed the covalent hydrogen bond of the hydrocarbons is the active string. Only the hydrogen atoms are providing an energy exchange with the nuclei of the molecule.

The experimental energy loss due to reduced temperatures is calculated "per hydrogen string" of the molecule:

$$\frac{\text{Total molecular wt.}}{\text{hydrogen units}} = \text{effective nuclear wt. of each hydrogen string;}$$

$h =$ Planck's Constant

$$E = \left(\frac{1}{239}\right)\left(\frac{\text{Total molecular wt.}}{\text{hydrogen units}}\right)(\text{mass of proton})(\Delta\text{Temperature})$$

$$E = (h)(\text{Total molecular wt.})$$

$$(h)(\text{Total molecular wt.}) = \left(\frac{1}{239}\right)\left(\frac{\text{Total molecular wt.}}{\text{hydrogen units}}\right)(\text{mass of proton})(\Delta\text{Temperature})$$

$$h = \frac{(\text{mass of proton})(\Delta\text{Temperature})}{239(\text{hydrogen units})}$$

Planck's Own Formula

$$\text{light frequency} = f = \frac{C}{\lambda} \quad ; \quad h = \text{Planck's Constant}$$

$$E = nh(f)$$

"N" is the number of oscillator's, according to Planck. The hydrogen units in the N-radiation experiment were the "Oscillating strings" exchanging radiant energy with the nucleus.

Let n= hydrogen units. (absorbing oscillating string-bonds)

$$nh = \frac{(\text{mass of proton})(\Delta\text{Temperature})}{239}$$

Calculating Plank from experimental data
Experimental result:
$h=\dfrac{(0.0057189542\Delta\text{temp per second})(1.6726e\text{-}27 \text{ kilograms } \textit{mass of proton})}{(60 \text{ electron strings})(239)}$ $= 6.6705177548e\text{-}34$
Actual value of $h=6.6260755e\text{-}34$
experimental as percentage of actual Planck's Constant=1.006707176. There is less than a 1% variance (.006) (Bohr had 7% for electron voltage)

Planck's constant is the change in temperature for one electron string oscillator per molecule for one string vibration per second. Multiply by frequency (in Hertz) to get absorbed energy of the frequency. It is the measure of the energy exchange between the electron string and the nucleus.

$$\text{Planck Energy} = (f)\frac{(\text{mass of proton})(\Delta\text{Temperature})}{239}$$

$$= (f) \ (6.9983263598\text{e-}30 \text{ kg})(\Delta\text{Temperature})$$

The change in temperature is a function of the number of string oscillator's per molecule and the rate of change in temperature is a direct function of frequency of vibration. This is the actual mathematics underlying Planck's Constant.

The Effects of N-Radiation Upon Gamma Emissions from U-238

Below are the results of the n-irradiation of approximately 5 grams of ore powder which contained 3% uranium oxide (isotope 238). Gamma emission were measured by a Gieger-Mueller meter sensitive to tenths of milli-Roentgens per hour and calibrated to Gamma radiation using Cesiuim-137. One "click" of the meter represented one Gamma radiation event which was was graphically stamped by a computer into one second units of a time line. The number of Gamma energy events during a three second interval became the test unit of measure.

60 such units of measure were sampled for both N-irradiated and non-irradiated Gamma emissions. The position of the meter in relation to the ore was the same for both samples.

The raw data, organized by number of Gamma "hits" for the time unit is the following
Gamma hits per 3 second intervals (in ascending order)

radiation	non radiation		radiation	non radiation
0	2		6	8
2	3		6	8
3	3		6	8
3	4		6	8
3	4		6	8
3	4		6	8
3	4		6	9
3	4		7	9
4	5		7	9
4	5		7	9
4	5		7	9
4	5		7	10
4	5		7	10
4	5		8	10
4	5		8	10
4	6		8	10
4	6		8	10
4	6		9	10
4	6		9	11
5	6		9	11
5	6		10	11
5	6		10	12
5	6		10	12
5	7		10	13
5	7		10	14
6	7		10	14
6	7		11	14
6	7		11	15
6	7		12	15
6	7		15	16

Non-Irradiated U 238

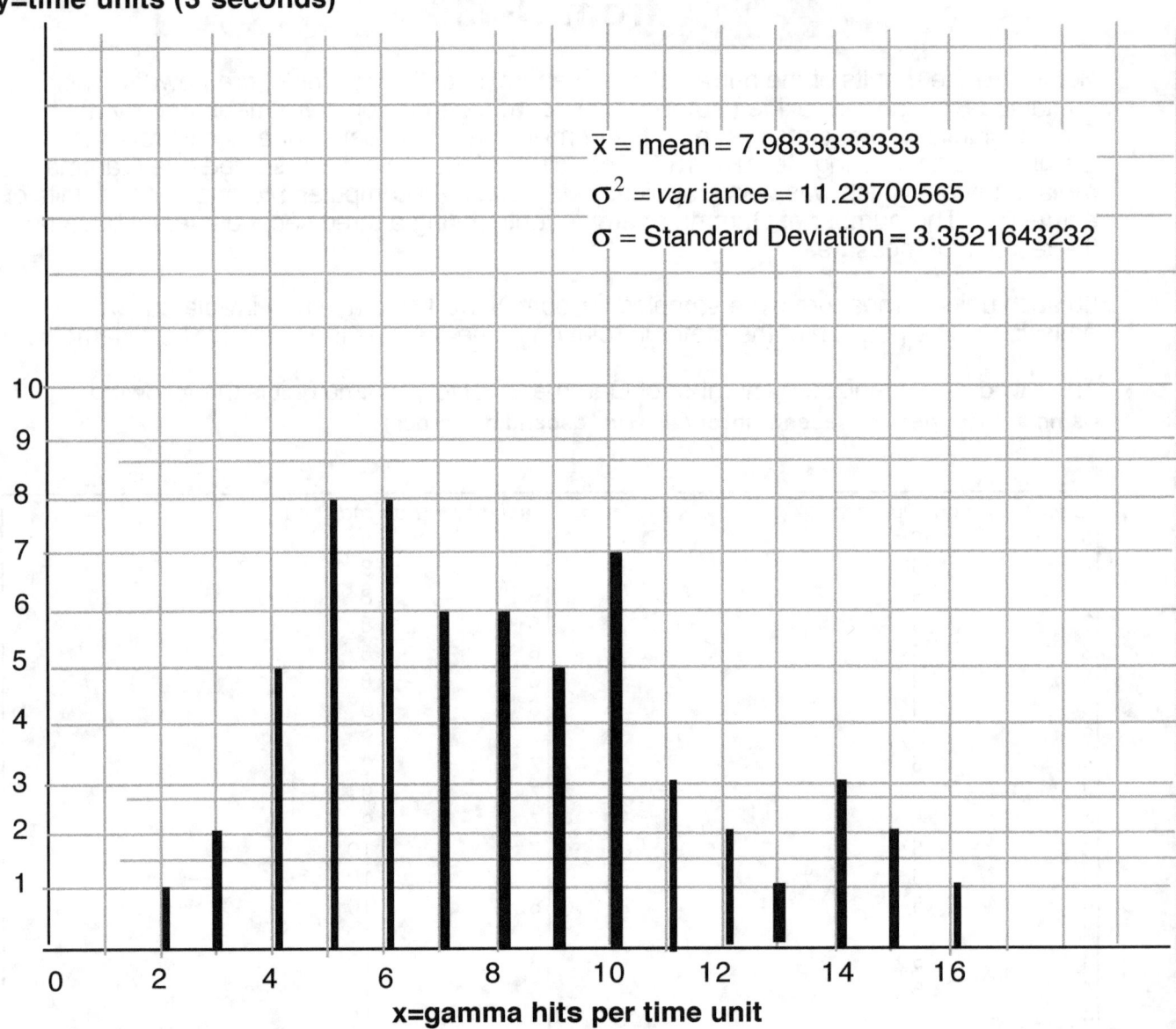

y=time units (3 seconds)

$\bar{x}$ = mean = 7.9833333333

$\sigma^2 = var$iance = 11.23700565

σ = Standard Deviation = 3.3521643232

x=gamma hits per time unit

Note the graph's mean is skewed away from the greatest frequency of hits per unit (5 and 6). This is due to a greater frequency of high end occurrences which produces a higher variance for the population.

Irradiated U 238

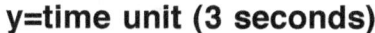

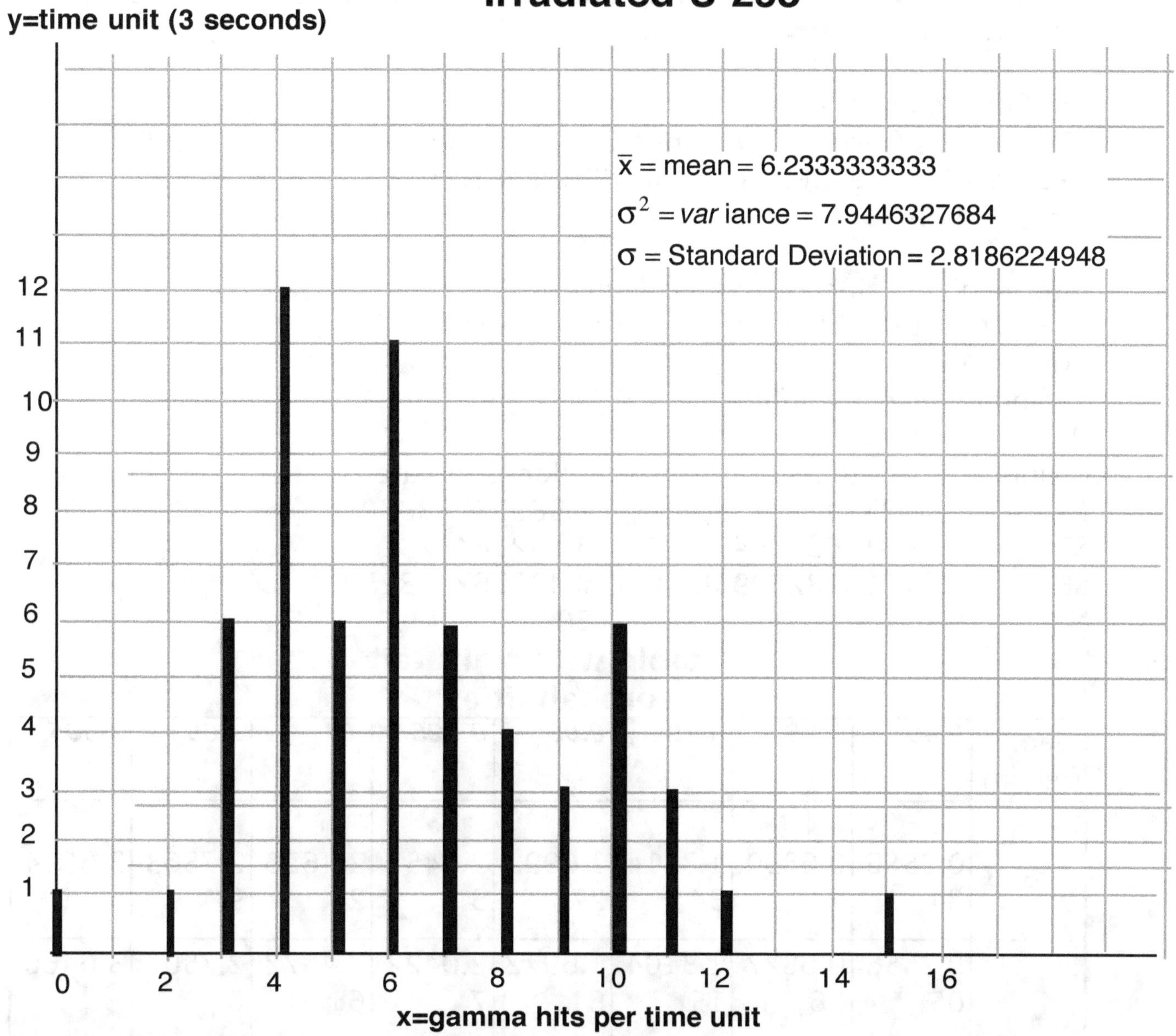
y=time unit (3 seconds)
x̄ = mean = 6.2333333333
$\sigma^2 = variance = 7.9446327684$
σ = Standard Deviation = 2.8186224948
12
11
10
9
8
7
6
5
4
3
2
1
0
2
4
6
8
10
12
14
16
x=gamma hits per time unit

Note the graph is less skewed towards the high end which produces a lower variance. This is due to less frequent high end occurrences (compared to non-radiated Gamma emissions) and more frequent low end occurrences. The difference in mean and variance between the two samples is statistically significant at the .0025 level meaning the differences in means and variance would only occur by chance 2.5 times out of every 1000 such comparisons. For the statistical test confirming this see below.

Difference between the two populations is statistically significant.

Unpaired t test calculations:

P value and statistical significance:

The two-tailed P value equals 0.0025 (significance <.01)

By conventional criteria, this difference is considered to be very statistically significant.

Confidence interval:
  The mean of Group One minus Group Two equals –1.750000000000
  95% confidence interval of this difference: From –2.869675959570 to –0.630324040430

Intermediate values used in calculations:
  t = 3.0951 (NOTE: see table below for probabilities)
  df = 118
  standard error of difference = 0.565

Review of data:

Group	Irradiated	Non–irradiated
Mean	6.233333333300	7.983333333300
SD	2.818622494800	3.352164323200
SEM	0.363882599387	0.432762553250
N	60	60

t table with right tail probabilities

deg. freedm	0.40	0.25	0.10	0.05	0.025	.01	.005	.0005
29	0.255684	0.683044	1.311434	1.699127	2.04523	2.46202	2.75639	3.6594
30	0.255605	0.682756	1.310415	1.697261	2.04227	2.45726	2.75000	3.6460
inf	0.253347	0.674490	1.281552	1.644854	1.95996	2.32635	2.57583	3.2905

NOTE: "Infinite" df must be used.   t=3.0951;   2.57583<t<3.2905; Therefore, for this t value, .005>probability> .0005 Probability calculated at .0025 by "two tail" test.

Methodological Problem

Because the Gieger-Mueller Meter is an analogue tube without the instantaneous reaction time of solid state circuitry, it was often difficult to distinguish between two "hits" very close together and a higher energy single "hit." The problem was much greater for the "non-radiated" sample than for the "radiated" sample. This was true because the "non-radiated' sample simply had a significantly greater number of closely spaced "hits" than did the "radiated" sample.

If the computer identified two "peaks" to a hit, it was counted as two "hits." If the computer only identified one "thick" peak, it was counted as a single "hit."

Discussion

A surface powdered with uranium oxide ore was irradiation at close distance with negative frequency light. The data shows that this resulted in the suppression of Gamma radiation emitted from the ore. When the data is corrected for background radiation, gamma emission were reduced by 25.06%. The reduction is statistically significant. There is only .0025 probability that the reduction was due to chance. This is well below the accepted scientific standard of .01 probability of chance.

Why N-radiation should suppress gamma emissions cannot be understood without an understanding of negative radiation. In turn, negative radiation cannot be fully understood without familiarity with of the four-dimensional model of the atom which emerged from systematic quantum mathematics[208]

The four dimensional model of the atom is sufficient to explain all measurable energy events on the atomic level. It is a "stand alone" model to be evaluated by the data and mathematics alone. It cannot and should not be forced into comparisons with standard three dimensional models of the atom. The conventional Bohr/quantum mechanics model of the atom could not have predicted these results.

Negative Radiation

In general, negative radiation frequencies are frequencies outside the orbital capacity of quantum electron strings to impede directly. The electron string can only impede the light wave in its "negative radiation pressure" troughs. In this condition, the energy of the nucleus must power the electron string to accelerate towards the source of radiation (negative radiation pressure[209]).

The most practical example of negative radiation is black light. Black light has a wavelength of between 365 and 380 nanometers. The highest quantum orbit available to the electron in the visible light range is Rydberg-formula 388.9 nm. The black light frequencies are just above the 388.9 nm quantum orbit. Black-light frequencies must be impeded in the negative pressure troughs of the wave which impels the nucleus to power the electron string to accelerate towards the source of the light.

Matter can only absorb negative radiation (N-radiation) by increasing the energy state of the electron at the expense of the energy state of the nucleus. The best and most easily

---

[208] See my paper "Dawson's Theorem —Differentiation by Euclidean Subdivision vs. Quantum Negation of Subdivision " for an introduction to systematic quantum mathematics.

[209] Negative radiation pressure exerted on kinks;" Péter Forgács, Árpád Lukács, Tomes Roma ń czukiewicz; Phys. Rev. D77:125012,2008

understood example of this is the "glowing" of hydrocarbons when bathed in black light. The Snake River N-Radiation Lab has demonstrated that the temperature of the hydrocarbon falls as it "glows" in black light and that this temperature fall is predictable using Planck's Constant and the atomic weight of the molecule. Negative radiation subtracts Planck's "frequency-only" energy packet from the nucleus.

The four-dimensional atomic model identifies the electron as a quantum particle. It is bonded to the nucleus by a field and a string. The energy state of the bound electron can be expressed as string vibrational energy or as field capacitance, that is, as the electric permittivity of the string length[210].

Each quantum string length (orbital radius) determines the tension on the string. In turn, the frequency of string vibration is then determined by this tension. String vibration outputs radiation at frequency. For negative radiation, string tension (a function of electron voltage) is reduced by the investment of energy from the nucleus. Electron capacitance/ field permittivity (the electrostatic field strength of the electron) is conversely increased.

This is the reason that hydrocarbons "glow" under the influence of n-radiation. Hydrocarbons are covalently bonded by a set of hydrogen atoms. Each shared hydrogen electron becomes an "oscillator" for the hydrocarbon molecule. Since hydrogen electrons are valenced and can acquire new orbital energy levels which directly impede n-radiation frequencies, they can therefore "glow."

However, non-organic molecules can also absorb n-radiation. Since they are composed of more complex atoms with fixed electron orbital levels they do not "glow." However, it is postulated that the loss of nuclear heat energy is invested in losses of electron voltage and increases of electron capacitance by quantum string harmonics[211]. Energy loss from the nucleus was invested in increased electric field permittivity for absorbing electron orbits.

A drop in temperature has been measured for U-238 under black light irradiation. . The reduction in gamma emissions was discovered to be accompanied by a drop of 2.3° F during a five minute period.   .

Any material which absorbs negative radiation must reduce the electron voltage of the strings impeding the negative frequencies. These string-bonds are the energy exchange mechanism between the electron and the nucleus. Since the emission of of beta radiation is the ejection of an electron from its orbit, it was postulated that that the reduction of string electron voltage should make this more difficult for the nucleus to accomplish. The data proved this to be the case.

A second conclusion must also be reported. It was postulated that gamma radiation is really the emissions produced by the breaking of the string when the electron is forcibly ejected from the atom. While this is not conventional theory, conventional theory (gamma as matter/antimatter destruction) is unproved esotericism. In point of fact, when U-238 emission samples where compared gamma was proven to be roughly 1/ 2 gamma plus beta emissions. The use of gamma as an indication of electron ejection activity was completely justified and further confirmation of the electron string theory.

---

[210] See *"Quantum/Electromagnetic Field Equations"* in Appendix
[211] See *"Quantum String Harmonics"* in Appendix

The Simplified Quantum Dimension
and the Definition of the Quantum Squared

This fourth quantum dimension I wish to present is dissimilar to any of the extra-dimensional speculations made by contemporary math and physics. It is not a "superior" dimension of mysterious content which can be measured only as an intersection with our known space (Riemann et. al.).

The quantum dimension is the "substructure" of all space. It is the "substructure" of our three dimensional space in the sense that all Euclidean dimensions are derived from and dependent upon the quantum dimension. The line and plane of our conventional Euclidean geometry are constructs of the quantum squared. Further, empty volume or vacuous space is also constructed by the quantum squared. The only authentic three dimensional Euclidean volume is mass and it is derived from the square of the quantum squared (Q^4). The quantum is thus the unrecognized substructure to all geometric space.

We cannot proceed further, however, without defining the quantum. The geometric quantum is not a vague concept; not a word of subjective and indistinct meaning. It has an exact mathematical definition.

> *DEFINITION: The quantum is an unit of distance which cannot be differentiated by subdivision, but must be differentiated by the negation of subdivision.*
>
> *The quantum is distinguished from a conventional Euclidean distance. The Euclidean distance is composed of a continuum of points and can be differentiated by subdivision without restriction.*

This definition is not as intimidating as it may first appear. It is actually composed of familiar concepts. "Differentiation of a unit of distance," for example, is used daily by school children. It is the set of subdivisional units on a child's ruler. Since most distances do not equal a whole number of inches or centimeters, some fraction or "partial" of the inch or centimeter must be used to measure them.

The unit of measure (the inch or centimeter) is "differentiated by subdivision" to provide this fraction. The inch subdivisional unit is achieved by dividing the inch by powers of "2." (produces 1/ 2 in, 1/ 4 in, 1/ 8 in. etc.)The centimeter subdivisional unit is achieved by dividing the centimeter by powers of "10" (1/ 10 centimeter=millimeter). [212]

The Negation of a Subdivision

Theoretically, any distance can be divided by a whole number "n" which produces exactly "n" equal subdivisions. Thus, if you divide an inch by "4" you get exactly "4" "1/4 inch" subdivisional units.

A quantum, however, cannot be subdivided. It cannot be subdivided because it is composed of only two points separated by "some" distance. If this unspecified distance —this "some" distance —were to be subdivided into two parts, an additional point would have to be supplied at the halfway measure of the distance.

---

[212] The inch subdivisional unit is achieved by dividing the inch by " 2^n." The centimeter subdivisional unit is achieved by dividing the centimeter by " 10^n ." "N" determines the resolution of the the subdivisional unit.

By the addition of this point the single quantum would become two quantums. To add points to the distance of separation —as required by subdivision — increases the number of quantums. Subdivision of the quantum distance multiplies the number of quantums within that distance.

Quantum distances, however, can be made smaller by subtracting space. They cannot be subdivided — there is no such thing as "half a quantum" —but the quantum can assume a distance value of one-half its original distance. The concept is similar to someone who loses a large amount of weight. We would not say he has become "half a person." We would say he is a person who has became smaller. It is the same with the quantum. The quantum cannot be divided into sub units, but it can be differentiated in the sense that the space separating the two points can assume a fraction of its former distance value.

This is not true of Euclidean distances which can be subdivided because the line composing the Euclidean distance is a continuum of points. All subdivisions, therefore, are possible for Euclidean distances because such subdivisions are constructed from a set of preexistent points.

Any subdivision of a distance, however, establishes two partials of the distance, not just one. The subdivision also establishes its negation which is defined as the distance *minus* the subdivision. This negation is the only differential of the distance available to the quantum.

$$\text{let the subdivision} = s_{ub} = \frac{1}{4}x$$

$$\text{negation} = x - s_{ub} = \left(1 - \frac{1}{4}\right)x = \frac{3}{4}x$$

Quantum distances are differentiated by the negation of subdivision. Since there are only two points defining the quantum, the quantum can assume a new distance value equal to the negation. In doing so, the quantum is still defined by only two points. But it has assumed a new distance of separation equal to the negation of the subdivision.

A quantum "ruler" might be composed of quantum units of distance. Any one unit of which can be differentiated by a series of such negations. Unlike the school child's Euclidean ruler (differentiated by subdivision), the quantum ruler must be differentiated by the negations of subdivisions.

For the purposes of illustration only, the quantum ruler units might assume distance values of: *Differentiation=(1-1/ n)Q, "n" being a whole number.*

Therefore, the differentiated quantum can assume the distance values:
(1-1/2)Q=.5Q; (1-1/ 3)Q=.67Q ; (1-1/ 4)Q=.75Q ; (1-1/ 5)Q=.8Q ; (1-1/ 6)Q=.83 Q; etc.

An Illustration of Quantum Differentiation

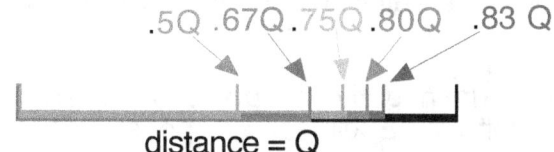

.5Q .67Q .75Q .80Q .83 Q

distance = Q

It might seem that differentiation by negation severely restricts the quantum "ruler" in the number of partials of any unit which it can measure. After all, differentiation of the Euclidean ruler by subdivision is unrestricted. Any one subdivision can be subdivided and subdivided again, increasing the resolution of the ruler with each subdivision. The quantum cannot do this as quantum differentiation is not unrestricted. Closer examination, however, shows this to be irrelevant.

The above illustration is single dimensional. But a single geometric dimension cannot exist autonomously. It can only exist as a component of a multidimensional unit of space— as a component of area or volume. So it is with the quantum dimension. It can exist only as a component of a quantum plane we have designated "the quantum squared."

A plane, however, is a two dimensional structure. Since there is only one quantum dimension and one possible quantum axis, the quantum plane must include an Euclidean dimensional axis:

$$Q^2 = (Q)(\epsilon\mu) \quad ; \quad \epsilon\mu = \text{Euclidean defined distance}$$

It is actually the quantum squared (Q^2) which is subject to the differentiation by the negation of subdivision. Quantum squared differentiation is negation of the subdivision of area:

$$d(Q_n^2) = \left(1 - 1/n^2\right)Q^2$$

"$1/n^2$" is the subdivisional unit for a unit of area. It is determined by squaring the subdivision of one of its dimensional components. Thus if a single dimensional side of a square unit of area is divided in half, the area unit is subdivided into four parts (2^2):

Sides subdivided by "2". Area Subdivided by "4" (2^2).

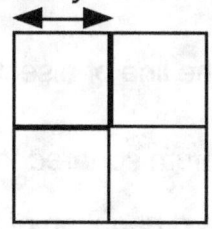

The differentiation of the quantum squared, therefore, is the negation of the subdivision of a unit of area or:

$$\text{"} d(Q_n^2) = \left(1 - 1/n^2\right)Q^2 \text{"}$$

Just as the quantum squared must be differentiated negatively, so it also provides a system of calculus "negative integration" which operates by subtraction rather than summation.

Negative integration requires much more information to explain fully. In the short version, the negative differentiation of the quantum squared (Q^2) provides for negative integration of the quantum. Since calculus integration is "anti-differentiation," negative quantum integration is "anti-anti-differentiation."

The negation of a negation is a positive. Quantum negative integration is the same as Euclidean differentiation. All possible point along the Euclidean component of the quantum squared can be identified by negative integration of the quantum squared.[213] Partials of the "ruler's" unit of distance are completely describable using the differentiated quantum squared while they cannot be by strict Euclidean subdivision.

The Euclidean "X" Axis as Established by The Quantum Squared

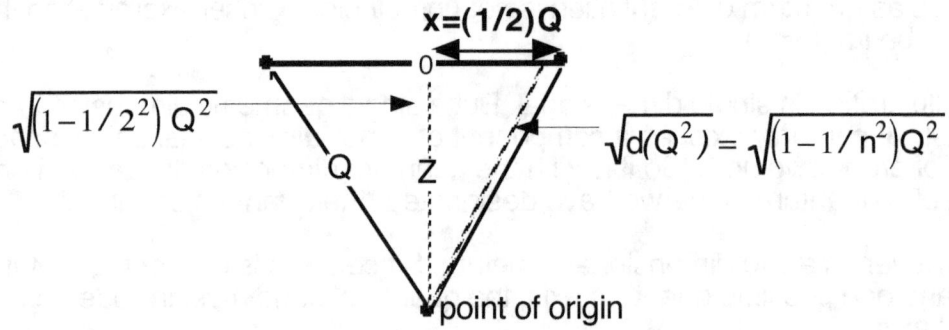

A completely measurable Euclidean line can be constructed from a single point by the quantum squared. The quantum squared is constructed by projecting two equal quantums from a point of origin at a 60° angle to one another. This will construct a Euclidean line the length of which is equal to the quantum. Henceforth this line will be referred to as "the x axis."

The x axis so constructed is Euclidean because a continuum of points along the line can be identified by quantum negative integration.

Quantum Negative Integration

Construct a third quantum line (Z) which bisects the angle of separation and the x axis. Let the point of intersection with the x axis be the "0" point. The value of x from "0" to either end points is "x=(1/2) Q."

By the Pythagorean Theorem, the square of the line of bisection is:
$$Q^2 - (Q/2)^2 = (1 - 1/2^2)Q^2$$
This is the negation of subdivision for the quantum squared for "n=2."

Construct a second line with a quantum-squared negation of subdivision value of "n>2" (designated as "$d(Q^2)$").

The value of "x^2" at the point which this second line intersects the x axis can now be determined as a quantum-squared negation of subdivision:
$$x^2 + (1 - 1/2^2)Q^2 = (1 - 1/n^2)Q^2$$
$$x^2 = (1 - 1/n^2)Q^2 - (1 - 1/2^2)Q^2 = (1/2^2 - 1/n^2)Q^2$$

---

[213] See Appendix "Differentiation by Subdivision and Quantum Negation of Subdivision."

"$\left(1/2^2 - 1/n^2\right)Q^2$" is the quantum-squared negation of the "$1/2$ subdivision."

X, therefore, is equal to the negation of the "1/2 subdivision" for every negation of the quantum squared.

If "$d(Q^2) = \left(1 - 1/3^2\right)Q^2$" then "$x^2 = \left(1/2^2 - 1/3^2\right)Q^2$" and so forth.

Every value of the negation of subdivision for the quantum squared will give a value for x squared.

The negation of subdivision for the quantum squared may be negatively integrated:

  let $Q = 1$;   let $w_n = n +$ any possible partial between "0" and "1"

  let $d(Q^2) = \left(1 - 1/w_n^2\right)Q^2 = \left(1 - 1/w_n^2\right)$

$$D(1 - \frac{1}{w_n^2}) = \frac{2}{w_n^3}$$

$$1 - \frac{1}{w_n^2} = \int \frac{2}{w_n^3}\, d(w_n)$$

$$\frac{1}{w_n^2} = 1 - \int_{w_n}^{0} \frac{2}{w_n^3}\, d(w_n)$$

Elsewhere I have proven that the negation of subdivision can be negatively integrated whereas subdivision itself cannot be integrated at all.[214] The problem is that the derivative of subdivision places the variable in the denominator and requires division by "0" for integration. In contrast, the derivative of the negation of subdivision allows the subtraction of the reverse integral from the whole to remainder the subdivision (see above integral).

Negative integration operates by the subtraction of indeterminate area from the whole of the area under the graph of the derivative (1 *minus* the reverse integration of the derivative from "w_n" to "0"). The subtraction of indeterminate area does not require division by "0" to determine. Therefore, the integral becomes possible.

Graph of the Derivative of $\left(1 - 1/w_n^2\right)$

$$D\left(1 - \frac{1}{w_n^2}\right) = \frac{2}{w_n^3}$$

Subtraction of Indeterminate area

remaindered area

$$\frac{1}{w_n^2} = 1 - \int_{w_n}^{0} \frac{2}{w_n^3}\, d(w_n)$$

$$x^2 = \frac{1}{2^2} - \frac{1}{3^2}$$

---

[214] See *"Dawson's Theorem, Proof of the Quantum Dimension"* in the Appendix.

In the above illustration, the area under the graph equal to "$1/2^2 - 1/3^2$" is shown. It is accomplished in the following manner:

$$\frac{1}{2^2} - \frac{1}{3^2} = \left(1 - \int_2^0 \frac{2}{w_n^3} \, d(w_n)\right) - \left(1 - \int_3^0 \frac{2}{w_n^3} \, d(w_n)\right) = \int_3^0 \frac{2}{w_n^3} \, d(w_n) - \int_2^0 \frac{2}{w_n^3} \, d(w_n)$$

The area under the graph equal to the integral "$\int_3^0 \frac{2}{w_n^3} \, d(w_n)$" is to the left of "$w_n = 3$."

It includes indeterminate area as "w_n" approaches "0" (requires division by "0" to determine). However, the indeterminate portion to the left of "$w_n = 2$" is being subtracted, remaindering determinate area between "$w_n = 2$" and "$w_n = 3$." The negative integral:

"$-\int_2^0 \frac{2}{w_n^3} \, d(w_n)$" subtracts the indeterminate area to the left of "$w_n = 2$."

The above graph illustrates the remaindered area.

The point of this rapid introduction to negative quantum integration is to demonstrate that the negation of subdivision for the quantum squared establishes an x axis composed of a continuum of points.

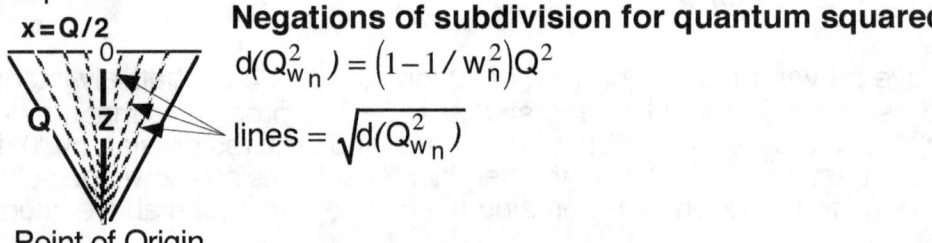

Negations of subdivision for quantum squared:

$$d(Q_{w_n}^2) = \left(1 - 1/w_n^2\right)Q^2$$

$$\text{lines} = \sqrt{d(Q_{w_n}^2)}$$

Point of Origin

In the above illustration, the lines indicated as "negations of subdivision" actually are a continuum of quantums intersecting the x axis. The intersections of this continuum of quantums compose the x axis as a continuum of points. Each one of these points will have a value along the x axis equal to:

$$x = \sqrt{\left(1/2^2 - 1/w_n^2\right)Q^2}$$

In the graph of the derivative above, it can be seen that an area value between "$w_n = 2$" and "$w_n > 2$" exists for every possible value of "$w_n > 2$." That area is equal to "x^2." The square root of all such "x^2" values compose an x axis.

The Euclidean axis of the quantum squared is therefore superior to a strictly Euclidean "ruler." Unrestricted subdivision of a strictly Euclidean "ruler" cannot identify all possible points as can the quantum "ruler."

The Quantum Squared Empirically Discovered

Differentiation by the quantum squared was empirically discovered by science at the turn of the twentieth century. The mathematician Janne Rydberg[215] discovered that the ultraviolet

[215] Born Nov. 8, 1854, died Dec. 28, 1919. Professor, Lund University, Sweden. Author of Rydberg radiation formula.

series of spectrographic lines emitted by energized hydrogen was predicted by a quantum-like differentiation of the highest frequency in the series. His model is exactly the negation of subdivision for the quantum squared:

"$(1/n'^2 - 1/n^2)\, Q^2$"

When the light emitted by heated hydrogen is defracted through a prism, it does not produce a gradated rainbow such as sunlight does. It produces bands of very specific spectrographic lines. Five separate bands of quantum radiological frequencies have been identified from energized hydrogen. For our immediate purpose, however, we are only interested in the band of highest frequency, the ultraviolet. Rydberg's equation was originally developed for the ultraviolet band which is called the "Lyman Series." The Lyman series consists of nine specific wavelengths in the ultraviolet range of light.

The Lyman Series

Rydberg: $\quad \dfrac{1}{\lambda} = \left(\dfrac{1}{1^2} - \dfrac{1}{n^2} \right) \dfrac{1}{\lambda_r} \quad ; \ 2 \ge n \le 9$

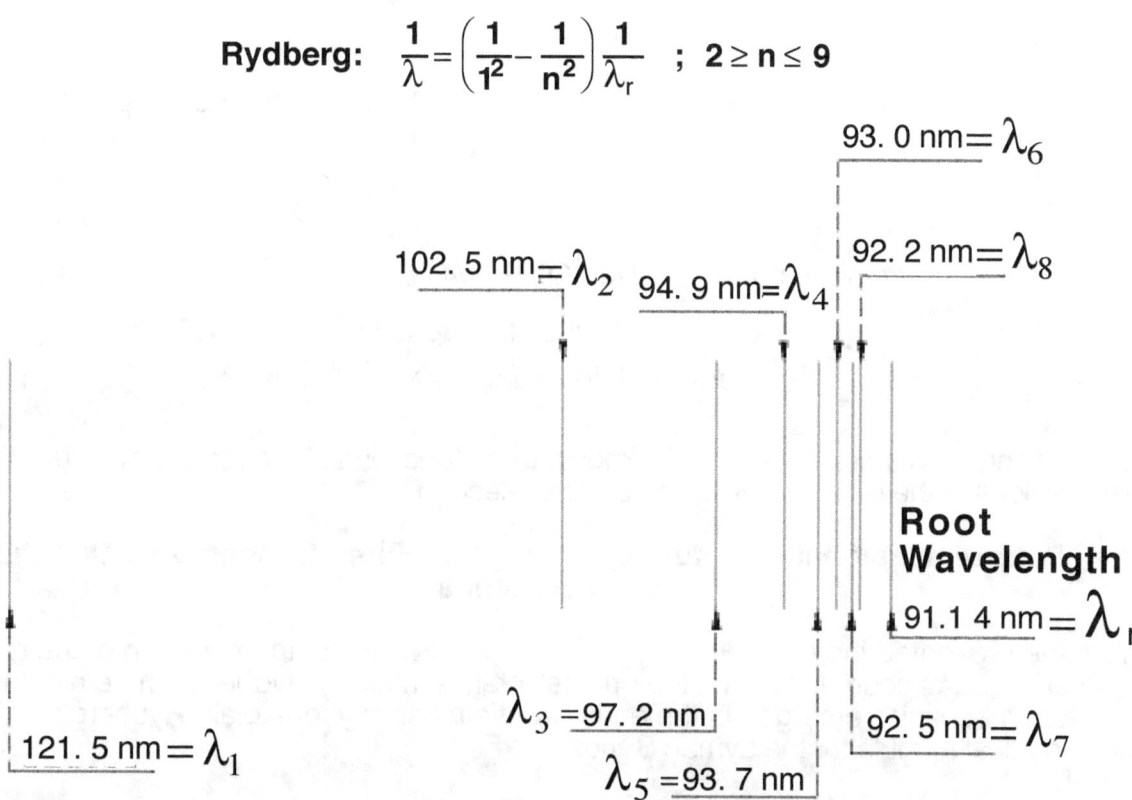

93. 0 nm$= \lambda_6$

102. 5 nm$= \lambda_2$ 94. 9 nm$= \lambda_4$

92. 2 nm$= \lambda_8$

Root Wavelength

91.1 4 nm$= \lambda_r$

$\lambda_3 = 97.\,2$ nm

92. 5 nm$= \lambda_7$

121. 5 nm$= \lambda_1$

$\lambda_5 = 93.\,7$ nm

Rydberg showed that all frequencies in the Lyman Series could be predicted as differentiations of the quantum squared. While Rydberg did not recognize his formula as the quantum negation of subdivision, the two formulas are identical.

The formula for the differentiation of the quantum squared is the same as the Rydberg formula. The differentiation of the quantum squared[216] (by negation of subdivision) is the following:

$$d(Q_n^2) = \left(1^2 - (1/n)^2\right)Q^2 = \left(1 - 1/n^2\right)Q^2 \ ;$$

"$d(Q_n^2)$" is the differentiation of the quantum squared at the "n" value.

---

[216] see Chapter 1 for an explanation of the differentiation of the quantum squared.

Is there simply a coincidental similarity between the Rydberg formula and the negation of subdivision for the quantum squared? Not if the Rydberg ultraviolet frequencies are a harmonic series related to the negations of subdivisions for quantum lengths — just as the sound frequencies or "notes" in music are harmonically related to the subdivisional lengths of vibrating strings.

If the Rydberg frequencies are the vibrational rates of tensioned "quantum strings" — if light is output by vibrating tensioned strings —then frequency change by differentiated quantum (squared) can constitute a quantum harmonic series. If string tension were determined by stretch and frequency were output by tension, then frequency would "fall" in direct proportion to string length.

string length $= Q^2$

tension produces frequency at $\lambda = $ 91.14 nm or "root" wavelength

string length $= \left(1 - 1\big/2^2\right)Q^2 = \left(3\big/4\right)Q^2$

tension produces frequency at $\lambda = $ 121.5 nm or

quantum differentiated at "n = 2."

The Rydberg frequencies are, in fact, known to be functions of a distance measure. They are functions of the radial length of the orbiting electron.

Four Dimensional Quantum Geometry vs. Three Dimensional Quantum Mechanics

We have presented the Rydberg formula as the equivalent of the negation of the quantum squared. This, of course, was not the original interpretation of Rydberg. Three dimensional quantum mechanics emerged in the early twentieth century to explain Rydberg's mathematical formula for the Lyman Series.

Both four-dimensional quantum geometry and three dimensional quantum mechanics agree that the radial distances of orbiting electrons are restricted to quantum values. That is, radial distances are functions of whole numbers. .

The disagreement between quantum geometry and three dimensional quantum mechanics occurs over the method by which those quantized radial distances output light frequency and this, in turn, is due to a dispute over the electron/ nuclear bond.

The Quantum Mechanical model of the atom has become the standard for all of physics. *(Hence forth, Quantum Mechanics will be abbreviated as "QM")*. The QM model, however, is seriously flawed. It was built in large part upon Niels Bohr's interpretation of "electron voltage."

Electron voltage is a energy state for a particular quantum radius of an orbiting electron per coulomb of charge for the particle. The electron voltage mathematics were developed by Niels Bohr from the Rydberg formula for the root frequency in the Lyman Series. Planck's Constant and the elementary charge of the particle were then applied using the formula "*Energy=(voltage)(charge)*." Specifically, electron voltage for any Rydberg frequency is the following:

λ_r = Rydberg root wavelength ; h = Planck's Constant ; e = elementary charge

eV = electron voltage ; c = speed of light ; (C / λ_r) = frequency

$$eV = \frac{(C / \lambda_r) h}{e} = 13.6$$

Minus this "13.6 eV" was said to be the potential energy possessed by the "primary orbit." Electron voltage is treated by Bohr as the equivalent of centripetal force as his model of the electron is a "little satellite" orbiting the nucleus. Centripetal force is modified by the square of the distance of the orbit. Potential Energy=$-13.6/ r^2$. "R" or radius in this case is restricted to quantum numbers becoming "Potential Energy=$-13.6/ n^2$."

A stable orbit of any body in a gravitational field possess potential energy. The potential energy is the balance of centripetal force (force toward the center) and centrifugal force (force away from the center). At greater distances, it takes less velocity (energy) to keep an object in orbit. It also requires less centripetal force and the bond of attraction (gravitational pull) possesses less potential energy. .

In the Bohr model, the electron "falls" from a higher quantum orbit (greater "n" value and weaker electron voltage) to a lower quantum orbit (lower "n" value and greater electron voltage.)

The greater "n" value orbit has less potential energy (less centripetal force or bond attraction and therefore weaker electron voltage). The lower "n" value orbit has greater potential energy (more centripetal force and greater bond attraction and therefore greater electron voltage). Energy is acquired during the "fall" from the lower potential energy state to the higher potential energy state.

For the purpose of Rydberg and the Lyman Series, the destination orbit of the "fall" is always the lowest orbit "n=1." The "falling" from "n" to "n=1" produces an increase in energy. Bohr argues that the increase in energy is output as radiation frequency.

Potential energy of the bond (centripetal force on the electron) is "electron voltage" and the energy value for "electron voltage" was calculated using Max Planck's formula for the energy in frequencies of light. *(More will be said about "Planck's Constant" later).*

Electron voltage is a function of electron's quantum radius. Bohr's formula is the following:

eV = electron voltage ; n= quantum radius of orbit

$$eV = \frac{-13.6}{n^2}$$

Electron voltage is "centripetal force" potential energy and is negative[217]. Electron voltage is divided by the square of the quantum radial distance. Therefore, higher "n" quantum orbits have less potential energy than lower "n" quantum orbits. Division by a greater number

---
[217] Consult a standard physics text for an explanation.

produces a lower numeric result for the division.

Rydberg is explained by the electron "falling" from orbit "n" to the first orbit (n=1). The potential energy of the destination orbit (n=1) is subtracted from the potential energy of the orbit from which the electron falls. This results in a positive, kinetic energy value:

$$\left(\frac{-13.6}{n^2}\right)-\left(\frac{-13.6}{1^2}\right)=\frac{13.6}{1^2}-\frac{13.6}{n^2}=\left(1-\frac{1}{n^2}\right)13.6 \text{ eV}; \; \textit{This duplicates Rydberg}$$

The electron is said to acquire "$\left(1-1/n^2\right)13.6$ eV" during the fall. This is a positive energy value which Bohr argues is output as ultraviolet light at the frequency of:

$$\left(1-1/n^2\right)13.6 \text{ eV} \; ; \; 13.6 \text{ eV}=\frac{(C/\lambda_r)h}{e}; \; \left(1-1/n^2\right)13.6 \text{ eV} = \left(1-1/n^2\right)(C/\lambda_r)(h/e)$$

That is, Bohr's electron volts are nothing more than the (Rydberg root frequency) *times* (Planck's Constant *divided by* the elementary charge). What is this value "(h/e)" which Bohr applied to the Rydberg frequencies?

The N-radiation studies of cotton revealed what Planck's Constant actually is. It is the energy exchanged by one hydrogen bond in a hydrocarbon molecule with one attached proton in the nucleus. It is that which is remaindered after frequency is subtracted by negative irradiation. Planck's Constant is a measure of exchange energy between the electron and nucleus. It is not a measure of orbital energy.

Niels Bohr's "little satellite" model of the atom is simply wrong. Planck's Constant *does not* identify "centrifugal force energy." It *is not* the energy possessed by an orbiting electron trying to escape from different radial distances. It *is not* the balance of escape-attraction forces.

Planck's Constant is the electromechanical capacity of the electron/proton to exchange energy with one another through four-dimensional space. The only adequate three-dimensional analogy for this four-dimensional process is the tensioned vibrating string. It can be simplified no further to accommodate restricted three-dimensional thinking.

If you place a heavy plumb on a string and let it drop, tension will be placed upon the string with a force value equal to "(mass of plumb) 9.80665 m/ sec 2." The tension thus established will produce a certain frequency of vibration for the string. If you pluck the string the energy of vibration will attempt to push the plumb into a pendulum motion of a certain period as determined by mass and length of string.

l = length of string ; g = gravity field = 9.80665 m/ sec^2 ; t = time

$$\text{frequency of pendulum motion}=\frac{1}{t}=\frac{1}{2\pi}\sqrt{\frac{g}{l}}$$

We can use Hook's Law[218] and the Dawson Tensor[219] for string frequency to determine a frequency value for the string

Hook's Law

k = force (tension) constant per unit of length ; Δx = change in length (stretch)

$F = mg$; m = mass of plumb; g = gravity acceleration = 9.80665 m/ sec^2; **F = k Δx**

---

[218] Robert Hook published in 1678 as *Ut tensio, sic vis*, meaning: " As the extension, so the force."
[219] See Appendix *"Dawson's Tensor"*

Dawson's Tensor

$$\text{frequency of string} = \frac{1}{t} = \sqrt{2}\, k(x_t - x_0) \; ; \quad x_t - x_0 = \Delta x$$

$$\text{frequency of string} = \sqrt{2}\, k\Delta x = \sqrt{2}\, mg$$

The frequency of the pendulum will always be much less than the frequency of the string vibration. Pendulum frequency is the square root of gravity reduced by factors of 2π and string length. String vibrational frequency is gravity multiplied by the mass of the plumb times the square root of 2 (1.414). Vibrational frequency is always greater than pendulum frequency.

Since frequency of the pendulum motion can never equal the frequency of vibration, motion interference will be established. Pendulum motion will always undergo forcible interference by string vibration. This interference will be resolved by the transference of pendulum energy into plumb spin. The energy of vibration for the string is converted to the energy of spin for the plumb. This is as close as three-dimensional physics can come in describing the energy exchange between the electron string and the attached proton.

Planck's Constant is analogous to a vibrating string energizing the spin of an attached plumb. It is the rate at which the electron string provides energy to the attached proton *per* cycle of vibration. The total potential energy which the electron string can supply the nucleus through this mechanism is Planck's Constant times the frequency of the string.

Frequency of string vibration is determined by string length. The greater the string length, the greater the potential energy exchange between nucleus and electron. Greater string length also increases the electron voltage. This leads us back to the problem with Bohr's application of electron voltage.

In the Bohr "little satellite" model, electron voltage is inversely related to the square of the presumptive radial distance of the orbiting electron. Electron voltage is thought of as the equivalent of gravity. The attraction of gravity is inversely related to the square of the distance between two bodies. For an orbiting body in a gravitational field, the lower the orbit, the higher the gravitation force, the higher the orbit the lower the gravitational force.

Bohr's model applies electron voltage in the same way. Electron voltage is inversely related to the square of the presumptive quantum orbital distance. The higher the orbit, the lower the electron voltage. The lower the orbit, the higher the electron voltage. Bohr's application of electron voltage is the equivalent of the gravitational pull upon a satellite.

Such argument by analogy, however, is not appropriate because gravity and voltage are not equivalent field forces. Bohr's electron voltage is determined by the elementary charge of the electron which is measured in coulombs (amps per second):

Energy=(watts)(sec)=(Voltage)(amps)(sec)=(Voltage)(coulombs or charge);
Voltage=Energy/ charge

Voltage, however, is not the measure of a field force which is inversely related to distance. The electrical field force which is inversely related to distance is capacitance. The actual field equation for the electron bond is the following:

$$C = \frac{Q}{V} \; ; \quad C = \text{capacitance} \quad V = \text{voltage} \; ; \quad Q = \text{coulombs} = e = \text{elementary charge}$$

$$E = \int_0^Q V \, d(Q) = \int_0^Q \frac{Q}{C} d(Q) = \frac{1}{2} \frac{Q^2}{C} = \frac{1}{2} \frac{e^2}{C} \quad ; \quad \text{let } 2C = eC = \text{ electron capacitance}$$

$$\frac{E}{e^2} = \frac{1}{eC}$$

E = (Planck's Constant)(root frequency)

Energy is equal to the integration of voltage over charge. The function of the derivative for this integral must be:

$$Q^2 / 2C \text{ or } e^2 / eC$$

Instead of Bohr's "eV=E/e " we have "1/ eC=E/e^2 ." That is, energy *divided by* elementary charge is electron voltage and energy *divided by* the elementary charge squared is the inverse of electron capacitance. We have not changed the values of anything. We have just changed the way they are displayed to better represent actual field equations.

In the first place, the electron/proton bond which these equations represent is the charge squared, that is, the charge of the electron *times* the charge of the proton. Since those charges are known to be equivalent, the product of multiplication is the charge squared.

In the second place, capacitance is a known field value of which voltage is only a component. The argument being made by Bohr is that electron voltage is a field value; the equivalent of a gravitational field for his "little satellite" model of the electron. Electron voltage is being made inversely proportional to the square of distance just as the gravitational force between two masses is inversely proportional to the square of distance.

The Quantum Geometric Alternative to Bohr's Electron Model

Quantum geometry identifies the relationship between orbital string length and electron voltage and capacitance differently than does the Bohr model of electron orbits.

Technically, electron field capacitance is the amount of "push" the negative electron charge has left at a specific string length. As the electron string gets longer and tighter, electron voltage increases and electron capacitance decreases proportionally. Both are forces partially initiated by the electron's charge.

The ± charge of a particle (its elementary charge) is measured in coulombs. We have demonstrated that the elementary charge has a distance value[220] consistent with four dimensional geometry. The charge is a projection into the extra dimension not conformed to the particle's own volume . The particle is "bonded" by attachment of this charge to the missing dimension contained within its opposite's volume geometry.

The covalent bond is the proton's positive charge (missing quantum dimension) *times* the electron's negative charge (missing Euclidean dimension). The charges multiply one another. The multiplication of a quantum by an Euclidean distance of equal value defines the quantum squared.

With three dimensional geometry, bonded particles could be pulled into contact with one another, but they would not be expected to merge. This is not so with four dimensional geometry. Both the quantum defined electron and the Euclidean defined proton have the

---

[220] See *Nuclear Capacitance: An Overlooked Quantum Field* in Appendix

capacity to define the same space simultaneously. There are two different geometric definitions competing for any volume.

Euclidean defined space expands and distorts any attempted quantum definition of the same space. Thus the Euclidean line is kinked by its quantum definition into a curved line and the Euclidean defined plane is kinked by its quantum definition into a curved plane.[221]

This would also be true of the attempt by a three-dimensional quantum defined electron to define the space occupied by an attached three dimensional 'Euclidean defined proton. The proton expands the electron in all directions, rotating the electron bond to 90°. Thus a force of contraction (the pull of the electron bond) is opposed by a force of expansion (kinking the rotated bond into curvature) creating tension upon the string/bond.

There are such oppositional forces clearly identified by the electron voltage equations. Science accepts Planck light energy as electron orbital energy ala the Rydberg distribution. Therefore, we have the Bohr equation:

$$\text{electron volts} = eV = \frac{(h)(f)}{\varepsilon} \quad ; \quad \varepsilon = \text{elementary charge}$$

As Planck energy, "$(h)(f)$" goes up, so does electron voltage. Voltage is defined as follows: *"Electrical tension (or voltage) is the difference of electrical potential between two points"*[222] As the distance between the two particles becomes greater, the potential difference between them (i.e. the amount of contested space) becomes greater. Electron voltage increases with electron energy. ALthough electron voltage is not synonymous with string tension, it is proportional to it.

On the other hand, capacitance is defined as follows: *"Capacitance is a measure of the amount of electric charge stored (or separated) for a given electric potential."*[223] Electron capacitance is the stored value of the electron charge at orbital distance. It is the measure of how much capacity the electron charge has to influence space outside itself measured in the force value "Farads." Its relationship to electron voltage is electrodynamically regular.

$$\frac{1}{eC} = \frac{(h)(f)}{\varepsilon^2} \quad ; \quad \frac{\varepsilon}{eC} = \frac{(h)(f)}{\varepsilon} \quad ; \quad \frac{\varepsilon}{eC} = eV \quad ; \quad eC = \frac{\varepsilon}{eV} \quad \textit{the standard formula for capacitance}$$

We have two interacting forces influencing the electron and its shell; the linear contraction force of electron voltage and the field expansion force of electron capacitance.

Electron voltage is the potential difference in the amount of space over which the contesting particles attempt definition. This is a linear force of tension directly related to string tension. Electron voltage attempts to close the distance between the particles and lower the potential difference of contested space. In this it acts like normal electrical voltage which seeks a current path to lower potential difference.

Electron capacitance is the remaindered portion of the electron charge which can be projected outside the electron as a field force. It represents the capacity of the electron charge to influence external space. It is an expansive force and may or may not be the expansive force contradicting the contraction force of electron voltage. That issue will need be resolved by further research.

---

[221] See Chapter 1.

[222] en.wikipedia.org/wiki/Voltage

[223] en.wikipedia.org/wiki/Capacitance

Conventional Electronic Capacitance

In electronics, capacitance is the measure of the induction force between two spatially separated plates. This induction force across the space of separation is called the "electrostatic field." Capacitance is inversely proportional to the distance between the two plates. The greater the distance of separation, the lower the capacitance.

The voltage is the "electromotive force" required to sustain the field. The greater the distance of separation between plates, the greater the "electromotive force" required to sustain the same amount of induction. Induction force is measured in coulombs or amps per second. For a constant induction force— as with the elementary charge of a particle— voltage becomes directly proportional to distance. The greater the distance, the greater the voltage required. Voltage operates in the opposite manner as that required by the Bohr model.

Electron voltage and electron capacitance are indirectly proportional to one another (eC=e/eV). Since the elementary charge "e" is a constant, electron voltage goes up as electron capacitance goes down.

It is electron capacitance which is inversely proportional to the square of the electron orbital distance (as identified by quantum geometry). It is capacitance which is known electronically as the electrostatic field value indirectly proportional to distance. The greater the distance between the two plates of a capacitor, the lower the capacitance. There is no such distance law governing voltage.

Returning the Bohr energy equation. Light frequency output is related to the orbital radius of the electron. As light frequency goes up energy also goes up *ala* Planck's Constant. As energy increases electron capacitance decreases and electron voltage increases. From electronics, distance is known to be indirectly proportional to capacitance and directly proportional to voltage. As energy increases, the radial distance of the electron must also increase. This is the opposite of the Bohr "little satellite" model.

Electron radial distances must increase with higher energy states, not decrease as Bohr proposes. The diameter of the atom must increase as the atom becomes hotter. Matter does indeed expand with increases in temperature. Real world data fits the electronic model better than the Bohr model.

We will demonstrate that this is not the only reason the Bohr "little satellite" model must be rejected in favor of the more accurate quantum string model.

The Bohr Atomic Model Cannot Identify "Root" Frequency

Niels Bohr was not completely wrong. He was a strong intuitions who discovered valid physical principles, even if his mathematical details proved inadequate[224]. Bohr intuitively applied a valid quantum mathematical principles to his model. The Bohr model works by negation and negation is the primary mathematical principle of the actual quantum (differentiation by negation of subdivision). The problem with the Bohr model is that it negates in the wrong direction.

In the Bohr model, the electron radius contracts (negates) *to* the highest energy state ("root" frequency), not *from* that state to lower orbits as the differentiated quantum must. Bohr

---

[224] He also intuitively identified the "even-odd" principle for atomic weights as the explanation of the differences in neutron capture resonances between uranium 235 and uranium 238. See Rhodes, "*The Making of the Atomic Bomb*".

explains all frequencies as being acquired energy in a "fall" — by subtraction from 13.6 electron volts, or the energy value of the root wavelength as determined by Planck's Constant divided by the elementary charge.

The greatest problem with the Bohr model is that the root frequency itself cannot be explained. The root frequency cannot be resolved by a "fall" to a whole number quantum orbit.

By the Bohr model, The root wavelength could only be explained by a "fall" to an orbit with a value of "$n = 1/\sqrt{2} = 0.7071067812:$ "

$$\frac{-13.6}{1^2} - \left(\frac{-13.6}{\left(1/\sqrt{2}\right)^2}\right) = \frac{13.6}{1/2} - \frac{13.6}{1} = (2)13.6 - 13.6 = 13.6 \text{ eV}$$

"$1/\sqrt{2}$" as a radial length is neither a whole number nor a quantum value. The Bohr model simply cannot explain the "root." This is a theoretical difficulty of major proportion.

ALL radiation wavelengths output by hydrogen are negations of the "root" wavelength. It is not just the ultraviolet Lyman Series. The root frequency is the basic quantum radial distance from which are other radiation frequencies are harmonically output by negation of subdivision. An inability to account for the root leaves the Bohr model completely bankrupt as an explanation of radiation emissions.

In addition to the Lyman Series of ultraviolet frequencies, four other bands of frequencies output by hydrogen have been discovered. All are derivable from the root frequency. These are the following:

Balmer Series (visible light frequencies). Formula: $\dfrac{1}{\lambda} = \left(\dfrac{1}{(2)^2} - \dfrac{1}{(n)^2}\right)\dfrac{1}{\lambda_r}$; $3 \leq n \geq 8$

Paschen Series (infrared spectrum). Formula: $\dfrac{1}{\lambda} = \left(\dfrac{1}{(3)^2} - \dfrac{1}{(n)^2}\right)\dfrac{1}{\lambda_r}$; $4 \leq n \geq 8$

Brackett Series (lower infrared spectrum). Formula: $\dfrac{1}{\lambda} = \left(\dfrac{1}{(4)^2} - \dfrac{1}{(n)^2}\right)\dfrac{1}{\lambda_r}$; $5 \leq n \geq 8$

Pfund Series (far infrared spectrum). Formula: $\dfrac{1}{\lambda} = \left(\dfrac{1}{(5)^2} - \dfrac{1}{(n)^2}\right)\dfrac{1}{\lambda_r}$; $6 \leq n \geq 8$

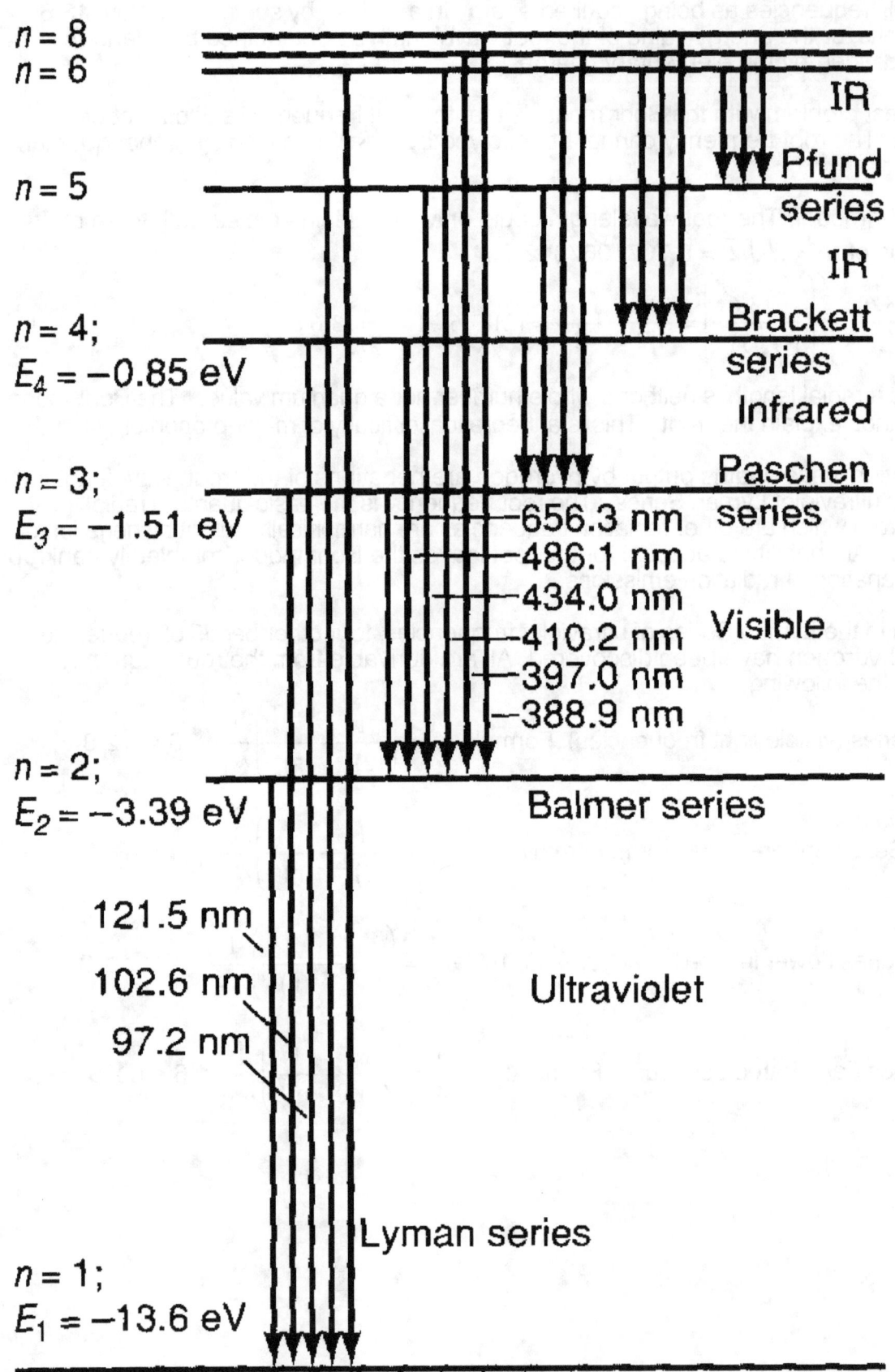

All four of the above are *"negations of quantum squared subdivisional units of the root frequency."*

The Balmer frequencies equal the quantum negations of the area subdivisional unit "$1/2^2$" That is, an area subdivision of 1/4 the whole area.

The Paschen infrared frequencies are the negations of the subdivisional unit "$1/3^2$" (area subdivision 1/ 9 whole area) and so forth.

All frequency series' are differentiations of a quantum squared subdivisional unit with the quantum squared equal to the root frequency. ALL radiation frequencies being put out by hydrogen are quantum squared differentiations of the root. The root itself is the quantum value (quantum squared).

The Geometric Construction of "Non-Kinked" Quantum Area (Q^2)
Quantum area is defined as the summation of all quantums made by the points along an Euclidean line and a quantum point outside that Euclidean line.

Quantum Area	Since there aren't two quantum dimensions, the quantum squared can only be defined by integrating the quantum values across the continuum of points along the Euclidean line. Specifically:
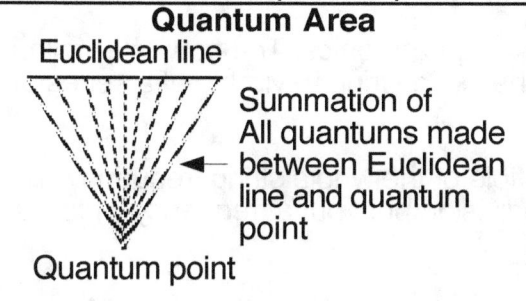 Euclidean line Summation of All quantums made between Euclidean line and quantum point Quantum point	*lim* $x = Q$; $x =$ Euclidean Line ; $Q =$ quantum $$\int_0^Q Q\, d(x) = \frac{Q^2}{2}$$

The actual area defined by the integral is the following:

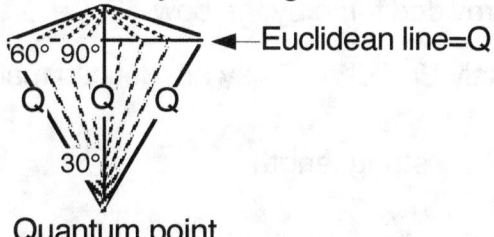

Euclidean line=Q

Quantum point

Total area is the area of two triangles each with a "base=Q" and a "height = Q/ 2."
Area= (2)1/2 (base)(height) = (Q)(Q/ 2) = $Q^2/2$
The quantum squared is geometrically constructed by "stretching" the mean quantum to an end point along the Euclidean line and returning it to its original position. Thus, the quantum squared is constructed in a manner similar to a vibrating string.

Quantum Harmonics by Negation of Subdivision
vs.
Euclidean Harmonics by Subdivision

The difference between the harmonics of sound and the harmonics of light is the difference between Euclidean differentiation and quantum differentiation. We have said that Euclidean distance is differentiated by subdivision and that quantum distance is differentiated by the negation of subdivision. This distinction is reflected in the difference between sound harmonics and light harmonics.

Sounds are harmonically related by frequency. These harmonics are Euclidean. Light is harmonically related by frequency energy. These harmonics are quantum. The impedance of harmonically related sounds by Euclidean strings are functions of frequency. The impedance of harmonically related light by electron strings are functions of frequency energy. .

A comparison of the tensioned strings of musical instruments and the string/bond of the electron illustrates the difference.

The violin operates upon the principles of Euclidean string harmonics. The bow of a violin, applies force to the string continuously. This causes the violin string to vibrate at a series of subdivisional lengths.

Each subdivision puts out a frequency which is a multiple of the whole string frequency. Each subdivision is $1/n$ of the whole. Each "nth" subdivision puts out a frequency which is "n" times the frequency of the whole string.

The bowed violin note is an Euclidean harmonic series . It is heard as a "richer," than the single, non harmonic, note given off by a "plucked" stringed instrument like the guitar.

Euclidean string harmonics are subdivisional. Subdivisional vibrations of the string occur naturally when the string is continuously provided force by the bow.

The string harmonics of sound are completely Euclidean. They are defined mathematically by the following:

$$\frac{SL}{n} = n\,(f) \quad ; \quad f = \text{frequency} \; ; \; SL = \text{string length}$$

An Euclidean tensioned string under continuous force application naturally vibrates subdivisionally. Each subdivision is found by dividing the whole by "n." Each one of these "nth" subdivisions produces a frequency which is "n" times greater than the frequency produced by the whole string.

The Energy in Vibrating Euclidean Subdivisions

The energy being put out by either a Euclidean or a quantum vibrating string is given by the following formula:

$m =$ mass of string ; $f =$ frequency of vibration ; $d =$ distance string is deflected

$$E = 8(m)(f^2)(d^2)$$

For the mathematical proof of this , see Appendix "Energy in Vibrating String"

The deflection of the string, "d," requires force to accomplish. Force is defined as mass *times* acceleration. That is, the force required for deflection is defined by the mass of the string *times* the acceleration across the deflection distance "d." The formula for the force of a deflected tensioned string is the following:

$$\text{Acceleration} = \frac{2d}{t^2} \quad ; \quad \frac{1}{t} = 2f$$

$$\text{Force} = m\left(\frac{2d}{t^2}\right) = 8(m)(d)\left(f^2\right)$$

If only "1/ n" of the string is vibrating, the mass value of the string portion vibrating is "(1/n)m." The frequency of the portion vibrating is "(n)*f* . " It can be shown geometrically that the deflection distance for the subdivisional unit is also reduced to "d/ n." That is, to "one nth" of the deflection provided the whole string by the bow force.

Applying these values to the force formula we can see that the required force of deflection is the same for the subdivisional unit as it is for the whole string.

$$\text{Force} = 8\left(\frac{m}{n}\right)\left(\frac{d}{n}\right)(n\ f)^2 = 8\frac{(m)(d)\left(f^2\right)n^2}{n^2} = 8(m)(d)\left(f^2\right)$$

The required force of deflection— " $8(m)(d)\left(f^2\right)$ " — is the same for the "nth" subdivision as it is for the whole string. It takes the same amount of force to deflect the "nth" subdivision to "d/ n" as it takes to deflect the whole string to "d."

This, however, is not true of energy. Substituting the "nth" subdivisional values into the energy formula, we get the following:

$$8\left(\frac{m}{n}\right)\left(n^2\ f^2\right)\left(\frac{d^2}{n^2}\right) = \frac{E}{n}$$

With the same amount of deflection force, the energy or "loudness" being put out by the "nth" vibrating subdivisional unit is only "one nth" as loud as the whole string. This is true even though the frequency or "pitch" is "n" times greater.

It takes the same amount of force to deflect a subdivisional unit as it does the whole string. This outputs a higher but softer (less energetic) harmonic pitch. The bowed violin provides a continuous multiplication of force to set the subdivisional units of the string vibrating. This composes a "note" of progressively higher but softer harmonic frequencies all of which are subjected to the same force of the bow.

The less energetic, higher frequencies are due exclusively to the shift in string-mass for subdivisional units. Only "one nth" of the string-mass is vibrating at the subdivision. This is not true of the quantum string/bond. Electron mass is constant for all values of differentiated string lengths.

The relationship between energy and frequency for Euclidean string harmonics is the inverse of the relationship which Max Planck discovered existed for light frequencies. Planck showed that the energy in light increases by a constant factor for increases in wave frequency.

Euclidean string harmonics are the inverse of quantum string harmonics. The Euclidean string is differentiated by subdivision. Frequency is increased by differentiation. Energy is inversely related to increase in frequency. The quantum string is differentiated by the negation of subdivision. Frequency is decreased by differentiation. Energy is directly related to decrease in frequency. Euclidean string harmonics and quantum string harmonics are the inverse of one another, just as the negation of subdivision is the inverse of subdivision.

Harmonics of the Quantum String

Some of the more important respects in which quantum string harmonics are the inverse of Euclidean string harmonics are the following: With Euclidean harmonics mass is changed by subdivision of the whole string and string tension remains constant, but with quantum harmonics, mass remains constant during negation of subdivision and tension changes; Euclidean frequencies are increased by differentiation while quantum frequencies are decreased by differentiation.

The greatest difference however is the way that frequencies are harmonically related.

The electron does indeed "fall" as Bohr suggests. It "falls" from the highest potential energy state and maximum extension of the electron radius (the root frequency) to a negation of that maximum. No matter what energy balance exists between the nucleus and the electron string; no matter what particular quantum orbit the electron may occupy, . that electron always has negative definition on the full range of string potential energies.

The potential energy state for any quantum string length has been identified by N-radiation experimentation as Planck's Constant *times* the string frequency. Frequency is determined by the string lengths which are produced by the negation of subdivision of the root frequency:

$h=$ Planck's Constant ; n= series or "band" number ; C = speed of light

$n' =$ series' element number; $n' > n$; λ_r = root wavelength (91.141 nm)

$$\text{Potential Energy} = PE = h\left(\frac{1}{n^2} - \frac{1}{n'^2}\right)\frac{C}{\lambda_r}$$

Any derived string length retains definition on the root string length by the following formula:

$$PE_{root} = h\left(\frac{1}{n^2} - \frac{1}{n'^2}\right)\frac{C}{\lambda_r} - h\frac{C}{\lambda_r} = \left(\frac{1}{n^2} - \frac{1}{n'^2} - 1\right)h\frac{C}{\lambda_r} = \left(\frac{n'^2 - n^2}{n^2 n'^2} - 1\right)h\frac{C}{\lambda_r}$$

always a negative number

String Potential Energy is the energy exchange rate between the nucleus and the electron. Even though the nucleus may not have enough energy to allow the electron to achieve the root orbit, the derivative orbit can always absorb or impede the root light frequency by virtue of the fact that the harmonically related light frequency can provide the string with enough electron voltage for impedance.

Harmonic impedance is a system of electron voltage exchange. The general principle is this: Any electron string length can impede a higher order harmonically related light frequency, but the amount of energy which can be absorbed by the string is restricted to its own potential energy (the Planck energy value for the string' s frequency).

196

The variance between string potential energy and Planck energy for the light frequency must be invested in increased electron voltage (string tension) to make the impedance possible:

f_l = frequency of light ; f_h = frequency harmonically related string

$\dfrac{h(f_l)}{\varepsilon}$ = string electron volts required for direct impedance of light

$\dfrac{h(f_h)}{\varepsilon}$ = string electron volts possessed by actual length

eV_{need} = Additional volts required = $\dfrac{h(f_l)}{\varepsilon} - \dfrac{h(f_h)}{\varepsilon} = \dfrac{h(f_l - f_h)}{\varepsilon}$

Electron strings can absorb higher order light frequencies in the harmonic series by applying a portion of the potential energy in the light frequency to needed additional electron voltage. Since the electron-light exchange is not established at the needed string length, additional voltage (tension) must be applied to the shorter string length.

The shorter string has negative potential energy with respect to the harmonically related string length which would directly impede the light frequency. The negative potential energy equation for an electron string length which is harmonically related to the root is the following: PE=(Planck's Constant)(frequency)[225]

Negative PE for Harmonic Partial of Root

PE_r = potential energy root = $h\dfrac{C}{\lambda_r}$

PE_s = potential energy string = $\left(\dfrac{1}{n^2} - \dfrac{1}{n'^2}\right)h\dfrac{C}{\lambda_r}$

$PE_{r/s}$ = potential energy for the string on root level = $\left(\dfrac{1}{n^2} - \dfrac{1}{n'^2}\right)h\dfrac{C}{\lambda_r} - h\dfrac{C}{\lambda_r}$;

always a negative number

The potential energy provided by the light must "fund" this negative difference in order for impedance to take place from a lower order string level. The variance between the string's own potential energy level and that of the light (the negative value above) must be made up (as additional electron voltage) so the string can impede the light. :

Energy Addition Needed to Allow Harmonic String to Impede Light Frequency

$-\left[\left(\dfrac{1}{n^2} - \dfrac{1}{n'^2}\right)h\dfrac{C}{\lambda_r} - h\dfrac{C}{\lambda_r}\right] = h\dfrac{C}{\lambda_r} - \left(\dfrac{1}{n^2} - \dfrac{1}{n'^2}\right)h\dfrac{C}{\lambda_r} = h(f_l) - h(f_h)$

Converted to Electron Volts

$eV_{need} = \dfrac{h(f_l) - h(f_h)}{\varepsilon}$

When the energy for the needed electron voltage is subtracted from the light's potential energy, the strings own potential energy is remaindered.

$h\dfrac{C}{\lambda_r} - \left[h\dfrac{C}{\lambda_r} - \left(\dfrac{1}{n^2} - \dfrac{1}{n'^2}\right)h\dfrac{C}{\lambda_r}\right] = \left(\dfrac{1}{n^2} - \dfrac{1}{n'^2}\right)h\dfrac{C}{\lambda_r} = string\ PE$

---

[225] By mathematics and by experimentation. See n-radiation study of cotton.

This is the only amount of light energy the lower order string can absorb from the higher order light frequency. That is, the total energy available to the harmonic derivative is less than the energy applied by the light.

The remainder of the light energy is used to fund the deficit between electron's own Potential Energy state and that of the root frequency orbit.

Energy is neither created nor destroyed, but is only exchanged. The energy difference did not simply disappear.

The light energy is split between absorption and managing the nuclear bond. The "electron voltage" of the string is increased relative to the nucleus when it absorbs higher frequency light in the harmonic series. That is, there is a portion of light potential energy which is no longer being exchanged with the nucleus, but is being applied to increased string tension to make impedance possible.

The electron string is reserving energy to itself. This increased string tension is not shared with the nucleus, but is retained solely by the electron. It offers the real explanation for the "photoelectric effect.[226] "

The Root Frequency: a Limit on a Vibrating String

The root frequency of 91.14 nm is actually a limit. It is the string-tension frequency at which maximum string vibration can reach the speed of light. The length of a string which can vibrationally accelerate to the speed of light (C) is determined by quantum mathematics. The length of the string must be 1/2 the frequency wavelength . Therefore, the length of the string/bond producing adequate tension to accelerate to "C" is 45.57 nm. The proof for this is the following:

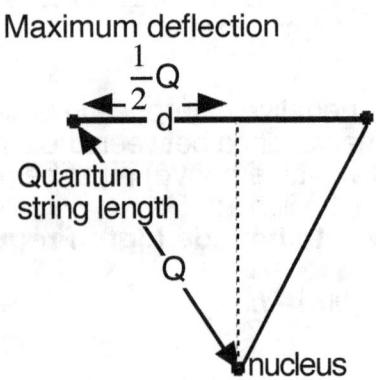

$t =$ time acceleration phase ; average velocity $= \overline{V} = \dfrac{d}{t}$; $d =$ deflection of string

frequency (per second) $= f = \dfrac{1}{2t}$;

final velocity $= V = 2\overline{V} = \dfrac{2d}{t}$; acceleration $= A = \dfrac{V}{t} = \dfrac{2d}{t^2}$

let $V = C =$ speed of light ; $\lambda_r =$ root wavelength $= 91.141$ nm

$$f = \dfrac{C}{\lambda_r} = \dfrac{1}{2t} \quad ; \quad \dfrac{1}{t} = \dfrac{2\,C}{\lambda_r} \quad ; \quad d = \dfrac{Q}{2}$$

$$V = \dfrac{2d}{t} = C = 2d\dfrac{2\,C}{\lambda_r} = \left(\dfrac{2Q}{2}\right)\dfrac{2\,C}{\lambda_r}$$

$$C = \dfrac{2\,CQ}{\lambda_r}$$

$$Q = \dfrac{\lambda_r C}{2C} = \dfrac{\lambda_r}{2} = 45.5705 \text{ nm}$$

Dawson's Theorem
(The Mathematical Proof of the Quantum Dimension)
Lawrence Dawson
Dawson's Theorem

Any Euclidean line in non-solid space is "kinked" into curvature by the intersection of the quantum dimension. The intersecting quantum dimension composes a linear plane with the Euclidean line which is also "kinked" into curvature. This structure composes a stand-alone unit of vacuous-volume designated as the quantum squared. The Euclidean line and its "kinked" curvature is a component of this discrete unit of quantum-squared space and the quantum imposes a unit of measure upon the Euclidean line equal to one half its own value. This unit of measure may be differentiated to "0" by negation of subdivision for the quantum squared at every possible point along the Euclidean line. A strictly Euclidean unit of measure must differentiated by subdivision and cannot be differentiated to "0." Only the intersecting quantum dimension allows rational calculus differentiation of the Euclidean line.

Traditionally, geometric space is represented on a Cartesian graph composed of two or three axii which are arrayed by units of measure and their subdivisions. The subdivisions of a unit of measure cannot identify all point values along the axis because there is always a space of separation between adjacent subdivisions.

The actual physical Euclidean dimensional axis, however, is composed of a continuum of points which can be differentiated to "0." That is there is no space of separation between adjacent point values along the axis. The Euclidean dimensional axis can only be accurately described by differentiation to "0" ($d(x)$) and by integration of the differential. Such differentiation, however, is not possible for an x axis which is arrayed by a unit of measure and its subdivisions.

DEFINITION: A dimensional axis which is differentiable to "0" ($\Delta x = 0 = d(x)$) and is therefore composed of a continuum of points will be designated a *strict Euclidean axis.*

A Strict Euclidean Axis Cannot Measure Distance

There is no space of separation between points along a strict Euclidean axis. Because there is no separation, distance cannot be measured by counting points between a beginning and an end point establishing distance along the axis. The strict Euclidean axis simply cannot determine distances.

Distances between pointal positions along a strict Euclidean axis can only be determined by arraying the axis with a unit of measure.

However, these units of measure cannot measure all possible points along the strict Euclidean axis by one-to-one correspondence. There must always be a space of separation between subdivisional units for any unit of measure.

The only way measure all possible distances contained within the continuum of points which compose the strict Euclidean axis is to integrate the derivative of the axis. That condition is established in the following manner:

Let y=f(x) = axis of measure
Let x=variable axis=the strict Euclidean axis

$$f(x) = x \quad ; \quad D(f(x)) = \frac{x + \Delta x - x}{\Delta x} = \frac{\Delta x}{\Delta x} = 1 \quad ; \quad f(x) = \int_{x=0}^{x} (1)d(x)$$

The derivative of the function is the unit value of the measured axis (unit=1). The variable axis "x" must be integrated *times* the unit value for the measured axis.

$$\mathbf{D(f(x)) = 1}$$

$$\mathbf{f(x) = \int_{x=0}^{x} (1)d(x)}$$

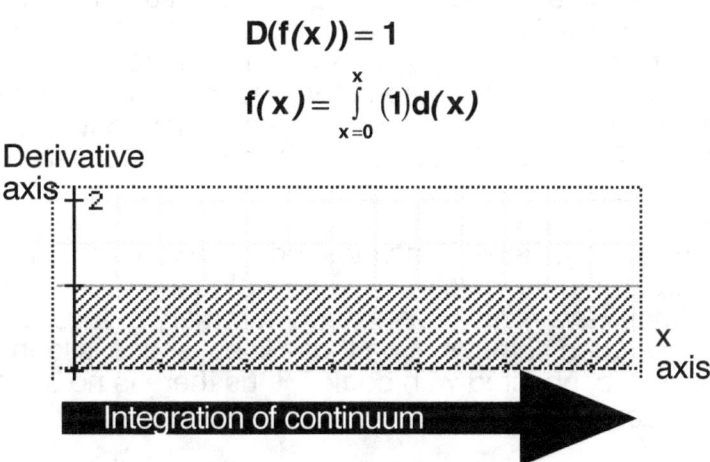

The value of "x" is the area under the graph.

Conceptually, this simple version of the measurement of a strict Euclidean length is satisfactory. The distance to any point along the x axis is measured by integration to the point *times* the unit of measure established by the derivative axis. The problem with integrating a unitized axis , no matter how small the unit of measure , is that the x axis cannot actually be integrated against it.

By finding the derivative of the function $f(x) = x$ we have forced the x axis to become a strict Euclidean axis. That is, we have eliminated the unit of measure from the x axis by differentiating it to "$\Delta x = 0 = d(x)$." No subdivision of the unit of measure can be differentiated to "0". That is, the distance between adjacent subdivisional units cannot equal "0." There must always be a space of separation between adjacent subdivisional units. Only the x axis as a strict continuum of points can be differentiated to "0."

The variable axis (the x axis) can no longer be defined by a unit of measure since Δx cannot equal "0" for any given subdivision of a unit of measure.

A strict Euclidean axis— one differentiated to "0" — cannot measure distance since integration of "$d(x)$" produces a continuum of points which are uncountable. Therefore, there is no possible distance value for "d(x.)" Multiplying the integration of "d(x) " by the unit of measure can give no distance measure.

The only possible distance value which can be applied to "$d(x)$" will be determined by the way the unit of measure is, itself, differentiated by subdivision. The formula by which a unit of measure is further subdivided supplies increasing numbers of countable points to the unit of measure. Adding levels of subdivisions increasingly differentiates the unit of measure by adding countable points to it. Differentiating a unit of measure makes the points within the unit approach the continuum of points defining a strict Euclidean length. However, the continuum can never be reached since there must always be a space of separation between adjacent subdivisional units.

The Concept of the Quantum as a Separate, Fourth Geometric Dimension

Before going on to the mathematical description of the quantum, it is necessary to pose the following question: "How is it possible that our observable three dimensional space can accommodate a fourth dimension regardless of what that fourth dimension might be composed?"The argument for quantum geometry proposes that the fourth quantum dimension does not reside outside of our knowledge and sight but exists in full view and is measurable at all times. For a possible answer we must turn to René Descartes and his modification by Albert Einstein.

The father of analytic geometry, René Descartes, postulated a view of the relationship between solids and empty volume which I will designated the "Cartesian controversy." Albert Einstein, in his book *Relativity,* gives a good summary of the Cartesian controversy: "Descartes argued somewhat on these lines: space is identical with extension, but extension is connected with bodies; thus there is no space without bodies and hence no empty space."[227]

The Cartesian viewpoint holds that space is not separate from the solid matter contained within that space. Rather, Descartes proposed that empty space is an extension of the solid. The fact that Einstein ultimately accepted the Cartesian viewpoint is contained in his *Note to the Fifteenth Addition* (June 9, 1952) of the book *Relativity.* This is the last entry Einstein penned before his death in 1955.

Einstein writes:
"I wished to show that space-time is not necessarily something to which one can ascribe a separate existence, independently of the actual objects of physical reality. Physical objects are not *in space,* but these objects are *spatially extended.* In this way the concept of "empty space" loses its meaning.(italics Einstein's)"

If space is an extension of a solid, then, Einstein believed, space itself must be a solid. If a solid is composed of a continuum of points in every direction, then the extension (empty space) must also be composed of a continuum of points in every direction. Einstein believed that empty volume or space was a solid and continuously refers to a line in space as a "stiff rod."

But what if space were an extension of a solid in a different sense than either Descartes or Einstein believed? I propose that empty volume is an extension of solids as Descartes and Einstein held. But space and the solid which extends that space have different geometric definitions. I propose that a solid is strictly three dimensional Euclidean. This three dimensional Euclidean solid extends or expands space by means of a geometric definition of the quantum.

Empty space, itself has only two Euclidean dimensions and acquires volume by a third quantum dimensional axis.

The difference between a solid and empty volume is the difference of one dimensional axis.

---

[227] Einstein, Albert *Relativity* ; Three Rivers Press, ISBN 0-517-88441-0 ; p. 156

The volume of a solid is defined by a third Euclidean axis. Empty volume is defined by a third quantum axis. The space we occupy, including both solid and empty volumes, is actually four dimensional. It is composed by three Euclidean dimension and one quantum dimension.

The concept then, is that a solid or "mass" and empty volume or "space" are completely geometrically distinguished. A solid is a solid because it is three dimensional Euclidean and is composed of a continuum of points in all directions. A solid is the only thing which is actually three dimensional in the usual Euclidean sense.

Space is empty volume because it is quantum in character and cannot define a solid. The volume of space is defined by the intersection of a series of Euclidean planes by the quantum dimension. It is only the Euclidean plane which can coexist with the quantum dimension in the same volume.

The analytic graph is a two dimensional Euclidean plane. It is, therefore, intersected by an unrecognized quantum dimensional axis. Both the "x" and "y" axii of the analytic graph are intersected by a quantum axis at 90°. A quantum plane is, therefore, formed with both axii of the graph. It is this quantum plane — or more accurately, the unit of area designated the "quantum squared" — which provides resolution to the problem of measurement for the Euclidean graph.

The Quantum Defined

The quantum may be thought of as two points separated by some distance. There are no further geometric points within this space of separation. This, however, is not its mathematical definition.

The mathematical definition of the quantum is the following:

A quantum is a unit of distance which cannot be subdivided, but must be differentiated by the negation of subdivision.

The quantum cannot be subdivided without adding points to the distance of separation. For example, to subdivide the quantum into 1/2 quantum units would require that an additional point be supplied at the half way mark of the distance of separation. The addition of this point would convert the quantum into two quantums. A quantum is two points separated by "some" distance and both halves meet that definition.

The subdivision of a quantum multiplies the quantum by the value of the subdivision. To subdivide a quantum into "n" subdivisional units multiplies the quantum by "n." This, of course, is not a multiplication of distance but a multiplication of the number of quantums contained within the distance.

The original quantum distance cannot be differentiated by subdivision, but must be differentiated by negation of subdivision. Any Euclidean length composed of a continuum of points may be differentiated by subdivision. Points already exist within the line composing the Euclidean length to accommodate such subdivisions. We have already established that a Euclidean unit of measure is differentiated by subdivision and such subdivisions cannot be differentiated to "0," as calculus requires.

Negation of subdivision, however, does allow the quantum to be differentiated to "0."

Negation of Subdivision

A quantum is composed of two points separated by "some" distance. This original quantum cannot be differentiated by subdivision without multiplying the quantum. The original quantum must be differentiated by negation of subdivision.

Any subdivision is a partial of the original which also provides its negation. This negation is a partial defined as the part of the whole which is left after the subdivision is removed:

$$\text{subdivision} = \frac{1}{n}(\text{length})$$

$$\text{negation} = \left(1 - \frac{1}{n}\right)(\text{length})$$

If "differentiation" is defined as a mathematically rational smaller unit of a whole, then a quantum can be differentiated by negation of subdivision. The two points establishing the quantum can assume a new distance value of "1- 1/ n" of the original distance of separation. The integrity of the quantum is thus preserved while assuming a mathematically rational partial of the original whole. The quantum distance of separation is thus "differentiated."

The quantum can be differentiated to "0" by negation of subdivision. If we allow "n=1" then the negation of subdivision is "(1— 1/1)(distance)=0." Unlike the Euclidean length which we have proven cannot be differentiated to "0," the quantum can be. The condition required by differential calculus can only be supplied by the quantum.

Differentiation by Negation and the Quantum Squared

We have said that the Euclidean plane of the standard analytic graph is intersected by an unrecognized quantum axis at 90° to both the x and y axii of the graph. Thus, the intersection of the x axis and the quantum axis produces a plane. The plane is defined by the intersection of two lines, the lines being the x axis and the quantum axis. The quantum plane, however, must define quantum area designated as "the quantum squared."

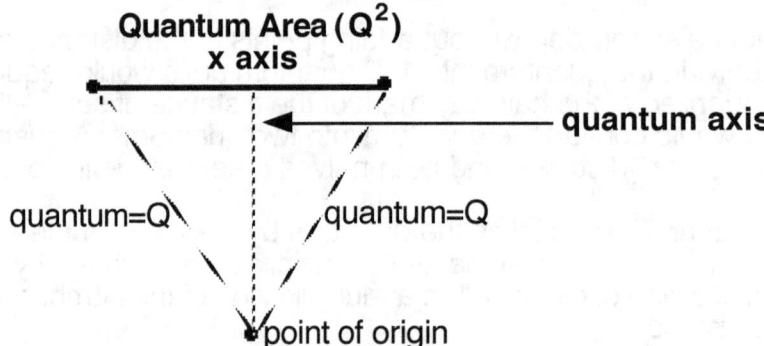

Quantum Area (Q^2)

The construction of quantum area requires the Euclidean axis (x axis). From a single point of origin, two quantum lines can be projected with an angle of separation between them. Such a projection will compose a quantum plane.

Constructing quantum area from the space bounded by the angle between the quantum lines and the secant which the quantum lines subscribe requires that the secant be a component of the x axis (composed of a continuum of points.)

Quantum area must be constructed as the summation of all possible quantums made by

the point of origin and a point along the secant. This requires that the secant be a Euclidean line composed of a continuum of points. A continuum is the only integrable line which provides all possible end points for the summation of the quantums as required to define area.

The quantum itself does not intersect the x axis at 90°. Yet the actual quantum axis must intersect the Euclidean x axis at 90°. Therefore, the actual quantum axis is a differentiated quantum; that is, a quantum which obtains its distance value by the negation of subdivision.

The angle of separation which must exist between the two quantums is mathematically determined. It must be 60°. A 60° angle is the only one which provides a unit of measure for the x axis which is exactly equal to the quantum distance. A 60° angle establishes an equilateral triangle.

Further, a 60° angle provides a quantum π value which is equal to "3." A whole number π value is compatible with the quantum which operates by the principle of whole numbers. The circumference of a quantum circle is equal to the summation of the secants made by the six 60° quantum angles defining the circle. Each secant is equal to the quantum which itself is the radius of the circle. Since π is the ratio of the circumference to twice the radius, quantum π = 6/2 = 3.

Finally, and most importantly, a 60° angle provides the means by which the Euclidean x axis might be differentiated and integrated using quantum principles. It allows for the differentiation of the x axis by negation of subdivision for area.

Negation of Subdivision for the Quantum Squared

Let us see how the 60° angle works to provide a rational negation of subdivision for area. The angle provides an equilateral triangle with the x axis bisected at a distance of 1/2 the quantum:

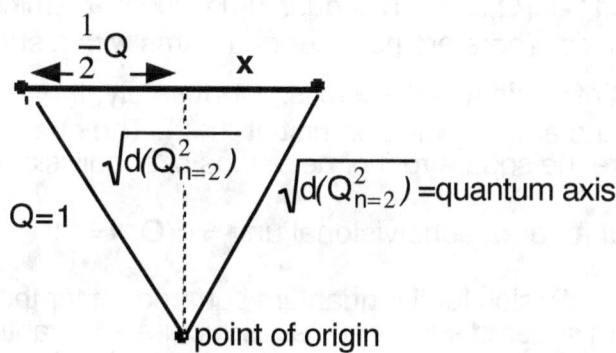

By the Pythagorean Theorem: $Q^2 - \left(\frac{1}{2}Q\right)^2 = d(Q_{n=2}^2)$

The Euclidean secant labeled "x" can be subdivided since it is composed of a continuum of

points. The differentiated quantum labeled "$\sqrt{d(Q_{n=2}^2)}$" intersects the x axis at the subdivisional point. "Subdivision= 1/2" is an exactly determined subdivision[228] .

---

[228] See my paper "*The Law of Exact and Inexact Subdivision.*" following.

The symbol "$d(Q^2_{n=2})$" represents the negation of subdivision for area (differentiation of the quantum squared) for "n=2." The negation of subdivision for a unit of area is unit of area minus the square of the linear subdivision:

$$d(Q^2_n) = 1^2 - (1/n)^2 = 1 - 1/n^2 \; ;$$

This is the negation of the subdivision of an Euclidean unit of area.

The subdivision of an Euclidean unit of area on a graph can be determined by squaring the subdivision of one of its sides. Thus, if one of the sides is subdivided into two parts, the unit of area is subdivided into quarters. The subdivision of a side by "1/2" produces four subdivisional units for the whole unit of area.

Thus a "1/2" linear subdivision produces a "1/4" area subdivision $\left((1/2)^2\right)$.

Quantum area cannot be sectored off or "subdivided" without the addition of new Euclidean lines to form the subdivisions. Such subdivisions are not possible since they would require new Euclidean lines which could never be mathematically constructed.

The quantum squared is mathematically differentiated by the negation of subdivision for Euclidean area. It is only a mathematical value. The differentiated quantum squared is not an actual unit of area, but a mathematical abstraction with strong mathematical application. The differentiated linear quantum value equals the square root of differentiated quantum squared.

The following formula is true both by the Pythagorean Theorem and the negation of subdivision for area in the above illustration:

$$Q^2 - \left(\frac{1}{2}Q\right)^2 = \left(1 - \frac{1}{2^2}\right)Q^2 = d(Q^2_{n=2})$$

True by Pythagorean Theorem. Also the negation of subdivision for Q^2 at "n=2".

The differentiated linear quantum is the square root of the negation of subdivision for area. It is therefore labeled "$\sqrt{d(Q^2_{n=2})}$." This quantum value is the minimum quantum distance provided by the x axis. There are no other quantums with a shorter distance of separation between the point of origin and the x axis. Conversely, the square of "$\sqrt{d(Q^2_{n=2})}$" is equal to the negation of the area subdivision unit at "n=2". The quantums differentiated by points along the x axis are the square root of negations of subdivisions for area:

$$\text{negation of an area subdivisional unit} = d(Q^2_n) = \left(1^2 - \left(\frac{1}{n}\right)^2\right)Q^2 = \left(1 - \frac{1}{n^2}\right)Q^2$$

Why negation of subdivision for the quantum squared rather than for the linear quantum (1-1/ n)? The quantum squared supplies a "substructure" of quantum differentiation to the Euclidean x axis which linear differentiation would not do. Ultimately, negation of subdivision for the quantum squared allows for a new quantum form of integration for the x axis, integration which a strictly Euclidean x axis disallows.

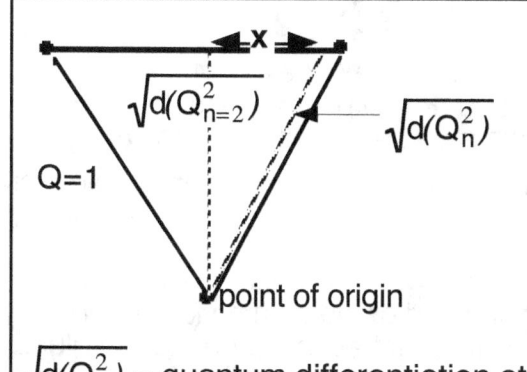

$$d(Q_n^2) = \left(1^2 - \frac{1}{n^2}\right)Q^2$$

$$d(Q_n^2) = x^2 + d(Q_{n=2}^2) \quad ; \quad d(Q_{n=2}^2) = \left(1^2 - \frac{1}{2^2}\right)Q^2$$

$$x^2 = \left(1^2 - \frac{1}{n^2}\right)Q^2 - \left(1^2 - \frac{1}{2^2}\right)Q^2$$

$$x^2 = \left(\frac{1}{2^2} - \frac{1}{n^2}\right)Q^2 \quad ; \quad n \geq 2$$

$\sqrt{d(Q_n^2)}$ = quantum differentiation at "n"

Notice how the differentiation of the quantum squared supplies a quantum differentiation to the subdivisional unit of the x axis.

The quantum is differentiated (in quantum squared units) at "n" as $d(Q_n^2) = \left(1^2 - \frac{1}{n^2}\right)Q^2$.

The x axis, however, is also differentiated by the quantum. $\quad x^2 = \left(\left(\frac{1}{2}\right)^2 - \frac{1}{n^2}\right)Q^2 \quad ; n \geq 2$

$\left(\frac{1}{2}\right)^2$ is an area value, representing the area of the subdivision (area is subdivided into quarters). "X^2" is therefore equal to the negation of the subdivision of the area-unit " $\left(\frac{1}{2}\right)^2$ " for all values of "n" greater than "2." Quantum differentiation has been superimposed upon the x axis.

At "n=2," x is equal to "0." "X" can now be differentiated to "0" because of the superimposition of a quantum differentiation upon the x axis. As we have shown, this is not possible with a strict Euclidean x axis which is differentiated by subdivision alone.

The x axis differentiates to "0" at every value of "n."

The value of "x^2" changes in direct relationship to "n." As "n" increases so does the value of "x^2." The formula is the following:

$$\Delta x^2 = \left(\frac{1}{2}\right)^2 - \frac{1}{(n+1)^2} - \left(\left(\frac{1}{2}\right)^2 - \frac{1}{n^2}\right)$$

$$\Delta x^2 = \frac{1}{n^2} - \frac{1}{(n+1)^2}$$

"$1/n^2$ " is a unit of area (albeit a subdivisional unit of area like "$\left(\frac{1}{2}\right)^2$"). It is a unit of area which is being differentiated by the quantum as a negation of subdivision. The subdivisional unit being used to negate the unit of area can also equal "n":

$$\Delta x^2 = \frac{1}{n^2} - \frac{1}{n^2} = 0 = d(x^2)$$

All values of "n" along the quantum defined x axis can be differentiated to "0."

Negative-Reverse Integration of $d(Q_n^2)$

While all subdivisional units ($1/n^2$; $n \geq 2$) of the quantum-defined x axis can be differentiated to "0," the x axis cannot be differentiated to "0" for all points along the continuum. The axis can only be differentiated at "n" points .. Therefore the x axis cannot be conventionally integrated using quantum differentiation ($\Delta x^2 = d(x^2) = 0$ is not a value for all points along the axis) .

However, the quantum provides an alternative system of integration which operates by negation and which can use the negative differentiation. Instead of the summation of known values along the x axis, quantum integration operates by the subtraction of unknown values along the x axis. This system will be identified as "negative-reverse integration."

Negative-reverse integration provides a solution to a general problem for Euclidean based calculus. That problem is identified by the way that a unitized x axis must measure distance. Any Euclidean length can be measured by a whole number of units of measure along the axis plus a partial of one unit of measure.

Only the partial need be integrated since it is only the partial which is "ambiguous." The whole number of units contained in any distance measure are precise. Any measure of distance, therefore is "n" units plus a partial unit.

Since Newton's day it has been known that the partial of a whole cannot be directly integrated but must be estimated by the derivative series postulated by Newton's protégé, Colin Maclaurin. While the Maclaurin derivative series is a mathematical complexity which can be ignored for the moment, it can be easily demonstrated why it is necessary.

A partial of a whole can be defined by " $f(x) = 1/x$" where x is any number greater than or equal to "1." While the value of x can be restricted to "≥1" the x axis itself cannot be. The axis must originate at the "0" point. The derivative and integral for the function are the following:

$$f(x) = y = \frac{1}{x}$$

$$D(y) = -\frac{1}{x^2} \; ; \; d(y) = -\frac{1}{x^2}\, d(x)$$

$$\frac{1}{x} = -\int_0^x \frac{1}{x^2}\, d(x)$$

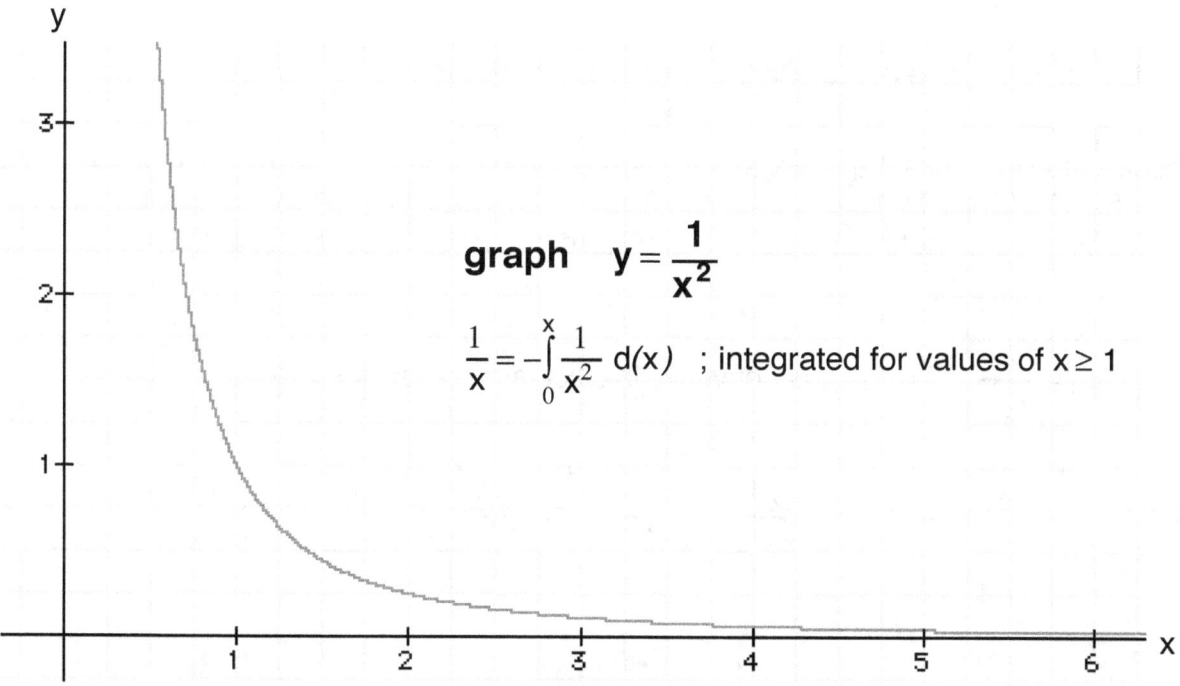

graph $y = \dfrac{1}{x^2}$

$$\frac{1}{x} = -\int_{0}^{x} \frac{1}{x^2}\, d(x) \quad ; \text{ integrated for values of } x \geq 1$$

The derivative variable "x^2" is in the denominator. To integrate the derivative from "0" to "x" requires division by "0" which is impossible. This is the reason that the integral of a partial cannot be calculated and must be estimated by the Maclaurin series. I will demonstrate that the quantum changes this significantly and that the partial can, in fact, be integrated by quantum methods.

Maclaurin has shown that the derivative series for a partial alternates between the partial and its negation[229] .

The primary example for our purposes is the fact that the second derivative of a partial equals the first derivative of the negation of the square of the partial:

$$D(\frac{1}{x}) = -\frac{1}{x^2} \;\; ; \;\; D^2(\frac{1}{x}) = \frac{2}{x^3}$$

$$D(1 - \frac{1}{x^2}) = \frac{2}{x^3}$$

$$D^2(\frac{1}{x}) = D(1 - \frac{1}{x^2}) = \frac{2}{x^3}$$

It can be seen that the negation of the square of the partial is similar to the negation of the quantum squared.

negation of the square of the partial $= 1 - \dfrac{1}{x^2}$

negation of the quantum squared($d(Q_n^2)$) $= 1 - \dfrac{1}{n^2}$

---

[229] See my paper *"An Exact Quantum Formula for the Periphery of an Ellipse and the Detection of Systemic Error in the Maclaurin Derivative Series"* Master's Thesis, The Virtual University.

They in fact can be made equal with the following modification: $1 - \dfrac{1}{x_n^2}$

This modification indicates that the x axis is intersected by a quantum axis and that the quantum is supplying the unit of measure to the x axis (Q=1 unit). The partial, then, can be defined by— but is not restricted to — $d(Q_n^2)$ or the negation of the quantum squared.

$$D\left(1 - \frac{1}{x_n^2}\right) = \frac{2}{x_n^3} \; ; \; Q = 1; \; \text{This derivative is calculated below for x alone.}$$

$$\frac{-\Delta\left(\frac{1}{x^2}\right)}{\Delta x} = \frac{\frac{1}{(x - \Delta x)^2} - \frac{1}{x^2}}{\Delta x} = \frac{x^2 - x^2 + 2x\Delta x - \Delta x^2}{(x - \Delta x)^2 \left(x^2\right)\Delta x} = \frac{\Delta x(2x - \Delta x)}{(x - \Delta x)^2 \left(x^2\right)\Delta x} = \frac{(2x - \Delta x)}{(x - \Delta x)^2 x^2} = \frac{2}{x^3}$$

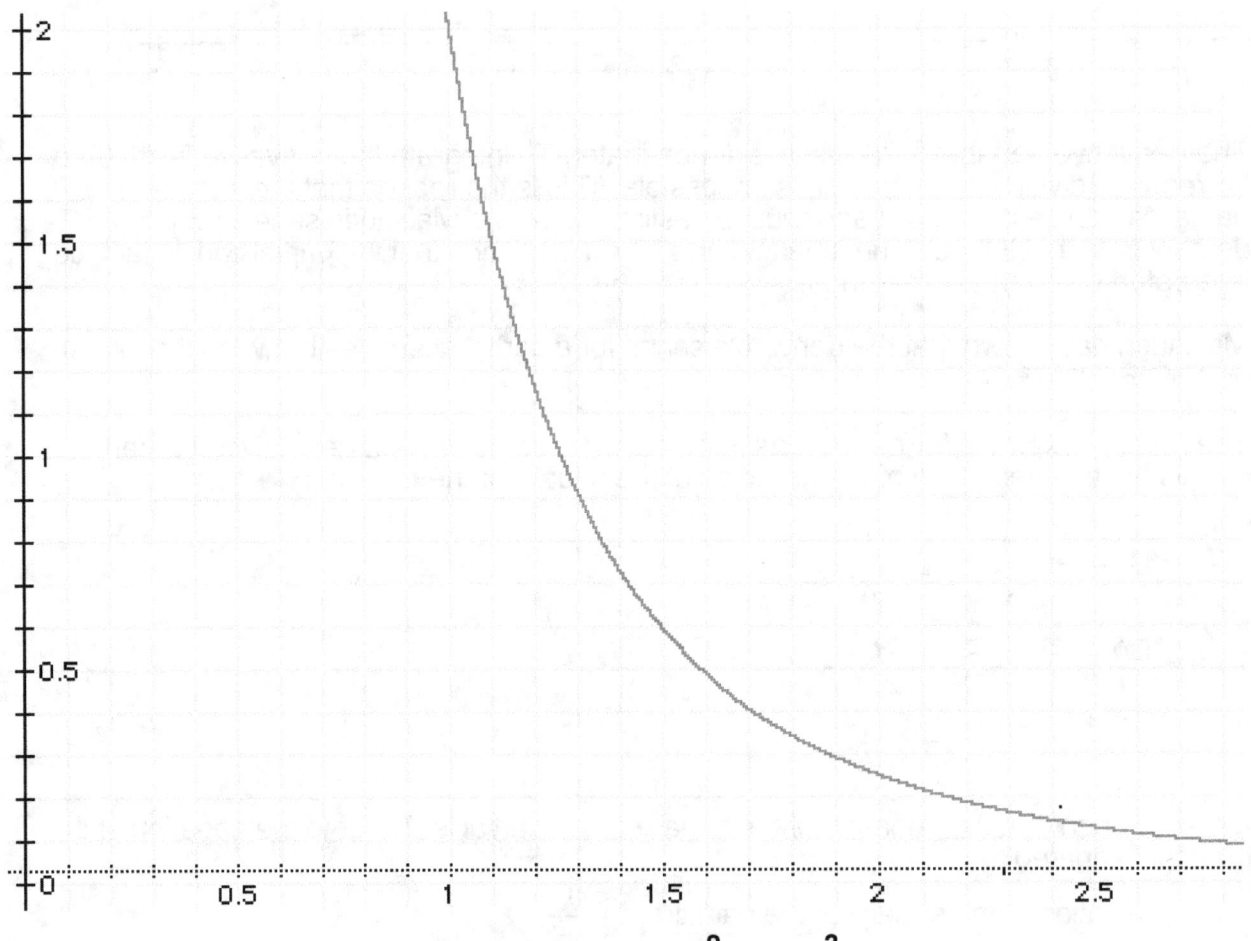

Graph of $D(1 - 1/x^2) = 2/x^3$
The Negative Integral

Let us first consider the integral for the x axis alone (without an intersecting quantum). It will demonstrate both the principle of negation and why a strict Euclidean integral cannot be resolved.

$$1 - \frac{1}{x^2} = \int\limits_0^x \frac{2}{x^3}\, d(x)$$

$$\frac{1}{x^2} = 1 - \int\limits_0^x \frac{2}{x^3}\, d(x)$$

"1" in the above equation signifies the whole of the area under the graph of the derivative. Therefore, "$1/x^2$" is equal to the whole of the area under the graph *minus* the integral of the derivative from "0" to "x."

It is crucial to understand the distinction between the "1" in the above formula and "x=1." The "1" in the formula represents the whole of the area under the graph. "X=1" is the point for which the area representing the whole unit being differentiated begins. For all partials, "x" must be restricted to "x≥1."

This integral, however, is inoperative. It requires division by "0" and therefore cannot be calculated. The non-quantum integral cannot be resolved and can provide no solution. The partial can only be estimated by a version of the Maclaurin derivative series. (The actual formula based upon the Maclaurin derivative series is unknown).

Quantum Negative-Inverse Integration

Consider what occurs when a quantum axis intersects the x axis and the unit of measure becomes the differentiated quantum squared ($d(Q_n^2)$). The intersection of the x axis by the quantum axis is designated by the symbol "x_n." This symbol is used to denote that the intersecting quantum can only assign value to the x axis at whole number intervals.

$$1 - \frac{1}{x_n^2} = \int\limits_{x_n}^0 \frac{2}{x_n^3}\, d(x_n)$$

$$\frac{1}{x_n^2} = 1 - \int\limits_{x_n}^0 \frac{2}{x_n^3}\, d(x_n)$$

The whole numbers along the x axis now become subdivisional values of the quantum squared. The only area under the graph of the derivative in which we are interested is that which is to the right of "$x_{n=1}$." This area represents the whole of the quantum unit which we are differentiating.

Notice that the definite integral has been reversed. Instead of integrating from "0" to "x," we are now subtracting the integral from "x" to "0." Inverse integration is necessary to accommodate the quantum function which is imposed upon the x axis and which is operating upon a quantum unit of measure.

Since there are no points between "n" and "n+1" for the quantum, the area under the graph of the derivative between those values is indeterminate (with respect to the quantum "n" but not with respect to x_n).

Negative-inverse integration operates by the subtraction of indeterminate area from the whole by means of inverse integration. Inverse integration reverses the direction of the integral. Integration proceeds from the x_n value towards the "0" point.

Thus the value of "$1/x_n^2$" is the whole of the area under the graph *minus* the subtraction of the indeterminate area as integrated from "x_n" towards "0."

The removal of indeterminate area *must* proceed in a reverse direction. Only knowable and determinable area can be subtracted from the whole by non reverse integration.

Since we are subtracting indeterminate area, division by "0" is no longer a problem. The "0" point is also "indeterminate" and simply a part of the area being subtracted.

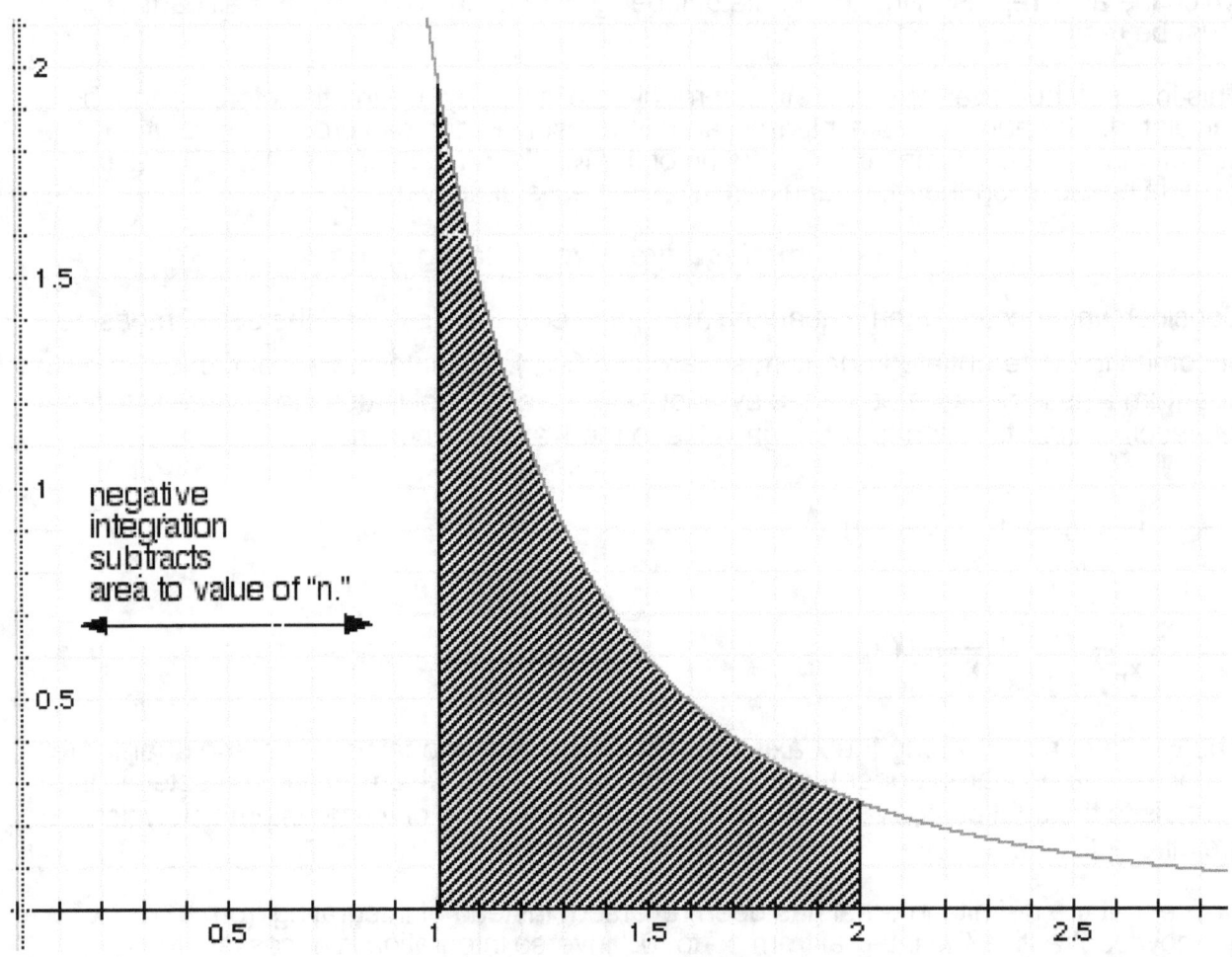

Negative-inverse integration is a requirement of the quantum.

The differentiated quantum can only take values equal to "$d(Q_n^2)=1-1/n^2$." The space between "n=1 and n≥1" is geometrically undefined relative to the quantum and area cannot

212

be determined from within that space. The "x" values within that space is integrated negatively – through subtraction of indeterminate area inversely integrated across the continuum of points constituting the x axis.

We can now determine the way negative-inverse integration operates upon the differentiated quantum ($d(Q_n^2)$). The following are the negative-inverse integrals for "x=n=1" and "x=n=2."

$$\frac{1}{x_{n=1}^2} = 1 - \int_1^0 \frac{2}{x_n^3} d(x_n) \qquad ; \qquad \frac{1}{x_{n=2}^2} = 1 - \int_2^0 \frac{2}{x_n^3} d(x_n)$$

T
he area to the right of "$x_{n=1}$" is the value of $1/1^2$. The area to the right of "$x_{n=2}$" is the value of $1/2^2$. If we subtract the area to the right of "$x_{n=2}$" from the area to the right of "$x_{n=1}$," the shaded area in the illustration above is remaindered.

This area is the partial of the whole quantum equal to the differentiated quantum squared for "n=2":

$$d(Q_{n=2}^2) = \frac{1}{x_{n=1}^2} - \frac{1}{x_{n=2}^2} = 1 - \frac{1}{2^2} = \left(1 - \int_1^0 \frac{2}{x_n^3} d(x_n)\right) - \left(1 - \int_2^0 \frac{2}{x_n^3} d(x_n)\right)$$

$$d(Q_{n=2}^2) = \int_2^0 \frac{2}{x_n^3} d(x_n) - \int_1^0 \frac{2}{x_n^3} d(x_n) = \int_2^1 \frac{2}{x_n^3} d(x_n)$$

The value of $d(Q_{n=2}^2)$ is the remaindered area when one subtracts "$1/2^2$" from "$1/1^2$". It is produced by an inverse *positive* integration. The negation of a negation has remaindered a positive integral under a graph which cannot be integrated positively by conventional methods.

Negative-inverse integration is the inverse of conventional integration in a second sense. It is not the summation across a continuum of "0" differential points. Rather, negative-inverse integration is subtraction across undefined quantum distance to reach the quantum squared "0" differential point ($d(x_n^2) = 0$). All "n" points may be differentiated to "0" as "n2" which must be reached by negative-inverse integration from "(n+1)2".

Comparisons between the strict Euclidean graph and the quantum defined graph are difficult at best. The strict Euclidean graph is a two dimensional plane, while the quantum defined graph is volume created by the intersection of the Euclidean plane by the quantum axis. The x axis is a component of quantum area and linear x values along the axis are quantum squared values for quantum area. For this reason, the quantum "$d(x_n^2)$" must be considered the equivalent of the strict Euclidean "$d(x)$."

With a quantum intersected x axis, the area under the graph of the derivative is negatively-inverse integrated to reach "$d(x_n^2) = 0$" as established by the quantum. In contrast, a strict Euclidean x axis integrates as the summation of a continuum of "$d(x) = 0$" to establish area under the graph of the derivative.

The second derivative from the Maclaurin series is the derivative of the quantum squared. The claim that this derivative can be integrated by quantum negation and inverse principles must be proven.

The area value of "1-1/2^2" equals "3/ 4." That is, the shaded area in the above graph is 3/4 of the area for the whole quantum. The area of the whole quantum is the area under the graph to the right of "x=1." This can be shown to be true if the area to the right of "x=2" can be shown to approach "1/ 2^2 = .25 (1/4)" as a limit.

The value of "1/ $(x = n)^2$" is the area to the right of "x=n" under the graph of the derivative. Since "x=n" is a nonrestrictive value (i.e. can increase indefinitely) the area to the right of any "n" value increases perpetually. That is the value of "1/ n^2 " is always inexact and irrational as predicted by the Law of Inexact Differentiation of Measure. The integral can be checked by summation with the following formula:

$$\sum_{n=n'}^{n} \frac{2n+1}{n^2(n+1)^2} \xrightarrow{lim} \frac{1}{(n')^2}$$

"$\frac{1}{(n')^2}$" is the limit of the summation of the units of area under the graph, unit being the area between "n" and "n+1."

$$\text{"area} = \frac{1}{n^2} - \frac{1}{(n+1)^2} = \frac{2n+1}{n^2(n+1)^2}\text{"}$$

sum check from "n=2" to "n=20" $\sum_{n=2}^{20} \frac{2n+1}{n^2(n+1)^2} = 0.2477324263 \xrightarrow{lim} \frac{1}{2^2} = .25$

As "n'" increases towards infinity, the sum will approach the limit ".25." The area under the graph to the right of "n=2" approaches the limit of .25 of the whole.

The reverse integral , "$\int_{2}^{1} \frac{2}{x^3} d(x) = 3/4$" correctly predicts the area it integrates.

By quantum negative integration, a set of positive integrals are established for a graph which cannot be integrated at all by conventional methods.

Discussion

The quantum axis is successfully reverse integrated between "n=2" and "n=1." The shortest differentiated quantum is the "Q axis" below. It is defined by "1-1/ 2^2." The differentiated quantum formula, 1-1/x_n^2 for 1 ≥ x_n ≤ 2 produces something less than the 3/4 which "1-1/ 2^2" produces. That is, the differentials between "1" and "2" reduce the "Q axis" to a limit of "0." The "Q axis" is successfully reverse integrated to the "0" differential at the point of origin between n=2 and n=1.

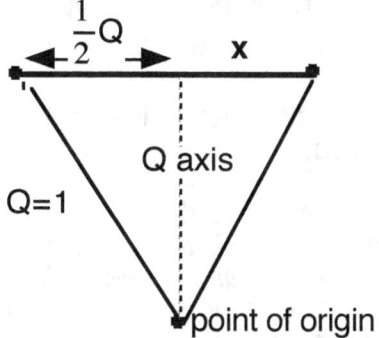

It is, however, the x axis, not the quantum axis which is significant. The quantum squared supplies the x axis with the following values:

$$x^2 = \frac{1}{2^2} - \frac{1}{n^2} \quad ; \quad n \geq 2$$

$$\Delta x^2 = \frac{1}{n^2} - \frac{1}{(n+1)^2} = \int_{n+1}^{n} \frac{2}{x_n^3} d(x_n) = 0$$

$$d(x^2) = \frac{1}{n^2} - \frac{1}{n^2} = -\int_{n+1}^{n} \frac{2}{x_n^3} d(x_n) = 0$$

Negative inverse integration from "n+1" to "n" produces " $1/n^2 - 1/n^2$" which is the same as the differential " $d(x^2)$ " for x at the value of "n." The difference between the change in x^2 and the differential of "x^2" is the difference between the negative and the positive integration from "x_n" to "x_{n+1}." Negative inverse integration of "x_n" (not the x axis itself, but the x_n axis) subtracts the area from "n+1" to "n", differentiating "n" to "0."

This is the most important principle in quantum mathematics. The negative quantum integral of "x_n" produces the positive Euclidean derivative of "x." Integration is known as "anti-differentiation." The negation of "anti-differentiation" is the positive differential. Negative quantum integration equals Euclidean differentiation.

On the other hand, positive inverse integration from "n+1" to "n" integrates the x axis between the points:

$$\frac{1}{n^2} - \frac{1}{(n+1)^2} = \left(1 - \int_{n}^{0} \frac{2}{x_n^3} d(x_n)\right) - \left(1 - \int_{n+1}^{0} \frac{2}{x_n^3} d(x_n)\right) = \int_{n+1}^{0} \frac{2}{x_n^3} d(x_n) - \int_{n}^{0} \frac{2}{x_n^3} d(x_n)$$

$$\int_{n+1}^{0} \frac{2}{x_n^3} d(x_n) - \int_{n}^{0} \frac{2}{x_n^3} d(x_n) = \int_{n+1}^{n} \frac{2}{x_n^3} d(x_n)$$

$$\Delta x^2 = \int_{n+1}^{n} \frac{2}{x_n^3} d(x_n)$$

Δx^2 is the inverse positive integrated distance between "x_n" and "x_{n+1}" for the x axis as a component of the quantum squared. After negative integration of the two components, the remaindered positive area under the graph is equal to the change in x^2 along the x axis.

The sum of the change in x^2 for all values of n>2 completely integrates the unit of measure for the x axis as a component of the quantum squared.

Law of Quantum Calculus

When the x axis is intersected by a quantum axis at 90° to the plane of the graph and the x axis becomes a component of the quantum squared, any partial of the unit of measure which is imposed upon the x axis is fully integrable by the second derivative of the partial. The second derivative is also the derivative of the quantum squared. Quantum integration requires the use of the method of negative-inverse integration which constructs the x axis as composed of a continuum of points.

The Law of Exact and Inexact Subdivision.
The impossibility of strict Euclidean subdivision

THE NUMERIC PROOF:
Any Euclidean Unit of Measure Must Be Exactly Subdivided

The Exact Subdivision: $1/2^n$

Can geometric construction identify points contained within an Euclidean unit of measure? Is differentiation of the unit by division possible? Yes it is.

One point contained within any Euclidean length can be identified geometrically. If one bends the line made by any two points such that the two original points coincide with one another, one will have identified a third point at exactly 1/2 the unit of measure. The bend identifies a point along the axis separated from the original points by distance exactly 1/2 the unit of measure.

By this geometric construction, one will have identified the midpoint or mean value of the unit of measure. If the process is repeated on a half unit, one will have identified a sub-unit equal to 1/4 the unit of measure. If repeated again, 1/8 the unit of measure will be identified and so forth.

Euclidean lengths can be exactly subdivided by the formula $1/2^n$ and only by this formula. Any real and rational number divided by "2^n" gives a rational and exact product of division.

Using our ten digit system, however, measured lengths can also be subdivided by "10^n and 5^n" to give rational (exact) products of division.

This, however, can be shown to be an artifice of the decimal number system itself. Division by "10^n" is irrational in the binary system.

10^{-n} (decimal) an Infinite Digression in the Binary Number System
10 decimal units = 1010 binary units

$$\frac{1}{1010 \text{(binary)}} = .00011001100110011+ \text{ (infinite digression)}$$

reiterated 25 place to right of "binary division point"

$$= 0.099999994 \text{(decimal)} \xrightarrow{\text{limit}} .1\text{(decimal)}$$

When the binary number system is employed, the only whole-number denominator which will provide an exact and rational product of division for any rational numerator is "2^n (decimal notation)"

"5^n" is explained below:

$$5^n = \frac{10^n}{2^n}$$

$$10^n = \left(5^n\right)\left(2^n\right) \qquad \text{Division by "} 5^n \text{" is division by "} 10^n \text{" } \textit{times} \text{ "} 2^n \text{."}$$

$$\frac{x}{5^n} = \frac{x}{10^n/2^n} = \frac{\left(2^n\right)x}{10^n}$$

When the natural, two digit binary system is used, all such products of divisions become irrational.

The binary system is composed of two digits which is the minimum needed for counting. Those digits (0,1) represent "existence" and "nonexistence," thus providing a physical foundation for the numbering system. It should be used in preference to the artificial decimal system for numbers theory.

The Inexact Subdivision: 1/ n

In the binary system, division of lengths by all values of "n" —which are not equal to 2^n — will give inexact and irrational products of division. An irrational subdivision is generally produced by the formula 1/ n.

(The only exception is: $n = 2^n$)

Consider the first number which is a non-exponential value of "2." This is the number "3." Subdivision by thirds is numerically inexact and irrational. Numerically, it approaches the fraction "1/3" as a limit. One cannot find a subdivisional point within the length which, if multiplied by three, produces the whole of the length.

Numeric division by three produces an irrational number (one which digress infinitely to the right of the decimal point).

The irrational product of division reflects our inability to identify a point along the axis which is exactly "1/ 3" of the unit of measure. The "1/ 3" point will reside somewhere in the space between $3\left(10^{-n}\right)$ and $4\left(10^{-n}\right)$. The "1/ 3" proportion of any Euclidean length has no exact value and we cannot intersect or identify an exact pointal position. The numeric "1/ 3" value for any unit of measure is irrational and only approaches the fraction as a limit which it never reaches.

This fact is proven by the derivative of the negation of subdivision for the quantum squared. All "n" values under the graph of the derivative produce inexact values:

$$1 - \frac{1}{(x = n)^2} = \int_x^0 \frac{2}{(x = n)^3} \, d(x)$$

$$\frac{1}{(x = n)^2} = 1 - \int_x^0 \frac{2}{(x = n)^3} \, d(x)$$

$$\sum_{n=n'}^{n} \frac{2n+1}{n^2(n+1)^2} \xrightarrow{lim} \frac{1}{(n')^2}$$

218

Planck's Constant: an Electron-Nuclear Energy Exchange
and the Einstein-Bohr Misapplication of Planck

Twentieth century science ignored the implication of the Rydberg formula — that all hydrogen radiation emissions are harmonic differentials of a root frequency. The second researcher to encounter a quantum phenomenon fared no better than Rydberg. He was the German physicist Max Planck[230] .

In 1900, Planck discovered that black bodies absorb light radiation in units of energy he called "quanta." A "black body" is a mass which absorbs all light frequencies and converts them to heat. With black bodies, none of the radiant energy is released back as light, black being the absence of all chromatic light frequencies.

From this discovery, Planck developed his famous constant which may be one of the most important single contributions in the history of physics. Planck's Constant (designated by the symbol "h ") provides an energy formula measured in joules. It is a monochromatic formula in that a single frequency *times* Planck's Constant *times* a whole-number of oscillators at the frequency equals energy:

$E = n(h)f;$ f = frequency ; n = whole number of oscillators at the frequency

h = Planck's Constant = $6.6260755(10^{-34})$ joules

The energy absorbed by black bodies from radiation of a certain frequency was a whole-number function of frequency *times* a constant.

Early twentieth century physics took Planck's discovery and built an irrational quantum science upon it. They did so because they failed to recognize what Planck's Constant actually was; a measurement of a the energy exchanged between nucleus and electron along a "string bond" between them.

In 1905, Albert Einstein used Planck to postulate a particle theory of light which he used to explain the "photoelectric effect."[231] To Einstein, Planck's "quanta" implied a mass-like light particle with frequency the equivalent of the velocity of the particle. Electrons were supposedly released by a momentum exchange with these light particles he called "photons."

However, I will show that Einstein's 1905 equation for the photoelectric effect did not, in fact, prove his particle theory of light. Einstein's photoelectric equations are the following:

h = Planck's Constant ; f = frequency of light ; f_0 = threshold frequency

$$hf = \phi + E \quad ; \quad \phi = hf_0 \quad ; \quad E = \frac{mv^2}{2} = hf - hf_0 = h(f - f_0)$$

$\phi = hf_0$ is the work function, min. energy needed to remove electron from surface

f_0 = threshold frequency for the photoelectric effect to occur ; m = mass of electron

The energy in the electron released by light is equal to the Planck energy value of the light *for one second* minus the minimum Planck energy value for release of the electron from

---

[230] Planck, Max Karl Ernest Ludwig; 1858-1947; Universities of Kiel, Munich and Berlin.
[231] Einstein did so in the German physics journal *"Annalen der Physik"* in 1905.

orbit *for one second.* It is Planck energy devoid of time as a continuous factor.

In 1916, Robert Millikan experimentally proved Einstein's mathematics to be correct. Despite this fact he made the following observation: "Einstein's photoelectric equation...cannot in my judgment be looked upon at present as resting upon any sort of a satisfactory theoretical foundation.[232] " He was referring to Einstein's "photon" explanation.

Robert Millikan's scientific reputation suffered because of this opposition to the Einsteinian "photon." Millikan was a "mere experimentalist" whose prestige could never equal the "theoretician" Einstein who had become a cultural icon after his general relativity field equations had been confirmed by astronomical observations during two solar eclipses.

Nonetheless, quantum electron string harmonics and the data generated by negative radiation studies is proving Millikan to be right and Einstein wrong.

Quantum String Harmonics Duplicate the Einstein Photoelectric Formula

The potential energy state for any quantum string length has been identified by N-radiation experimentation as Planck's Constant *times* the string frequency[233] . This is the potential energy which can be exchanged by the electron and nucleus. Frequency is determined by the string lengths which are produced by the negation of subdivision of the root frequency:

$h =$ Planck's Constant ; $n =$ series or "band" number ; $c =$ speed of light

$n' =$ series' element number; $n' > n$; $\lambda_r =$ root wavelength (91.14 nm)

$$\text{String Potential Energy} = PE = h\left(\frac{1}{n^2} - \frac{1}{n'^2}\right)\frac{c}{\lambda_r} \; ; \; by\ N\text{-}radiation\ data\ ^{24}$$

Any derived string length is harmonically related to root string by the following formula:

$$PE_{r/d} = PE_{deriv.} - PE_{root} = h\left(\frac{1}{n^2} - \frac{1}{n'^2}\right)\frac{c}{\lambda_r} - h\frac{c}{\lambda_r} = \left(\frac{1}{n^2} - \frac{1}{n'^2} - 1\right)h\frac{c}{\lambda_r}$$

always a negative number

Any harmonically related string length has *negative potential energy* definition on any higher harmonic frequency. Lower frequencies strings impede higher frequencies as follows:

(harmonics for root frequency demonstrates general principle)

$$\text{Total energy} = E_t = (\text{Mechanical Impedance Energy} = E_i) + PE$$

$$E_i = h\frac{c}{\lambda_r} \quad ; \quad PE_{r/d} = \left(\frac{1}{n^2} - \frac{1}{n'^2}\right)h\frac{c}{\lambda_r} - h\frac{c}{\lambda_r}$$

$$E_t = h\frac{c}{\lambda_r} + \left[\left(\frac{1}{n^2} - \frac{1}{n'^2}\right)h\frac{c}{\lambda_r} - h\frac{c}{\lambda_r}\right]$$

$$E_t = \left(\frac{1}{n^2} - \frac{1}{n'^2}\right)h\frac{c}{\lambda_r}$$

$$\left(\frac{1}{n^2} - \frac{1}{n'^2}\right)h\frac{c}{\lambda_r} = PE\ (\text{of derivative string})$$

Lower Frequencies Must Harmonically Impede at Own String Potential Energy

[232] Physical Review, 22 April, 1999.
[233] See *"The N-Irradiation of Cotton Derives Planck's Constant"* in the Appendix.

The above equations demonstrate that a lower frequency, harmonically related string, can impede or absorb higher frequency light but only at the string's own potential energy level. There is a variance between the Planck energy in the light and the Planck energy absorbed by the harmonically related string. The available energy is greater than can be absorbed by the string:

let $h(f_1) =$ energy of light ; let $h(f_h) =$ harmonically related string potential energy.

let E = available energy

$$E = h(f_1) - h(f_h)$$

But this harmonics formula is a duplication of the Einstein formula if we allow "$h(f_h) = h(f_0)$"

$$hf_1 = \phi + E \quad ; \phi = hf_0 \quad ; E = \frac{mv^2}{2} = hf_1 - hf_0 = h(f_1 - f_0)$$

$$h(f_h) = h(f_0)$$

$$E = \frac{mv^2}{2} = hf_1 - hf_h = h(f_1 - f_h)$$

The Rydberg frequencies are a quantum harmonic distribution of string lengths. They are similar to the Euclidean harmonic distribution of string lengths except that Euclidean harmonic lengths are determined by subdivision and quantum harmonic lengths by the negation of subdivision for the quantum squared. Both can vibrate sympathetically or in resonance with frequencies higher in the harmonic series.

The above equations demonstrate that the amount of energy any electron string can absorb in resonant vibration with higher frequencies is the string's own potential energy. String potential energy has been experimentally shown to be Planck's Constant times the string's frequency of vibration.[234] There is a variance, however, in the energy in the higher frequency light (Planck's Constant times light frequency) and the lower frequency string potential energy (Planck's Constant times string frequency). Einstein's photoelectric equation shows that this variance can be applied to aiding an electron's escape from the atom and into a current.

In summary, the photoelectric effect is shown to be a phenomenon of the fixed orbits of light-absorbing conductors, fixed orbits which do not allow the electron to acquire the orbital level of higher frequency harmonically related light. it does not require a light particle or photon to explain — as Millikan correctly anticipated.

The photoelectric effect is due solely to the variance between the energy in light and the ability of a fixed orbit — but harmonically related — electron string to absorb the light energy. The excess energy is invested in emitting the electron from the atom, as Einstein's formula demonstrates. The experimental scientist Millikan, was right. The cultural icon, Einstein, was wrong. The "photon" is not needed to explain the photoelectric effect.

Particle momentum-exchanges (Einstein's photon) and wave impedance (electron string behavior) have completely different energy signatures. This is due to differing applications of the time variable.

In a momentum exchange, time determines the rate of acceleration/deceleration and therefore the force component of the energy equation. Time is the period covering the acceleration/deceleration.

---

[234] *"The N-Irradiation of Cotton Derives Planck's Constant"* The Snake River N-Radiation Lab; in Appendix

For a waveform, however, energy is the number of energy "peaks" which pass a point in space during a period of one second. Energy is a function of frequency and the frequency of the wave is measured *per second.* Planck energy is the energy *per second.* The Constant *times* frequency produces the number of energy "hits" per second.

For an orbiting electron, "$h(f)$" must be multiplied by the number of seconds the electron string is vibrating to equate to energy. Time is a continuous factor, not a specific period determining a single energy event.

In the case of a photoelectric event, the period covering the "event" (the ejection of the electron from the atom) is also determined by frequency. The period of the "event" is framed by the time factor for the frequency. The relationship between time and frequency is the following: $t = 1/f$. The time which one vibration takes (in seconds) is equal to one divided by frequency.

This translates into the following energy equation: $E = h/t$. Planck's " joules *times* frequency" are now expressed in a small fraction of a second equal to "$1/f$ " — or the time required by one vibration. Planck's Constant can be stated in smaller amounts of time which multiplies total energy. The time period determines the expression of energy. The expression of the energy ejecting the electron from the atom requires the following amount of time:

f_l = frequency light ; f_h = frequency harmonic string

$$\text{photoelectric energy} = E_p = \frac{h}{t_p} = h(f_l - f_h) \quad ; \text{ photo electric time period} = t_p$$

$$t_p = \frac{1}{f_l - f_h}$$

The time taken to eject the electron from the atom; the "period of the event." will be:

$$t_p = \frac{1}{f_l - f_h}$$

The time will be one *divided by* the frequency of light *minus* frequency of impeding string.

Time covering the ejection of the electron is determined by the frequencies of the carrier light wave and the frequency of the impeding electron string, not by the momentum exchanged on impact with an alleged :photon.

Acceleration/deceleration of impacting particles is a one-time energy event and time is a set period determined by relative momentums. Frequency is a multi-energy event over time. To confuse the two is, to say the least, irrational thought.

In 1913, Niels Bohr tried to explain Rydberg's and Planck's quantum discoveries with Einsteinian photons. To do so, Bohr had to treat Planck's "packets of energy" as stripped of their time component.

Specifically, Bohr explained Rydberg frequencies as photons output by electrons falling between quantum orbits. These "falls" were one-time events and supposedly resulted in one-time releases of a photon.

Bohr proposed that electron orbits were restricted to quantum distances. Each electron quantum orbit possessed a potential energy value which he identified as "electron voltage." He calculated electron voltage from the Planck energy value for the Rydberg root frequency as divided by the elementary charge.

Bohr's formula was the following:

Rydberg root frequency $= \dfrac{c}{\lambda_r}$; Planck's Constant (*in joules*) $= h$

elementary charge of electron (*in coulombs*) $= \varepsilon$; electron volts $= eV$

Energy $= (\text{charge})(\text{volts})$; *from the equation for an electrical capacitor field.*

$$h\dfrac{c}{\lambda_r} = (\varepsilon)(eV)$$

$$(\text{electron volts}) = h\dfrac{c}{\lambda_r} \Big/ \varepsilon = h\dfrac{c}{\lambda_r \varepsilon} = 13.6 \;\; joules\; per\; coulomb$$

(Planck energy)/(elementary charge)= Electron Volts

This calculated to 13.6 electron volts which Bohr defined as the base or highest energy state of the electron at quantum orbit number "1."

The orbit's potential energy was defined as "$-13.6/\,1^2$ eV."

Each subsequent quantum orbit would have a potential energy defined as "$-13.6/\,n^2$ eV." Orbits could only acquire quantum values defined by the whole number "n."

When the electron fell from a *lower* potential energy quantum orbit (greater "n" value) to a *higher* potential energy orbit (lower "n" value), it gained energy. This acquired energy was emitted as a "photon."

$$\left(\dfrac{-13.6}{n^2}\right) - \left(\dfrac{-13.6}{1^2}\right) = \dfrac{13.6}{1^2} - \dfrac{13.6}{n^2} = \left(1 - \dfrac{1}{n^2}\right)13.6 \; eV = \text{energy emitted as "photon"}$$

The alleged photon had an energy value determined by the "static " value of Planck's Constant. It equaled Planck energy for one second of time only. Bohr had used the static Planck's Constant to calculate electron volts.

However, Einstein's "static" Planck energy allegedly possessed by "photons," were frequency values which kicked an electron out of its orbit. Static Planck values were never applied to orbiting electrons which were continuously emitting light.

I have shown above that Planck's Constant distributed into the time demarcated by frequency, establishes the period of electron ejection. Static Planck energy can be used to mathematically describe the photoelectric ejection of electrons.

It is true that static Planck energy might also describe change in orbit as Bohr is arguing. The period covering the change in state can be described by frequency demarcated time as with the Einstein equation. Any energy gain by this change in orbital state, however, could never be output as a "burst of light" (Einstein's photon). The energy thus gained would be needed to sustain the new orbit *over time.*

Any electron state continuously emitting light must be described by *non-static Planck energy.* An orbit itself cannot be restricted by time. The vibrational frequency of that orbit is restricted by time, but the string must be free to absorb and emit the light at frequency for an unresolved amount of time.

Any static energy gained by change in orbit must be applied to string potential energy. In fact, Bohr's electron "fall" to a higher potential energy electron state would be, mathematically, the reverse of the Einsteinian photoelectric equation as modified by Rydberg string harmonics:

$$eV = \frac{E}{\varepsilon} = \frac{hf_l}{\varepsilon} = \frac{h(f_l - f_h)}{\varepsilon} + \frac{hf_h}{\varepsilon}$$

This is based upon the reverse of the Einsteinian photoelectric equation and applies to electrons which are not restricted to fixed harmonic orbits (such as the hydrogen electron and possibly, the covalent hydrocarbon bond).

The electron can acquire a higher energy state when it impedes a higher frequency within the harmonic series. By the rules of harmonic impedance, it can only absorb energy values equal to the potential energy of its current orbital string length. The variance between string potential energy and light energy can be applied to "jumping" to the higher and more energetic orbit. Planck light energy equals variance plus string potential energy.

Change in orbit is described by variance energy which is a static Planck value. It must be added to current string potential energy to gain the string length equal to the Planck potential energy for the carrier light wave. The period of transition between orbital string lengths is given by the time value for light frequency *minus* original string frequency. It is only the transition or change in state which can be described by static Planck energy.

In summary, quantum electron string harmonics allows for a change in static Planck energy being applied to transitions between orbital energy states as Bohr proposes. However, this change in static Planck energy cannot be applied to the output of a fictional photon. The whole of the change in static Planck energy must be invested in the new energy state.

The Einsteinian-Bohr misunderstanding of the use of frequency-based time for transitional static energy had serious consequences. A virtual Gordian Knot of mysticism was wrapped about light radiation, leading ultimately to Heisenberg's uncertainty principle and its absurd proposition that time and space are no longer connected on the atomic level.[235]

It is pointless and futile to attempt to unravel these layers of time-irrational mysticism. It is better to sever the knot in one blow, with the time-correct interpretation of the Rydberg and Planck discoveries.

This correction begins with the recognition that time and space are not as we imagine them to be. Our geometry is four dimensional, not three and the extra dimension is the quantum dimension. Both the electron and vacuous space are defined by this unrecognized fourth dimension. They operate by principles we have failed to identify.

---

[235] This is the only conclusion one can draw from the "uncertainty principle" which asserts that, if the position of a particle (its spacial measurement) is known, the particle's momentum (its measurement over time) is uncertain. Space and time are disconnected.

Further, time itself is quantum. On the most minute level, time is composed of "jumps" of 3.34×10^{-24} seconds[236] — much too small for us to measure. We can measure time in "nanoseconds" (one billionth of a second), but there are 3,340 trillion quantum time units in a single nanosecond.

Time originates quantum space because time is composed of two separated values. One value is 3.34×10^{-15} nanoseconds behind the other and this time variance must be spatially separated into "time points" at a set distance apart (the α space). The time differential between the two divergent points is a force which compels quantum spacial separation. Time-force constructs four-dimensional matter and provides an unrecognized characteristic to vacuous space which allows it to conduct light radiation as waveform energy.

---

[236] Determined by the speed of light across the basic quantum (the α space) which must equal the radius of a proton and the shortest wavelength. Both measures are the same order of magnitude at 10^{-15} meters.

The Electron Field
and Nuclear Capacitance

Four-dimensional quantum geometry identifies a completely different form of potential energy than that proposed by Niels Bohr and Quantum Mechanics. The potential energy of the bond is the potential energy possessed by a tensioned string.

Quantum mathematics identifies a four dimensional atom with the bond between the electron and proton obeying four-dimensional principles. That bond is a "field and a string." It is not a gravity-like attraction between two masses.

When electron and proton are bound together to produce a hydrogen atom, that atom has a four dimensional structure. This four-dimensional structure requires an electron/proton bond of a type which is unrecognizable using three dimensional presumptions.

The bond produces a "quantum-squared string" and a quantum "expanded field" which surrounds the nucleus and which contracts back upon that nucleus .

The nucleus oscillates within the envelope of this field. At higher energy states, the oscillating nucleus expands the field — as well as its string — to a new quantum distance. The expansion stretches the field/string to new longer quantum radii and the "stretching" increases string tension.

The Electron Field

In an analysis of the determination of the elementary charge, I mathematically demonstrated that charge has a distance value.[237] It has a distance value because the charge is a geometric projection onto the fourth dimension relative to the particle. Energy is divided by the charge squared (e^2) to give capacitance ($1/C$).

The square of charge-as-distance is charge-as-area. The charge-as-area is the field value of the elementary charge. Capacitance is the energy expressed by the electron distributed into area. Let us review our modification of Bohr's mathematics to identify how electron field capacitance might operate:

$$(\text{voltage})(\text{charge}) = eV(e) = \text{Energy} \quad ; \quad f = \text{frequency} \quad c = \text{speed of light}$$

$$E = f(h) \quad ; \quad f = \left(\frac{1}{n^2} - \frac{1}{n'^2}\right)\frac{c}{\lambda_r} \quad ; \quad 1 \leq n \leq 7 \quad ; \quad 8 \geq n' > n$$

$$\text{capacitance} = \text{charge} / \text{voltage} = eC = e/eV$$

$$eC = \frac{e}{eV} = \frac{e^2}{E} = \frac{e^2}{f(h)}$$

$$f = \frac{e^2}{eC(h)}$$

It can be seen from the above that as the energy goes down, electron capacitance goes up. Energy goes down as frequency goes down. Frequency is a function of the negation of subdivision for the quantum squared. A simple hypothesis identifies what the root string length must be and this can be used to determine the relationship between string tension,

---

[237] See *A Review of Millikan's Determination of the Elementary Charge* in the Appendix.

electron voltage and electron field capacitance.

Hypothesis: *The root frequency is determined by a string vibration, the maximum velocity of which is restricted to the speed of light.*

Using the strict Euclidean or linear quantum plane as the plane of vibration, root quantum string length must be the following by the above hypothesis:

Maximum Deflection=Maximum Velocity=2d/t
(by law of string acceleration)

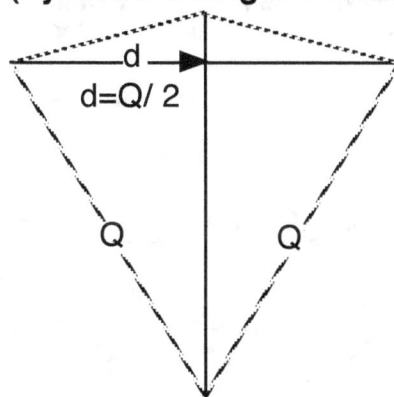

$$\frac{1}{t} = 2f \quad ; \quad f = \text{frequency} = \frac{c}{\lambda_r} \quad ; \quad c = \text{speed of light} \quad ; \quad 2d = Q$$

$$\frac{2d}{t} = c \quad ; \quad \textit{maximum velocity is speed of light}$$

$$Q\left(2\frac{c}{\lambda_r}\right) = c$$

$$Q = \frac{\lambda_r}{2}$$

Dawson's Tensor[238] can then be used to determine what the tension constant "k" must be.
For quantum squared:

$$f = \frac{1}{t} = \sqrt{2}\,k(x_t - x_0)^2 \quad ; \quad (x_t - 0)^2 = \left(\frac{1}{n^2} - \frac{1}{n'^2}\right)Q^2$$

$$\frac{e^2}{eC(h)} = \sqrt{2}\,k\left(\frac{1}{n^2} - \frac{1}{n'^2}\right)Q^2 \quad ; \quad Q^2 = \left(\frac{\lambda_r}{2}\right)^2$$

$\dfrac{e^2}{eC(h)} = \sqrt{2}\,k\left(\dfrac{1}{n^2} - \dfrac{1}{n'^2}\right)\left(\dfrac{\lambda_r}{2}\right)^2$ $\sqrt{2}\,k = \dfrac{4c}{\lambda_r^3} \quad ; \quad k = \dfrac{4\,c}{\sqrt{2}\lambda_r^3} = \dfrac{f_r}{\sqrt{2}\,SL_r^2}$	$eC = \dfrac{e^2}{\left(1/n^2 - 1/n'^2\right)c/\lambda_r\,(h)} = \dfrac{e^2\lambda_r}{\left(1/n^2 - 1/n'^2\right)c(h)}$

f_r = root frequency ; Root String Length = $SL_r = \lambda_r/2$ $k = 1.1200\ 10^{30}$ hz / m^2
The tension constant "k" is given in hertz (frequency measure) per quantum squared (in

---
op.cit..

meters). This indicates that the electron string is not linear, but a unit of the quantum squared. It can also be given in quantum electrodynamic units.

$$f_r = \frac{(\text{Planck Capacitance})}{eC_r} \quad ; \ \text{Planck Capacitance} = \frac{e^2}{h}$$

$$k = \frac{(\text{Planck Capacitance})}{\sqrt{2}\ eC_r\left(SL_r^2\right)}$$

The tension force for any string length is always frequency of the string divided by the square root of two. Using Planck's energy formula, and electrodynamic values, "k" becomes a function of :

$$F = \frac{f}{\sqrt{2}}$$

$$f_r = eV_r\frac{e}{h} \quad ; \ eV_r = 13.6 \text{ eV} \quad ; \ \frac{e}{h} = \frac{\text{Planck Capacitance}}{e} = \frac{C_P}{e} \ ;$$

$$k = \frac{f_r}{\sqrt{2}\ SL_r^2} = \frac{eV_r(e)}{\sqrt{2}\ SL_r^2(h)} = \frac{eV_r C_P}{\sqrt{2}\ SL_r^2\ e} = \frac{1}{\sqrt{2}\ SL_r^2}\left(\frac{\Delta e}{e}\right) = 5.103 \ 10^{22} \ (\text{units } e)/m^2$$

$$eV_r C_P = \Delta e$$

$$\Delta e + e = eV_r C_P + eV_\lambda\left(eC_\lambda\right)$$

k is a function of % increase in electron elementary charge

(from negative radiation) divided by root string length squared.

The string tension constant in quantum electrodynamic units is the number of elementary charges per meter squared. This constant number of elementary charges per meter squared is the tension factor which converts string length to frequency. The tension on the string produces frequency of vibration:

$$f = \text{frequency at string length} \quad ; \ SL = \text{string length}$$

$$f = \left(\frac{1}{n^2} - \frac{1}{n'^2}\right)\frac{c}{\lambda_r} \quad ; \quad SL = \left(\frac{1}{n^2} - \frac{1}{n'^2}\right)\frac{\lambda_r}{2}$$

$$f = \sqrt{2}\ k(SL) = \left(\frac{1}{n^2} - \frac{1}{n'^2}\right)\frac{(\sqrt{2}\ k)\lambda_r}{2} \qquad \textit{reader can confirm by calculation}$$

$$SL = \frac{f}{\sqrt{2}\ k} = \left(\frac{1}{n^2} - \frac{1}{n'^2}\right)\frac{c}{(\sqrt{2}\ k)\lambda_r}$$

Planck Capacitance and String Tension

What is Planck Capacitance (PC)? It is the upper end limit of string capacitance just as the root wavelength establishes the upper end limit for frequency. PC is the upper limit which any one electron string can acquire. It is defined by capacitance without frequency.

$$\text{String Tension Force } (F_s) = \frac{\text{Planck Capacitance (PC)}}{\text{Electron Capacitance (eC)}}\left(\frac{1}{\sqrt{2}}\right) = \frac{\text{Frequency } (f)}{\sqrt{2}}$$

$$\frac{PC}{eC} = f = \left(\frac{1}{n^2} - \frac{1}{n'^2}\right)\frac{c}{\lambda_r}$$

$$PC = \frac{e^2}{h} = 38.74 \ \mu F \ \text{(micro Farads)}$$

$$eC = \frac{PC}{f} = \frac{38.74 \ \mu F}{f}$$

$$eC \ (\text{root}) = \frac{38.74 \ \mu F}{3.2893620584\left(10^{15}\right)} = 1.1777 \left(10^{-14}\right) \ \mu F$$

Root frequency we know is the maximum radial extension of the electron. For the other frequencies in the Rydberg series (the quantum differentiated orbits) the formula for electron capacitance is the following:

$$f = \left(\frac{1}{n^2} - \frac{1}{n'^2}\right) \quad f_r = \left(\frac{1}{n^2} - \frac{1}{n'^2}\right) 3.2893620584\left(10^{15}\right)$$

$$eC = \frac{38.74 \ \mu F}{f} = \left(\frac{n^2 n'^2}{n'^2 - n^2}\right) 1.1777 \left(10^{-14}\right) \ \mu F$$

Proof of Equality

$$\frac{f(h)}{e^2} = \frac{1}{eC}$$

$$f = \left(\frac{1}{eC}\right)\frac{e^2}{h}$$

$$\frac{1}{PC} = \frac{h(1)}{e^2}; \ \text{Planck's Constant times frequency of "1"}$$

$$PC = \frac{e^2}{h} = 38.74 \ \mu F$$

$$f = \frac{PC}{eC}$$

$$F_s = \frac{PC}{eC}\left(\frac{1}{\sqrt{2}}\right) = \frac{38.74 \ \mu F}{eC\sqrt{2}} = \frac{27.39 \ \mu F}{eC}$$

The Electron Capacitance Field

To understand what these capacitance formulae mean, we must review the operational principles of the electron-proton bond. The theory holds that particle charges are distance projections into the fourth dimension relative to the particle's three dimensional volume. The attached bond between proton and electron is a Euclidean projection from the quantum electron *times* a quantum projection from the strictly Euclidean proton. The quantum *times* an Euclidean length equal to the quantum distance is the definition of the quantum squared. The bond, therefore, is the quantum squared.

Literally, the quantum squared must be $Q^2/2$ because intersecting quantums originating from a single point must form the plane . Parallel quantums between two parallel Euclidean lines could not form a quantum plane since quantum distance could not be restricted to the shortest distance between the parallel Euclidean lines.

That the quantum squared electron string may actually be a plane rather than a linear string is irrelevant to the tension force on the string. It was demonstrated above that Dawson's Tensor reduces the quantum squared to the exact tension force as the formula would for a linear tensioned string. That tension force equals the frequency of vibration divided by the square root of two.

The quantum-squared value for the electron bond is equal to the radial length which outputs the root frequency. All other Rydberg frequencies are negations of subdivision of this root quantum squared. The root frequency establishes the quantum by the limitations of the speed of light. At root frequency string length, the maximum acceleration of the string reaches the speed of light.

Although longer string lengths are possible for the electron, they are "escape" distances from which the electron can no longer orbit the nucleus. The radiation output at these escape distances are wavelengths greater than the root ultraviolet wavelength of 91.14 nanometers, and are primarily gamma and x-ray. Their radiological signature is short bursts or pulses rather than the continuous output of the orbiting electron.

The point of this brief review is to demonstrate that the hydrogen electron bond is stretched or given tension force by the energy state of the nucleus and that the maximum extension of the electron is the field value "$Q^2/2$." That is, the tensioned electron bond-string is always quantum area which distributes tension over a field, rather than along linear distance. The electrostatic field force of "capacitance" is the appropriate equivalent of the mechanical tension which is distributed into area.

It is important to recognize the difference between electromagnetic field force and electrostatic field force. Electromagnetic force is a force of attraction. Amperage is defined by the force of attraction between two parallel wires at set distance carrying current in the same direction. The force is "magnetic" in the sense that the two wires are attracted to one another by a set amount of Newtons as determined by the amount of current and the distance of the wires. Amperage is the force of attraction per unit of area.

Electrostatic field force, on the other hand, is a force of repellence. Capacitance is the field force which can "push" a current in the opposite direction for a set amount of time. Since time is a factor it is clearly related to frequency as the common capacitance device, "the relaxation oscillator" so aptly demonstrates.

In a relaxation oscillator, a capacitance field is charged with voltage until it reaches a threshold at which it pushes a current for a set amount of time as determined by resistance. The capacitor is then said to be discharged over this period of time and the process starts over. The capacitor "oscillates" as determined by the time value. Oscillations per unit of time is the definition of frequency.

Electron capacitance is defined above as Planck capacitance *divided by* frequency which is the same as Planck capacitance *times* time. The time for one oscillation of a vibrating string is equal to 1 *divided by* frequency. Therefore: "eC= PC/ f = PC(t)."

Planck Capacitance

Planck capacitance is the "base line" electron capacitance. It is the highest capacitance because it is the amount required to "push" the elementary charge for the full second which is the unit of time against which all frequency is measured. Frequency is defined as the number of oscillations *per* second. As frequency increases, time goes down. As time goes down, the amount of capacitance field strength required also goes down. Electron capacitance equals baseline capacitance (PC) *times* the amount of time required to discharge. Time to discharge is the inverse of the frequency of string vibration.

Frequency of string vibration determines the amount of field capacitance the electron field is expressing. The longer the string, the greater the frequency and the less the less capacitance expressed by the field.

Negative Radiation and Electron Field Capacitance

In February of 2008 an article in the Journal of Physical Review theoretically predicted the existence of negative radiation[239] .

Abstract: The interaction of a kink and a monochromatic plane wave in one dimensional scalar field theories is studied. It is shown that in a large class of models the radiation pressure exerted on the kink is negative, i.e. the kink is *pulled* towards the source of the radiation. This effect has been observed by numerical simulations in the ϕ^4 model, and it is explained by a perturbative calculation assuming that the amplitude of the incoming wave is small. Quite importantly the effect is shown to be robust against small perturbations of the ϕ^4 model. In the sine-Gordon (sG) model the time averaged radiation pressure acting on the kink turns out to be zero. The results of the perturbative computations in the sG model are shown to be in full agreement with an analytical solution corresponding to the superposition of a sG kink with a cnoidal wave. It is also demonstrated that the acceleration of the kink satisfies Newton's law.

A "kink" is an attempt on the part of strict Euclidean geometry to describe the quantum squared, although its originators are not aware of that fact. A kink is defined as a "1+1" dimensional object which is integrable or partially integrable. Kinks are often likened to clothespins attached to a twisted rope[240] . Only the clothespins are not separate objects, but "kinks" in the geometric dimension defined by the rope. The "kink" is then described as a "soliton" or single wave with "vacuum" or "emptiness" on either side of it.

A kink does not establish two dimensional space but only a "protrusion" into two dimensional space. It is treated as a soliton because there is only emptiness, vacuum or "nothingness" outside of the protrusion. The kink as thus conceptualized could have no area, so it is provided with a field value to make it a true soliton.

Essentially, a "kink" is a strict Euclidean attempt to identify the quantum squared.

---

[239] *"Negative radiation pressure exerted on kinks;"* Péter Forgács, Árpád Lukács, Tomes Romań czukiewicz; Phys.Rev.D77:125012,2008

[240] *"Solitons"* Sascha Vongehr, 1997: available on internet in PDF format.

Vongehr likens "kinks in 1+1 dimensions" to clothespins attached to a twisted rope[241] . The "kinks" are projections into vacuum along a single dimension but a dimension which is under tension (the twists on the rope to which the kinks are attached). Essentially, the amount of twist on the rope is differentiating kink-influenced vacuum by tension. An untensioned vacuum is likened to the rope with no twisting torque and all clothespin "kinks" pointing downward. The greater the twist on the rope the more tension the "kinks" are applying to vacuum. The 1+1 dimensional kink is a conceptual attempt to supply a field value to vacuum.

The kink so described is the quantum squared turned on its head. Quantum geometry identifies quantum squared vacuum as applying tension to the Euclidean dimension (Vongehr's rope). It is not the rope which is applying tension to vacuum. In fact, it is the tension being applied to the "rope" which is causing it to "kink."

The quantum squared is a kink into quantum space along the x axis. It is a discrete unit of area projecting into quantum space. The quantum dimension must be defined by such "kinks" along the Euclidean x axis, since an intersecting Cartesian quantum axis could not restrict the quantum to vertical values. Kink geometry is the only known mathematical interface between four dimensional quantum physics and three dimensional physics.

"Negative radiation pressure" on an impedance string which is bent or "kinked" into the quantum dimension is mathematically shown to be robust. "The interaction of a kink and a monochromatic plane wave in one dimensional scalar field theories is studied. It is shown that in a large class of models the radiation pressure exerted on the kink is negative, i.e. the kink is *pulled* towards the source of the radiation.[242] "

In the summer of 2008, the predicted "negative radiation pressure" was demonstrated by the Snake-River N-Radiation Lab. The Snake River N-Radiation Lab studies of cotton irradiated by black light (370 nm) revealed a very curious fact about frequency of string vibration and electron field capacitance. Negative frequency light as a carrier wave can sustain string vibrational frequency while simultaneously reducing the energy or tension in the string.

That is, Tomes Roma ń czukiewicz's, "negative radiation pressure" was empirically confirmed. The negative radiation pressure extracted energy from the nucleus, while simultaneously removing tension from the electron string. Since electron field capacitance is an inverse function of electron string tension, the capacitance of the field increased as the tension on the string was reduced.

That capacitance under N-irradiation increases was proven by using bare wire electrical leads. When bare wire leads were irradiated by black light, the measured capacitance increased between two and five times. By controlling the dielectrics of the space separating the leads, this increase in capacitance was shown to be strictly a function of irradiation. The amount of increase in capacitance could only be controlled by shading of the irradiation source and by changing distances from the irradiation source.

The increase in capacitance was not accompanied by a decrease in the frequency of reflexive light output. Change in string tension was not accompanied by a decrease in string frequency. This was proven by the cotton studies. Irradiated cotton "glowed white" as temperature-drop measurements demonstrated that each hydrogen bond acquired Planck

---

[241] *"Solitons"* p.5 op. cit.
[242] op. cit.

Capacitance of 38.74 μF, or the absolute upper limit on electron string capacitance.

The color of the "glow" did not change in frequency from the natural reflective color of the cotton. The glow remained brilliant white, indicating that there was no downward shift in frequency of the hydrogen bonds which were outputting the "glow."

The full implications of this increase in electron field capacitance under N-irradiation must await further research, a part of which is being conducted by the Snake River N-Radiation Lab —the current author being the director of research.

One thing is clear. The increase in capacitance due to nuclear cooling is not restricted to the hydrocarbons which "glow" under N-irradiation. The negative radiation study had identified a new model of the nucleus which was used to guide research into unstable elements.[243] Negative radiation was shown to impact the radiation emissions of U-238. An initial study of N-irradiation of the 238 isotope of uranium conducted by the Snake River N-Radiation Lab has shown that emissions of beta and gamma are significantly reduced by increasing electron field capacitance relative to the nucleus.[244]

---

[243] Not covered in this work.

[244] "*The Effects of N-Radiation Upon Gamma Emissions from U-238* ; The Snake River N-Radiation Lab; See appendix

Nuclear Capacitance: an Overlooked Quantum-Field Definition

Millikan's elementary charge is a quantum discovery

Niels Bohr explained Rydberg frequencies and Planck energy values as a function of the elementary charge of the electron. Bohr derived an energy constant for Rydberg using Planck's energy formula. He then set this Rydberg/Planck energy constant equal to a proposed electron field using the following formula:

$$C = \frac{Q}{V} \; ; \; C = \text{capacitance} \; ; \; Q = \text{charge (in coulombs)} \; ; \; V = \text{voltage}$$

$$E = \int_0^Q V \, d(Q) = \int_0^Q \frac{Q}{C} d(Q) = \frac{1}{2} \frac{Q^2}{C} = \frac{1}{2} C \, V^2 = \frac{1}{2} V \, Q$$

Energy=(voltage) (charge) ; voltage=(Energy)/ (charge)

"Electron volts" were this Rydberg/ Planck energy constant divided by the "elementary charge" of the electron; the charge value which had been determined by Robert Millikan three years earlier.[245] The method that Bohr used to derive his "electron volts" is given by the following example for an orbital fall from "n" to "n=1":

$$C / \lambda = \text{frequency} \; ; \; C = \text{speed of light} \; ; \; h = \text{Planck's Constant}; \; E = (C / \lambda) h;$$

$$E_r = \text{Rydberg energy}; \; RC = \text{Rydberg Constant}; \; \varepsilon = \text{elementary charge of electron}$$

Energy = (Potential Energy orbit of origin) - (Potential Energy orbit of destination)

$$\text{Planck light energy} = E = \left(-\frac{E_r}{n^2} \right) - \left(-\frac{E_r}{1^2} \right) = \left(1 - \frac{1}{n^2} \right) E_r = (C / \lambda) h$$

$$\frac{1}{\lambda} = \left(1 - \frac{1}{n^2} \right) \frac{E_r}{C(h)}$$

Rydberg :

$$\frac{1}{\lambda} = RC \left(\frac{1}{1^2} - \frac{1}{n^2} \right) \qquad ; \; \therefore RC = \frac{E_r}{C(h)} \; ; \; E_r = (RC)(h)(C)$$

$$E_r = (RC)(h)(C) = 2.1795301543 \left(10^{-18} \right) \text{ joules} ; \quad \textit{the Rydberg / Planck constant}$$

$$eV = \frac{E_r}{\varepsilon} = \frac{2.1795301543 \left(10^{-18} \right) \text{ joules}}{1.60217733 \left(10^{-19} \right) \text{ coulombs}} = 13.6035513263$$

Bohr failed to recognize that the "energy constant" which he developed for the Rydberg formula ($E_r = 2.1795301543 \left(10^{-18} \right)$ joules) was nothing more than the Planck energy for the root frequency in the Rydberg formula:

$\lambda_r =$ root wavelength (91.141 nm)

$$E_r = (RC)(h)(C) \; ; \; \text{Rydberg's Constant } (RC) = \frac{1}{\lambda_r}$$

---

[245] R.A. Millikan, *A new modification of the cloud method of determining the elementary electrical charge and the most probable value of that charge*, Phys. Mag. XIX, 6(1910), p. 209.

$$E_r = \left(\frac{1}{\lambda_r}\right)(h)(C) = \left(\frac{C}{\lambda_r}\right)(h) \quad ; \quad \frac{C}{\lambda_r} = \text{root frequency}$$

E_r = (root frequency)(Planck's Constant) ;

The Rydberg energy constant is the Planck energy value for the root frequency.

Once again, the Rydberg root frequency was ignored, lying hidden in non-comprehension. The possibility that the electron bond is a quantum string exhibiting quantum wave harmonics was beyond the knowledge base of the period. It was an age without quantum geometry and with the most primitive of wave physics.

The Elementary Charge is a Distance Value as Predicted by Quantum Geometry

Bohr's "13.6 eV" is the voltage factor of the electrodynamics equation producing the Rydberg/ Planck energy constant. Energy is produced by multiplying the elementary charge by this voltage. Actually, Bohr's electron volt identifies electron string tension.

Electron voltage is defined by the Planck/ Rydberg light-energy value (assumed to be possessed by the electron) *divided by* the elementary charge of the electron. The formula used is "Energy=(voltage)(charge) ; voltage=(Energy)/ (charge)."

Since the charge is a constant, as the Energy increases, the voltage increases proportionally. Energy increases only by frequency (E=Planck's Constant *times* frequency). Therefore, voltage increases with frequency.

In string theory, frequency is a function of string tension. Frequency increases directly by increases in string tension. The electron/proton bond is a string which is stretched to acquire higher quantum orbits. Tension and frequency are directly proportional to the orbital distance. Frequency increases directly proportionally to tension. Frequency also increases directly proportionally to voltage. tension=f(voltage). Tension is a function of voltage.

Elsewhere, I have determined what this function is: tension=(1.7098 10^{14}) eV[246] .

The proposition that "tension=f(voltage)" and the Bohr formulation can be employed to demonstrate that the elementary charge is a distance value, just as four-dimensional quantum geometry proposes. Above, the charge was defined as " *a potential attachment to a dimension which is extraneous to the volume of the particle possessing the charge.*" That is, the charge is a distance in a dimension which the particle's volume cannot address.

The proof is built upon the following modification of the Bohr electrodynamics formula:

$$E = \int_0^Q V \, d(Q) = \int_0^Q \frac{Q}{C} d(Q) = \frac{1}{2}\frac{Q^2}{C} \quad ; \quad Q = \text{elementary charge} = e$$

$$\frac{E}{e^2} = \frac{1}{2C} = \frac{1}{eC} \quad ; \quad 2C = eC = \text{electron capacitance}$$

Conventional electron/nuclear bond theory recognizes that the bond is the result of the electron's elementary charge *times* the proton's elementary charge. The elementary charge of the electron *times* the elementary charge of the proton is the charge squared .

---

[246] See *"The Electron Field and Electron Capacitance"* in the Appendix.

Using "e^2," as the energy denominator instead "e" produces the following:

$$\frac{E_r}{e^2} = \frac{eV}{e} = \frac{1}{eC} \quad ; \quad \text{. } \textit{This is "per string."}$$

The exchange of energy between a single nuclear proton and its attached electron is equal to the electron voltage divided by the elementary charge which equals the inverse of the electron's field capacitance. Capacitance is the proper measure of the electron field[247].

The geometric value of the elementary charge is deduced below.

$$Amp = \frac{(2 \ 10^{-7})\text{Newtons}}{\text{meter}^2} \quad \text{a field force in Newton's spread over area of a square meter.}$$

$$\text{Newton} = (1 \text{ kilogram})(1 \text{ meter per second per second of acceleration}) = kg \ \frac{\text{meter}}{\text{sec}^2}$$

$$\text{Coulomb} = (\text{amp})(\text{second}) = \frac{\text{Newton}}{\text{meter}^2}(\text{sec}) = \frac{\text{Newton}}{(\text{meter}/\text{sec})(\text{meter})} = \frac{F}{v(d)}$$

$$\text{Impedance} = \frac{F}{v} = \text{the amount of force absorbed per velocity of impeded object.}$$

$$\frac{F}{v(d)} = \frac{\text{impedance}}{d} = \text{the amount of force absorbed per velocity per unit of distance.}$$

$$f(\textit{voltage}) = \text{tension} = eV \ \frac{e}{\sqrt{2h}} = eV(1.7098 \ 10^{14}) \quad \textit{See "Electron Field" Appendix}$$

$$\text{capacitance} = \frac{\text{coulombs}}{\text{voltage}} = \frac{\text{impedance} \ (1.7098 \ 10^{14})}{\text{tension}(d)}$$

$$\frac{E}{e^2} = \frac{1}{\text{capacitance}} = \frac{\text{tension}(d)}{\text{impedance}(1.7098 \ 10^{14})}$$

$$E = \frac{\text{tension}(d)}{1.7098 \ 10^{14}}; \quad \text{tension}(d) = \left[\text{voltage}(1.7098 \ 10^{14})\right](e) \quad\quad e = 1.60217733e\text{-}4\alpha$$

$$e = d$$

$$e^2 = \text{impedance} = d^2$$
is the linear value of charge

"Charge" is a fourth dimensional distance value just as determined by quantum geometry. Energy is the tension divided by the voltage constant ($1.7098 \ 10^{14}$) *times* this fourth dimensional distance value. The square of this fourth dimensional distance value is impedance. It is impedance because "the charge squared" is the only charge value which can have definition in three dimensional space (since it is a fraction of the quantum squared). The charge must have physical dimensionality in order to provide impedance.

---

[247] Proven by N-radiation studies reported in Appendix.

The electron volt — no matter how important a measure to nuclear physics — is still an incomplete description of the energy exchange between nucleus and electron string. The charge is a fourth dimensional projection. But it must have dimensional reality in three dimensional space in order to impede or absorb light energy. It is only the charge squared, that is the proton's "quantum charge" times the electron's "Euclidean charge" which can provide that dimensional reality.

The charge squared is a fraction of the quantum squared. As such, it can provide definition to the "field of separation" between the Euclidean proton and the quantum electron. The quantum squared is a soliton field as shown in chapter 1.

Quantum geometry cannot accommodate the mental restrictions of three dimensional thought any more than a spherical earth can accommodate flat-earth beliefs. All vacuous space or empty volume is defined by the quantum squared. The only precedent to explain vacuous volume as the distortion of a two dimensional field is soliton geometry (as applied to four-dimensional quantum space)..

The quantum squared constructs vacuous volume because the line of intersection between a quantum plane and an Euclidean plane is curved to become two dimensional. Further, the quantum plane itself is curved to produce volume. Since there is only one quantum dimension, the quantum squared must employ an Euclidean axis which forces curvature of line and plane and the construction of volume[248]

The electron is a particle because it is the only three dimensional, strictly quantum form in existence. It is the derivative of the square of the quantum space which it occupies.

$$\text{space} = \frac{Q^2}{2}$$

$$\left(\frac{Q^2}{2}\right)^2 = \frac{Q^4}{4}$$

$$\text{electron} = d\left(\frac{Q^4}{4}\right) = Q^3\, d(x) \; ; \; d(x) = \text{charge}$$

Derivative cannot be integrated because four dimensional space cannot exist. However, "d(x)" can have a "projected" distance value of 1.60217733 (10^{-4})α.
The electron may be thought of as a hollow sphere which contains at least one Euclidean plane. Just as the circumference of the Euclidean circle is the derivative of the circle's area, so the quantum particle is the derivative of the quantum squared, squared.

As stated above, Planck's Constant *times* frequency of the "root" string *divided by* the elementary charge squared equals the inverse of the electron's smallest possible field capacitance. Root frequency is the highest possible energy value (for any electron still in orbit) and therefore produces the lowest possible capacitance. Capacitance is inversely proportional to Planck energy.

eV = electron volts ; ε = elementary charge

$$eV = \frac{\varepsilon}{eC} = 13.603564223$$

1.1777629037(10^{-8}) pico Farads is lowest possible capacitance for orbiting electron.

---

[248] For a more complete explanation, see *"The Theory of Time Enforced Space"* Chapter 5.

Electron field capacitance is a valuable unit of measure for whole atoms or molecules. The mathematical components of field capacitance —string energy exchanges and total string charges — vary molecularly by the state of ionization (or antionization) and by atomic weight. To understand the full relationship between electron field capacitance and the molecule we must turn to Millikan's oil drop experiment.

The Millikan Discovery:
the Elementary Charge is Quantum with Respect to the Molecule

 How was Millikan's oil drop experiment able to filter out a single electron charge from the millions of molecules which compose a drop of oil? The experiment was able to do so because the elementary charge proved to be a uniform quantum for all of the molecules composing the drop. These millions of molecules generated a single electron field which increased in force by quantum steps.[249]

Millikan discovered that the electrostatic field is contiguous with respect to all molecules within a mass and the force value of that field is quantized by elementary charges per molecule.

The Millikan Apparatus

The droplets   **The atomizer**

The microscope

The parallel plates
generate a voltage field

The X-ray tube

Millikan determined the elementary charge by isolating and suspending a small droplet of oil between two electrically charged plates. When a single droplet was isolated and suspended, Millikan researchers would spend hours testing the velocities of the drop under a variety of conditions. The droplet would be observed through the microscope as it rose and fell between two lines 10.2 millimeters apart. The time it took to cover this known distance determined velocity.

Two types of velocities were observed; a "fall" velocity with the field turned off and under gravitational influence; and a "rise" velocity caused by the voltage field. These velocities determined the "drag force" which the atmosphere was placing upon the droplet.

Any object being accelerated through an atmosphere reaches a terminal velocity when the

---

[249] See *"A Review of the Millikan Determination of Elementary Charge"* in Appendix.

drag force equals the force of acceleration. The drag force could be determined by terminal velocity using Stokes Law[250] and the known factors of air density, air viscosity and density of the oil.

The drag force on the falling object equaled the mass of the droplet *times* gravity(its weight). The drag force on the rising object equaled the force of the electrostatic field; or charge *times* voltage *divided by* the distance between the plates.

The Millikan experiment proceeded by holding the voltage of the field constant while changing the charge of the oil drop. The charge was varied by x-ray ionization of the oil droplet. (see illustration above).

Multiple experimentations with a single drop were used. The ionized drop was accelerated to terminal velocity by the voltage field. The terminal velocity was measured by the time it took to move between the two lines of known distance.

The field was turned off and the drop fell at gravitational terminal velocity while being re-ionized by the x-ray. The voltage field was again applied and the drop rose to a new field terminal velocity. The data consisted of a set of terminal velocities produced by different ionization states under the same voltage. The charge of the oil droplet was calculated from these terminal velocities.

Mathematically it can be shown that the variance in the charges between any two ionized drop tests is equal to a drag force constant *divided by* the known voltage *times* the terminal velocity of the first ionized state *minus* the terminal velocity of the second ionized state:
k = drag force constant ; ε = charge ; $\Delta\varepsilon$ = change in charge ; υ = voltage field strength

v_1 = terminal velocity ionization 1 ; v_2 = terminal velocity ionization 2

$$\Delta\varepsilon = \frac{k}{\upsilon}(v_1 - v_2)$$

Without going into the complex mathematics involved, Millikan discovered that the *change* in charge between any two different ionization states always produced a whole number ("n") multiple of a basic charge measured in coulombs. A whole number minus a whole number always equals a whole number.

At no time during extensive experimentation with hundreds of individual drops did Millikan find a *change* in charge which was a fraction of the basic charge:
  e = fundamental charge

  $$\Delta\varepsilon = n_1(e) - n_2(e) = \pm n'(e) \quad ; n' = \text{whole number} \geq 0$$

The fundamental charge always produced the measured charge for any ionized state. Regardless of the size or mass of the drop, its various ionized states possessed charges which were always "n" times the fundamental charge. This experimentally determined fundamental charge was identified as the "elementary charge" of the electron or proton.[251]

How is this possible? How can a molecular weight (mole) possess the charge of a single

---

[250] In 1851, George Gabriel Stokes derived " Stokes' law," for the frictional force — also called drag force — exerted on very small particles in a continuous viscous fluid.
[251] Millikan's measurement of the elementary charge was .4% too low because he used the wrong viscosity value for air which the Stokes formula for drag force requires.

electron multiplied by its mass value? In the excitement of identifying a charge value for the electron, science missed this larger significance of the Millikan experiment.

The Millikan Experiment is the Third Detection of a Quantum Relationship between Light Radiation and Matter

Millikan's changes in ionization were produced by the photoelectric effects of x-ray irradiation — the same photo electric effects which Einstein had argued were caused by photon momentum exchanges. However, since Millikan's data proved that the same number of electrons were removed from every molecule within the oil mass, it is difficult to see how every molecule was impacted by the proposed light particle in exactly the same way. This is why Millikan became one of the strongest critics of the "photon" theory even while admiring and confirming Einstein's mathematics. [252]

The whole-number changes in elementary charge for the total mass of the oil drop can only be explained by the systematic removal of the same number of electrons from every molecule within that mass. The energy in the x-ray is a quantum value with respect to the forced ejection of electrons from any sized mass.

Einstein's photoelectric equation cannot account for this. Einstein's equation for the ejection of an electron from the atom is the following:

$$hf = \phi + E \quad ; \quad \phi = hf_0 \quad ; \quad E = \frac{mv^2}{2} = hf - hf_0 = h(f - f_0)$$

$\phi = hf_0$ is the work function, min. energy needed to remove electron from surface

f_0 = threshold frequency for the photoelectric effect to occur ; m = mass of electron

Einstein's equation for photoelectric ejection is light-energy "$h(f)$" per electron "m." The ionization of the oil droplet must be accounted "per electron." The Einstein formula cannot account for ionization "per unit of molecular mass" and, therefore, cannot account for the uniform ionization of the whole oil drop.

This, however, is not true of four-dimensional quantum mathematics which identifies the Millikan experimental results as strictly quantum phenomena.

Ionization by N-Radiation

The Millikan results are very similar to those obtained in the study of the irradiation of cotton by negative frequency light. The hypothesis that frequencies very close to Rydberg frequencies must be absorbed by negative impedance was tested. That hypothesis predicted that hydrocarbons glowing in the black light N-radiation bands were radiating off their nuclear temperatures. The hypothesis was confirmed.

It was discovered that the drop in temperature for the whole mass reflected a mathematically regular drop in temperature for each molecule. As with the Millikan data, changes in energy for the whole of the test subject accurately represented changes on the atomic/molecular level.

---

[252] "Einstein's photoelectric equation...cannot in my judgment be looked upon at present as resting upon any sort of a satisfactory theoretical foundation, even though it actually represents very accurately the behavior of photoelectricity." R. Millikan, *Physical Review*, Jan. 1916.

In the N-radiation study, changes in molecular energy as measured by drops in temperature over time derived Planck's Constant . The number of electron "oscillators" per molecule (hydrogen bonds) and the atomic weight of the molecule were factored by measured change in temperature to extract Planck's Constant.

Planck's constant is discovered to be the change in temperature per hydrogen bond *times* the weight (mass) of the proton attached to the electron string oscillator. Total energy (lost) is Planck's Constant *times* the number of string oscillators *times* the nuclear weight of the molecule[253] :

$$h = -\frac{(\text{mass of proton})(\Delta\text{Temperature})}{239(\text{hydrogen string oscillators})} \quad [254]$$

$$E = -(h)(\text{Total molecular wt.})$$

The difference between negatively absorbed radiation and positively absorbed radiation is frequency. Change in temperature is always *per string oscillator*. Positively absorbed radiation is determined by Planck's energy formula "E=n(h)(f)" where "n" is the number of string oscillators[255].

Negatively absorbed radiation "subtracts" frequency from the string value remaindering Planck's Constant. The amount of energy subtracted from the molecular nucleus is the frequency contribution to Planck energy for the string[256] :E=n(h)(1)

N-Radiation Produces a Quantum Field Capacitance

This principle of subtraction of frequency from the Planck energy value of a quantum electron string produces a quantum field capacitance value for any electron string which absorbs negative radiation. Previously we demonstrated that the "Rydberg energy constant" which Niels Bohr used to develop his electron volts standard was more properly an electron capacitance with the following formula:

$$\frac{C}{\lambda_r}(h)\left(\frac{1}{\varepsilon^2}\right) = \frac{1}{eC} \quad ; \quad eC = \text{electron field capacitance}$$

Bohr used the formula "Energy=(charge)(volts)" to determine electron volts. He did so by dividing both sides of the equation by "elementary charge."

However, the energy equation is also "Energy=(charge)^2 / (capacitance)." This equation develops the "electron field capacitance" formula above. It is more proper because the electron bond is produced by the elementary charge of the electron *times* the elementary charge of the proton or the "(charge)^2."

Negatively absorbed radiation substitutes "(h)(1)" for "(h)(f)" as the energy value for any electron string impeding the negative frequency. This establishes a new capacitance value for the field:

$$"\left(\frac{(h)(f)}{\varepsilon^2}\right)" \text{ becomes } "\left(\frac{(h)(1)}{\varepsilon^2}\right)"$$

establishing a constant quantum capacitance for the string. This is the maximum capacitance

---

[253] For a fuller discussion see below.

[254] *A note of caution:* Planck's Constant has a frequency value of "1." See below for full explanation.

[255] Planck himself identified "n" as the number of oscillators.

[256] This requires negative quantum integration: See *Quantum Mathematics* in Appendix.

which any string can possess and is a value restricted to negative radiation absorption.

$$\left(\frac{(h)(1)}{\varepsilon^2}\right) = \frac{1}{eC} \quad ; \quad eC = \frac{\varepsilon^2}{(h)(1)} = 3.874\left(10^{-5}\right) \text{ Farads}$$

Any electron string which absorbs negative frequencies of light— regardless of the frequency's magnitude — will acquire a quantum field capacitance equal to 38.74 µF [257].

The string will also acquire a quantum micro voltage of $h/\varepsilon = 4.1357\left(10^{-15}\right)$ eV. This micro voltage is well over a trillion times smaller than the lowest electron voltage identified by the bottom Rydberg far-infrared frequency (0.0048 eV).

An electron string absorbing any frequency of negative radiation will undergo a massive increase in field capacitance and an equivalent massive decrease in voltage. To understand the significance of this requires understanding the difference between a voltage "force" and a capacitance "field force."

A voltage force is produced by the potential difference between unlike charges. It is the force needed to neutralize the potential difference. In the case of the electron string, potential difference is increased by length and therefore voltage is increased by length. Since string tension is also increased by length, tension is a function of voltage. Under negative irradiation, energy in the nucleus neutralizes electron string tension, decreasing voltage. Voltage is a linear force value.

In contrast, the capacitance field is the force of potential difference between like charges projected into area. A "field" is a nonlinear projection of force. In electronics, capacitance is the force projected between two separated parallel plates (area values).

Capacitance is the electron projection of field force; the capacity to influence external matter. It is the "electrostatic field" of the atom — or the negatively charged force which the electron can project outside the atom. Negative radiation increases electron capacitance by reducing the potential energy of the string (neutralizes tension).

This increase in electron field capacitance under N-radiation can be routinely measured by a standard electrician's capacitance meter. The measured capacitance between two bare wires in a circuit will be increased between 2 and 5 times under black light N-irradiation. When the source of the black black light is turned off, the increase is lost.

Under normal atomic conditions (i.e. positive radiation impedance) electron capacitance is inversely proportional to the Planck energy being absorbed. Higher energy, lower capacitance.

Ionization of oil by x-ray could not be the photoelectric effect.
Earlier I demonstrated that Einstein's photoelectric effect equation could be derived by Rydberg frequency harmonics. Specifically, higher frequency light in the harmonic series could be impeded by lower frequency strings, but that such strings could only absorb energy at their own potential energy (string tension). This leaves an excess of energy which could be applied to ejection of the electron. The mathematical formula for such harmonic impedance exactly duplicated the Einstein formula for the photoelectric effect.

X-ray frequencies are not Rydberg harmonic frequencies. The photoelectric effect is restricted to fixed orbit electrons and the number of electrons ejected are dependent upon many variables per molecule. Ejection would not be uniform. Finally, ionization from photoelectric occurs within a closed electric circuit and not by total ejection of the electron.

Ionization must be by absorption of x-ray as negative radiation
The reason that the Millikan data identified whole number multiples of the elementary charge is due to a fortuitous accident. Millikan's choice of x-ray as the ionizer and a hydrocarbon (oil) as the target made the observations possible. In recent times, Millikan has been criticized for excluding data from his results. The data he excluded, however, were for non-oil droplets which simply did not give the "whole-number elementary charge" results.

X-ray is negative radiation with respect to organic molecules (hydrocarbons). The Rydberg root wavelength (high ultraviolet at 91.141 nm) is the highest frequency which the orbiting hydrogen electron can absorb. X-ray is just above that frequency (starts at 10 nm wavelength). The furthest orbit of the electron (C/ (91.141 nm) string frequency) can only absorb x-ray (C/ (10 nm) frequency) as negative radiation since the electron cannot acquire the string length necessary for direct absorption.

The oil hydrocarbon molecule makes negative absorption of x-ray possible. The hydrogen bonds are the electron "oscillators" which absorb negative radiation, as the N-radiation study of hydrocarbons proved.

From the furthest possible orbit at root wavelength, the absorption of x-ray increases capacitance from "1.1778 (10^{-14}) micro farads" to "38.74 micro farads." and decreases electron voltage from "13.6 eV" to "4.1357 (10^{-15}) eV ."

By negative impedance of x-ray— from the furthest possible orbit — the tension attaching the electron to the nucleus (electron voltage) is reduced to insignificance. Simultaneously, the force of external projection (capacitance) is increased significantly. It is the formula for the ejection of the electron from the atom.

Proof Oil Drop Will Change by Whole Number of Charges
The number of electrons ejected will be the same for all molecules within the oil drop. This will result in a single change in charge value for the whole drop. This is proven by the following set of equations:
If ionization is a ratio of positive to negative charges per molecule, then:

$$\text{ionization} = \frac{(x-1)e^-}{xe^+} \text{ where x=positive charges and "x-1=negative charges."}$$

If the ionization state is multiplied by "n" molecules, then:

$$\text{ionization} = \frac{n(x-1)e^-}{nxe^+} = \frac{(x-1)e^-}{xe^+} \text{ Ionization ratio doesn't change no matter how}$$

many molecules are added.
Change in ratio of Charges is:

$$\frac{e^-}{e^+} - \frac{n(x-1)e^-}{nxe^+} = \frac{xe^- - xe^- - e^-}{xe^+} = -\frac{e^-}{xe^+}$$

Multiply change in ratio of charges by positive charges to give change in charges:

$$\text{Change in charges} = -\frac{e^-}{xe^+}\left(xe^+\right) = -e^-$$

Ionization is the change in the ratio of negatively charged electrons to positively charged protons. In non-ionized material, the ration is "1 to 1" (e^- / e^+). Ionization is the process of changing this ration. The ionization of each molecule will be a whole number.

The above equation is for a change by loss of one electron: $(x-1)e^-/xe^+$.

The N-radiation studies proved that all molecules with a cotton mass uniformly changed in electron capacitance and in electron voltage. The derivation of Planck's Constant proved the measure of temperature change for the whole mass equaled the measure of temperature change for a single molecule. Change in temperature for the mass was the same as change in temperature for a single molecule. Since x-ray is a form of n-radiation which completely penetrates the oil mass, the change in ionization for each molecule may be presumed to be the same:

$$n(x-1)e^-/nx\,e^+ \qquad \text{where n is the number of molecules.}$$

The change in ratio for the whole mass is the same as the change in ratio for a single molecule:

$$\frac{e^-}{e^+} - \frac{n(x-1)e^-}{nxe^+} = \frac{xe^- - xe^- - e^-}{xe^+} = -\frac{e^-}{xe^+}$$

Since the change in ratio is the change in negative charges *per* positive charges, the change in negative charges can be found by multiplying the ratio by the positive charges which gives a single value for the whole mass;

$$\text{change in charge for whole mass} = -\frac{e^-}{xe^+}\left(xe^+\right) = -e^-$$

The change in charge for the whole mass is the same as change in charge for a single molecule which is exactly what the Millikan data found to be true.

A Review of the Millikan's[258] Elementary Charge

Millikan compared electrostatic field influenced velocities by measuring the time in which the drop covered the distance between two horizontal lines.

A Table of Results

Velocity using time measurements over a standard distance	difference in time values	Number of measurement
0,0417		
	+0,0003	1
0,0420		
	−0,0181	2
0,0239		
	−0,0090	3
0,0149		
	−0,0087	4
0,0062		
	−0,0001	5
0,0061		
	+0,0090	6
0,0151		
	−0,0001	7
0,0150		
	+0,0176	8
0,0326		
	−0,0176	9
0,0150		
	+0,0001	10
0,0150		
	+0,0187	11
0,0238		

[258] R.A. Millikan, *A new modification of the cloud method of determining the elementary electrical charge and the most probable value of that charge*, Phys. Mag. XIX, 6(1910), p. 209

Looking at the table one may see that in cases "1, 5, 7, 10" the drop charge didn't change. Numbers "3, 4 and 6" had identical changes in charge. The quantity of the change is the elementary charge - e. In cases "2 and 9" the quantity of the charge decreased by 2•e, and in cases "8 and 11" it increased by 2•e. The charge of the drop always changed by whole number units of "e." This observation Millikan confirmed experimentally with hundreds of drops. He never reported changes of charge equaling a fraction of e.

The Millikan Apparatus

The droplets

The atomizer

The microscope

The parallel plates
generate a voltage field

The X-ray tube

Method

Initially the oil drops are allowed to fall between the plates with the electric field turned off. They very quickly reach a terminal velocity because of friction with the air in the chamber. The field is then turned on and, if it is large enough, some of the drops (the charged ones) will start to rise. (This is because the upwards electric force F_E is greater for them than the downwards gravitational force W, in the same way bits of paper can be picked up by a charged rubber rod.) A likely looking drop is selected and kept in the middle of the field of view by alternately switching off the voltage until all the other drops have fallen. The experiment is then continued with this one drop.

The drop is allowed to fall and its terminal velocity v_1 in the absence of an electric field is calculated. The drag force acting on the drop can then be worked out using Stokes' law:

$$F_d = 6\pi \, r \, \eta \, v_1$$

where v_1 is the terminal velocity (i.e. velocity in the absence of an electric field) of the falling drop, η is the viscosity of the air, and r is the radius of the drop.

The weight W is the volume V multiplied by the density ρ and the acceleration due to gravity g. However, what is needed is the apparent weight. The apparent weight in air is the true weight minus the upthrust (which equals the weight of air displaced by the oil drop). For a perfectly spherical droplet the apparent weight can be written as:

$$W = \frac{4}{3}\pi r^3 \, g(\rho - \rho_{air}),$$

Now at terminal velocity the oil drop is not accelerating. So the total force acting on it must be zero. So the two forces F and W must cancel one another out.
F = W implies:

$$6\pi \, r \, \eta \, v_1 = \frac{4}{3}\pi r^3 \, g(\rho - \rho_{air})$$

$$r^2 = \frac{9\eta v_1}{2g(\rho - \rho_{air})}$$

Once r is calculated, W can easily be worked out.

Now the field is turned back on, and the electric force on the drop is

$$F_E = q \, E$$

where q is the charge on the oil drop and E is the electric field between the plates. For parallel plates

$$E = \frac{V}{d} \quad ; qE = \text{force} = \frac{(\text{coulombs})(\text{volts})}{\text{distance}}$$

real formulation; Energy = (force)(distance) = (coulombs)(volts)

where V is the potential difference and d is the distance between the plates.

One conceivable way to work out q would be to adjust V until the oil drop remained steady. Then we could equate F_E with W. But in practice this is extremely difficult to do precisely. Also, determining F_E proves difficult because the mass of the oil drop is difficult to determine without reverting back to the use of Stoke's Law. A more practical approach is to turn V up slightly so that the oil drop rises with a new terminal velocity v_2. Then

$$qE - W = 6\pi \, r \, \eta \, v_2 \quad ; W = 6\pi \, r \, \eta \, v_1 \quad ; 6\pi \, r \, \eta = \frac{W}{v_1}$$

$$qE - W = \frac{W v_2}{v_1}$$

$$qE = W + \frac{W v_2}{v_1} = W\left(1 + \frac{v_2}{v_1}\right)$$

The field force "qE" is a function of the Stokes formula and velocity.

e = elementary charge ; ε = field ; $q\varepsilon$ = field force

$$q_n = \frac{mgT_f}{\varepsilon}\left(\frac{1}{T_f} + \frac{1}{T_r}\right)$$

$$q'_n = \frac{mgT_f}{\varepsilon}\left(\frac{1}{T_f} + \frac{1}{T'_r}\right)$$

$$q'_n - q_n = \frac{mgT_f}{\varepsilon}\left(\frac{1}{T_f} + \frac{1}{T'_r}\right) - \frac{mgT_f}{\varepsilon}\left(\frac{1}{T_f} + \frac{1}{T_r}\right) = \frac{mgT_f}{\varepsilon}\left(\frac{1}{T'_r} - \frac{1}{T_r}\right)$$

$$q_n = ne \ ; \ q'_n - q_n = n'e$$

$$n'e = \frac{mgT_f}{\varepsilon}\left(\frac{1}{T'_r} - \frac{1}{T_r}\right)$$

$$e\frac{\varepsilon}{mgT_f} = \frac{1}{n'}\left(\frac{1}{T'_r} - \frac{1}{T_r}\right)$$

$$\frac{\varepsilon}{mgT_f} = k \quad \textit{a constant for the experimental drop}$$

$$k(n'e) = \left(\frac{1}{T'_r} - \frac{1}{T_r}\right)$$

The Energy Equation for the Voltage Field

$$C = \frac{Q}{V}$$

$$E = \int_0^Q V \, d(Q) = \int_0^Q \frac{Q}{C} d(Q) = \frac{1}{2}\frac{Q^2}{C} = \frac{1}{2}C V^2 = \frac{1}{2}V Q$$ [259]

The Energy Equation Used by Millikan

$$F_E = \frac{V}{d}Q \quad ; d = \text{distance of separation between plates}$$

$$F_E(d) = V Q \quad \text{(implicitely} = 2E \text{ or twice Energy)}$$

[259] Hammond, P, Electromagnetism for Engineers, pp44-45, Pergamon Press, 1965.

Quantum/ Electromagnetic Field Equations

Free-Space Magnetic Permeability: **Maxwell's Electromagnetic Waves cannot Escape Local Bond with Matter.**

Field Equations: **Electromagnetic Resonance with Vacuum Solitons Produce Light Waves**

The modern version of Maxwell's electromagnetic field equations is given in the 'Heaviside' form[260]. The electromagnetic field is a function of magnetic permeability in free space and electric permittivity in free space which is set equal to the speed of light:

c = speed of light ; μ_0 = magnetic permiability in free space

ε_0 = electric permittivity in free space

$$c = \frac{1}{\sqrt{\mu_0 \varepsilon_0}}$$

The interaction of the magnetic field component of free space (μ_0) and the electric field component of free space (ε_0) is said to produce the speed of light and thus represent an electromagnetic wave moving through vacuum. The values are measured in units of force and electrodynamics fields.

In this particular formulation, magnetic permeability (Heaviside's term) is a constant from which electric permittivity is calculated:

$$\mu_0 = 4\pi\left(10^{-7}\right)\frac{\text{Newton}}{(\text{amp})^2}$$

$$\varepsilon_0 = \frac{1}{\mu_0\left(c^2\right)} = 8.8541878176 \ 10^{-12} \ \frac{\text{Farads}}{\text{meter}} \textit{(Calculated value, not current SI value)}$$

In SI units, permittivity is measured in farads per meter (F/m). The constant value ε_0 is known as the electric constant or the permittivity of free space, and has the value $\varepsilon_0 \approx 8.854\ 187\ 817\ 10^{-12}$ F/m or
$A^2s^4kg^{-1}m^{-3}$ *in SI base units*.[261]
1.1126500561e-17 of seconds squared per meter squared of time force equals 5.5632502803e-11 Newtons for each second squared per meter squared.
However, Heaviside had to mathematically operate in four dimensional space in order to gain his units of measure for both magnetic permeability and electric permittivity. The actual geometry governing the field values have been hidden in transitional forms of the equation. Heaviside is described as impatient with systematic geometry.

"Heaviside was not interested in (geometric) rigor. His poorest subject at school had been the study of Euclid, a topic in which the emphasis was on rigorous proof, an idea strongly disliked by

---

[260] Oliver Heaviside; electrical engineer and mathematician who became chief interpreter of Maxwell's field equations. Born: 18 May 1850 in Camden Town, London, England. Died: 3 Feb. 1925 in Torquay, Devon, England.
[261] Wikipedia; *http://en.wikipedia.org/wiki/Permittivity*

Heaviside who later wrote:- ' It is shocking that young people should be addling their brains over mere logical subtleties, trying to understand the proof of one obvious fact in terms of something equally .. obvious.'"[262]

To Heaviside, geometry was intuitively "obvious" and he felt no compulsion to identify his maneuvers through it. He intuitively applied a four-dimensional model. His magnetic permeability unit of measure is actually a four-dimensional field:

$amp = \dfrac{2(10^{-7})\text{Newton}}{(\text{meter})^2}$ $\mu_0 = 4\pi(10^{-7})\text{Newton}/amp^2$	$\mu_0 = 4\pi(10^{-7})\dfrac{\text{Newton}}{\left(\dfrac{2(10^{-7})\ \text{Newton}}{(\text{meter})^2}\right)^2} = \pi\,\dfrac{\text{meters}^4}{(10^{-7})\ \text{Newton}}$

Magnetic permeability, when reduced to its field value, is geometrically related to the inverse of the amperage field except that the amperage force value (in Newtons) is distributed into four dimensions not two as with the amperage field.

$$\frac{1}{amp} = \frac{(\text{meter})^2}{2(10^{-7})\text{Newton}} \qquad : \quad \mu_0 = (2\pi)\,\frac{(\text{meter})^4}{2(10^{-7})\text{Newton}}$$

With the addition of the circularity factor "2π," magnetic permeability distributes the amperage force value "$2(10^{-7})$ Newtons" into meters to the fourth power. That is, Heaviside's free-space magnetic-permeability constant is a four-dimensional field.

Why is such a four dimensional field necessary? Heaviside was forced into four-dimensional geometry because the amperage field cannot define the magnetic field. Of necessity, the amperage field is two dimensional and the magnetic field is three. Further, the two dimensional amperage field cannot be converted into three dimensions because the geometric factors (legs) of the two dimensional amperage plane are not equivalent and cannot be identified by the square root of the field:

$$"m^2\sqrt{m^2}" \text{ cannot equal } "m^3" \text{ because } "\sqrt{m^2}" \text{ is impossible.}$$

In this, the amperage field is similar to the quantum squared unit of area because the factors of the quantum squared cannot be identified by the square root. The quantum squared is the product of a quantum distance and an equivalent Euclidean distance neither of which can be factored out by the square root.

Similarly, the "meter^2" of the amperage field is produced by nonequivalent factors which cannot be factored out by the square root. The amp is measured by two parallel wires carrying the same current which are set at one meter apart. One ampere is amount of current producing "2 (10^{-7}) Newtons" of force in one meter of distance along the parallel wires. The area field value is achieved by multiplying the meter of separation by the meter of distance along the wire.

The field area represents the product of two nonequivalent factors. The distance along the wire (a mass value) is multiplied by the distance of separation (a spatial value). Force can only be increased by multiplying field area along one leg. If distance of separation is increased, force goes down. If distance along the wire is increased the amount of force goes up. The equation allows for the multiplication of force by multiplication of area, but this is only true for one leg of the field. The inverse is true for the second leg.

[262] School of Mathematics and Statistics, University of St Andrews, Scotland; JOC/EFR © October 2003

The factors composing the geometric amperage field are not equivalent and cannot be addressed by taking the square root. Therefore, the required three dimensional vacuous space for the magnetic field must be achieved by squaring the amperage field:

$$\left(m^2\right)^2 = m^4$$

Only in quantum geometry can a field value be squared to produce rational geometric space and that space must incorporate both mass as well as vacuum. A very brief review of the squaring of the quantum squared using Einstein's equation will show why this is true:

$$E = mc^2 \; ; \; m = \frac{E}{c^2} \; ; \; Q^2/2 = F_t^2\alpha^2/2 \; ; \; E = \Delta T^2/2 = F_t^2\alpha^2/2 \; ; \; 1/c^2 = \Delta T^2/\alpha^2 = F_t^2$$

$$m = \left(F_t^2\alpha^2/2\right)F_t^2 = F_t^4\alpha^2/2 \; ; \; \left(Q^2/2\right)^2 = \left(F_t^2\alpha^2/2\right)^2 = F_t^4\alpha^4/4 = m\left(\alpha^2/2\right)$$

Without reviewing the details of the difficult quantum geometry involved, the squaring of the quantum time-field produces a rational four-dimensional space composed of mass times a unit of vacuous space:

$$\left(Q^2/2\right)^2 = F_t^4\alpha^4/4 = m\left(\alpha^2/2\right)$$

where "m" equals mass and "$\alpha^2/2$" equals a soliton vacuole[263] (a quantum unit of vacuous space).

Similarly, the squaring of the amperage field to produce a four dimensional field requires that mass as well as vacuum (free space) be a component of that four dimensional field. Heaviside's four-dimensional free-space magnetic field must incorporate mass as a component. This is consistent with the known characteristics of magnetic fields. They are force projections into vacuous space from an originating mass.

The attempt to supply vacuous space alone with magnetic permeability and electric permittivity, as the electromagnetic wave theory requires, is simply not possible. The disguising of four-dimensional field assumptions by reducing the free space permeability formula to amperage does not eliminate this fact. The magnetic permeability of free space must always originate from matter and cannot exist independent of matter. By this fact alone, the projection of electromagnetic waves across the vast distances of interstellar space is just not attainable.

This is not to say that the electromagnetic wave doesn't exist — as it clearly does. The magnetic permeability of vacuous space does receive the timed discharges of electric permittivity capacitance at the speed of light. Capacitance pulses at regular frequency are projected by the interacting fields through vacuous space.

Rather, the claim is being made that such fields are extremely localized because the fields are four-dimensional —as the atom itself is four dimensional — and therefore tied to the mass which originates the fields. These localized electromagnetic waves interface with quantum space which actually can conduct wave form radiant energy over vast distances. Heaviside's free space permeability-permittivity formulation demonstrates that quantum space and the electromagnetic wave do interact.

The Interface between Quantum Defined Space and Localized Electromagnetic Waves

Heaviside's permeability-permittivity formula for free space identifies exactly how localized electromagnetic waves interact with quantum space. To review, magnetic free-space permeability *times* electric free-space permittivity is equal to the inverse of the speed of

---

[263] See chapter 1

light squared:

$$c = \frac{1}{\sqrt{\mu_0 \varepsilon_0}}$$

$$c^2 = \frac{1}{\mu_0 \varepsilon_0}$$

$$\mu_0 \varepsilon_0 = \frac{1}{c^2}$$

Permeability "μ_0" is a measure of the magnetic field density available to free space. Permittivity "ε_0" is a measure of capacitance per meter of distance permitted to penetrate free space. The multiplication of the two "$\mu_0\varepsilon_0$" is equal to the time force sustaining the quantum-squared vacuum vacuoles of which free space is composed:

Time-Force Vectors and the Quantum-Squared Vacuole

$$T_1 - T_2 = -\Delta T \quad ; \quad c = \frac{\alpha}{\Delta T}$$

quantum $= \alpha$

$$\frac{\Delta T^2}{2} = \frac{F_t^2 \alpha^2}{2} \quad ; \quad F_t = \text{time force}$$

$$F_t^2 = \frac{\Delta T^2}{\alpha^2} = \frac{1}{c^2}$$

Sum of Vectors

$$\left(1 - \frac{1}{x_n^2}\right) F_t^2 + \frac{F_t^2}{x_n^2} = F_t^2$$

by Pythagorean Theorem

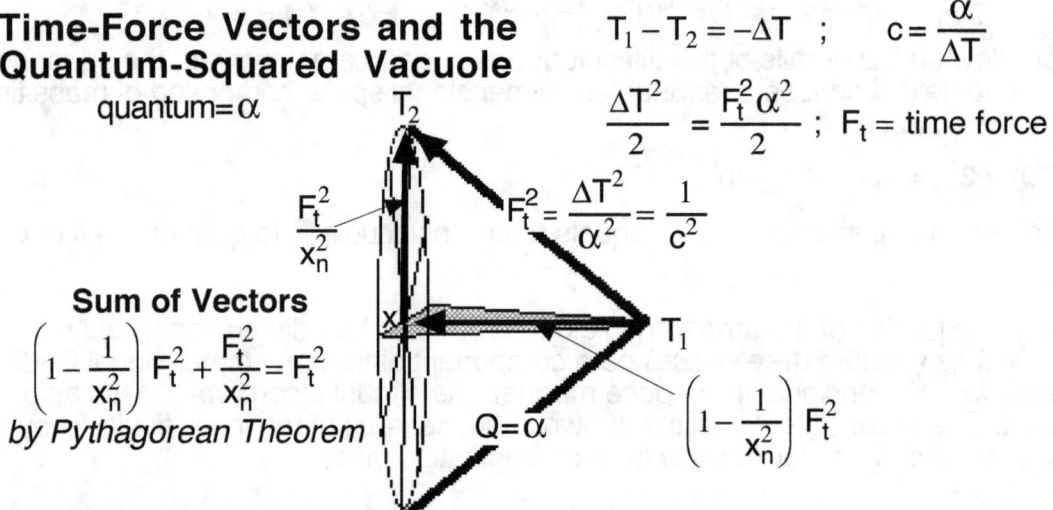

$$Q = \alpha$$

$$\left(1 - \frac{1}{x_n^2}\right) F_t^2$$

The electromagnetic equivalent

$$\mu_0 \varepsilon_0 = \frac{1}{c^2}$$

$$F_t^2 = \frac{1}{c^2}$$

$$F_t^2 = \mu_0 \varepsilon_0$$

Time-force is the force sustaining separation between two incompatible time values. It is measured by the variance between the time values and the amount of spatial separation:

$$\Delta T^2 = F_t^2 \alpha^2 \quad ; \quad F_t^2 = \Delta T^2 / \alpha^2 = 1 / c^2$$

Time force is not the direct equivalent of Newtonian force (F=ma). It is measured by Heaviside's electromagnetic field equation for free space. The multiple of free-space magnetic permeability by free-space electric permittivity in the equation factors out force (in Newtons), remaindering a time-force value. To demonstrate this we need the electric permittivity unit of measure which is determined to be the following:
The SI unit of measure for capacitance is the Farad "F." It may be calculated by the following:

1 Farad=1 amp-sec./1 volt ; 1 Watt / 1 Amps = 1 volt;

1 Farad=amp²(sec.)/1watt; 1 joule/sec.=1 watt;

1 joule=1Newton(1 meter); 1 Newton=1kg(1 meter/ sec²)

$$F = q/v \quad ; \quad v = \text{watt} / \text{amp}$$

$$F = \frac{amps(sec)}{watts\,/\,amps} = \frac{amps^2(sec)}{watts} = \frac{amps^2\,sec^2}{joules} = \frac{amps^2\,sec^2}{(Newton)(meter)} = \frac{\left[2\left(10^{-7}\right)\right]^2 Newton\left(s^2\right)}{meter^5}$$

$$\varepsilon_0 \text{ unit of measure} = \frac{F}{m} = \frac{\left[2\left(10^{-7}\right)\right]^2 Newton\left(s^2\right)}{meter^6}$$

The $\mu_0 \varepsilon_0$ unit of measure:

$$. = \varepsilon_0 \frac{\left[2\left(10^{-7}\right)\right]^2 Newton\left(s^2\right)}{meter^6}\left(\frac{\pi m^4}{\left(10^{-7}\right)Newton}\right) = 4\pi\varepsilon_0\left(\left(10^{-7}\right)amp-Newtons\,\frac{sec.^2}{m^2}\right)$$

$$\frac{4\pi\left(10^{-7}\right)^2 Newton}{\left(10^{-7}\right)Newton} = 4\pi\left(10^{-7}\right)\,amp\text{-}Newtons$$

$$amp\text{-}Newtons = \left(10^{-7}\right)Newton$$

$$amp-Newtons\,\frac{sec.^2}{m^2} = \frac{F_t^2}{4\pi\left(10^{-7}\right)\varepsilon_0} = \frac{8.8541878176e\text{-}12}{\varepsilon_0} = 1$$

One amp-Newton of force for a second squared of quantum time-variance per meter squared of quantum space is equal to time force squared ($F_t^2 = 1/c^2$) *divided by* $4\pi\left(10^{-7}\right)\varepsilon_0$.

Therefore, $F_t^2 = 4\pi\left(10^{-7}\right)\varepsilon_0 = 1.112650056 \quad 10^{-17}$ amp-Newtons of $\Delta sec.^2$ quantum time-variance per meter^2 of quantum space.

$$F_t^2 = \frac{1}{c^2} = \frac{\Delta T^2}{\alpha^2} = \frac{\left(3.34^{-24}\Delta sec.\right)^2}{\left(10^{-15}\,m\right)^2} = \frac{\left(3.34^{-9}\Delta sec.\right)^2}{m^2}$$

$$F_t^2 = 1.11265\left(10^{-17}\right)\,amp\text{-}Newtons\,of\,\frac{\Delta sec^2}{m^2}$$

" $1.11265\left(10^{-17}\right)$ **amp-Newtons of** $\dfrac{\Delta sec^2}{m^2}$ **Force"** **is the Cosmological Constant**

This is an important equation because it determines that the square of the quantum time variance per alpha space is exactly equal to amp-Newtons of force. The Heaviside formulation of Maxwell has demonstrated that quantum squared time force is the equivalent of conventional Newtonian force as denominated by half-amps. Einstein's cosmological constant is the tension on space which prevents it from contracting under gravitational influence. It is determined to be the force of the time-quantum (squared) sustaining the alpha-quantum squared spacial separation. Einstein's cosmological constant is a function of the time quantum and is measured in amp-Newtons as determined by the speed of light.

Final Note: Calculating Electrostatic Field Values Using Electric Permittivity

The unit of measure for electric permittivity "ε" is "Farads/ meter."

$$\text{Unit of "}\varepsilon\text{"} = \frac{F}{m} = \frac{\left[2\left(10^{-7}\right)\right]^2 \text{Newton}\left(s^2\right)}{\text{meter}^6}$$

Permittivity of free space : $\varepsilon_0 = 8.8541878176 \ 10^{-12}$

Applying permittivity to quantum geometric electron-string model (Rydberg) we get:

Permittivity of Rydberg Root String (Farads *per* meter)

sl=electron string length ; c = speed of light ; C=electron capacitance=$\dfrac{e^2}{f(h)}$[264]

e= elementary charge=$1.60217733 \ 10^{-19}$ coulombs; h=Planck's Constant

$$\varepsilon_r = \frac{C_r}{sl_r} = \frac{e^2 \lambda_r}{c(h)(\lambda_r / 2)} = \frac{2e^2}{c(h)} = 2.5844853661 \ 10^{-13} \text{ F/ m}$$

This is less than the permittivity of free space, therefore the atom cannot project an electrostatic field into free space from this root orbital level

Permittivity of Differentials of Root

$$\varepsilon_\lambda = \frac{e^2 \lambda_r / c\left(1/n^2 - 1/n'^2\right)h}{\left(1/n^2 - 1/n'^2\right)\lambda_r / 2} = \frac{2e^2}{c(h)\left(1/n^2 - 1/n'^2\right)^2}$$

$$\varepsilon_\lambda = \frac{\varepsilon_r}{\left(1/n^2 - 1/n'^2\right)^2} \text{ F/m}$$

Electric permittivity of differential is the increase of root permittivity by the square of the inverse of the differential (always less than "1"). Permittivity increases as electron orbital distance falls towards the nucleus by the inverse of the square of the distance.

Orbital Restrictions upon Projection of Electrostatic Field:

The visible band (Balmer Series) for the Rydberg distribution is defined by "n=2"
Visible Band

$$\frac{1}{\lambda} = \left(\frac{1}{2^2} - \frac{1}{n'^2}\right) \frac{1}{91.14 \text{ nm}}$$

Only the lowest frequency in the visible band(n'=3) does the permittivity exceed the

---

[264] See *"Electron Capacitance; an Overlooked Field Defintion."* above

permittivity for free space:

$$\frac{1}{\lambda} = \left(\frac{1}{2^2} - \frac{1}{3^2}\right) \frac{1}{91.14 \text{ nm}} = \frac{0.1388888889}{91.14 \text{ nm}}$$

$$\lambda = 656.2 \text{ nm}$$

$$\varepsilon = 1.3398 \ (10^{-11}) \ \text{F/m}$$

Permittivity of this wavelength is greater than free space permittivity. Therefore, the dielectric atom can project an electrostatic field into free space from this orbital level. Any frequency greater than Rydberg 656 wavelength cannot project an electrostatic field. Any Rydberg frequency lower than 656 can project such a field and the field will strengthen as the frequency decreases.
Rydberg "656" is the transition point for electrostatic fields generated by hydrogen bonds.

The Required Capacitance Measures in the Presence of Light if the Electromagnetic Wave Were the Carrier

Maxwell's equation for electric field flux is the following:

ρ = electric charge density in (coulombs)/m^3 ; ε_0 = electric permittivity of free space
∇E = electric field flux

$$\nabla E = \frac{\rho}{\varepsilon_0} \quad ; \quad \varepsilon_0 = 8.8541878176 \ 10^{-12} \ \frac{\text{Farads}}{\text{meter}}$$

$$\nabla E = \frac{\rho}{8.8541878176 \ 10^{-12} \ \text{F/m}}$$

Planck's Constant can be used to determine the energy value for any frequency of light:

Light Energy = $\frac{c}{\lambda}(h)$ *(in joules)* ; h = Planck's Constant

Using the capacitance formula which can be converted to "joules," an electrodynamic formula for light energy can be determined:

Light Energy = $\frac{c}{\lambda}(h)$ joules = $\frac{c}{\lambda}(h) \frac{\text{coulombs}^2}{\text{Farads}}$;*a fact. known by capacitance formula*[265]

Setting light energy (in electrodynamic units) equal to field flux we get the following:

$$\frac{c}{\lambda}(h) \frac{\text{coulombs}^2}{\text{Farads}} = \frac{\rho \ C/m^3}{8.8541878176 \ 10^{-12} \ \text{F/m}}$$

$$\frac{1}{\rho} \frac{\text{coulomb}}{\text{Farad}} \text{meter}^3 = \frac{\lambda}{1.758837891e\text{-}36} \quad ; \quad \frac{\text{coulomb}}{\text{Farad}} = \text{Volt}$$

$$\frac{1}{\rho}\text{Volts} \bullet \text{meter}^3 = \frac{\lambda}{1.758837891e\text{-}36}$$

By inversion of the equality:

ρ/ Volts·meters^3 = 1.758837891e-36/λ ; ρ/ Volts·meters^3 = Capacitance *per* meters^3

---

[265] See *"Quantum/Electromagnetic Field Equaitons"* in the Appendix.

When the "meters^3" are converted to light wavelengths, capacitance values in the nanofarads and picofarads are given. These capacitance values are easily measured using off-the-shelf meters. They should be measurable in the presence of light from distant sources, yet they never are. In the absence of such measurements, the electromagnetic wave as the carrier of light across interstellar distance is brought into question.

Light as a Quantum Transverse Wave
Failings of Special Relativity & Electromagnetic Wave Theory

During the early part of the nineteenth century, the English physicist, Thomas Young proved the wave characteristics of light. Using a slit experiment, Young had shown that two beams of light can "interfere" or cancel each other out. In this, light acted like water waves which can also cancel each other by opposition.

Prior to Young's observation, the wave theory of light was given little credence. Light transmits energy. The earth, after all, is warmed by the sun. While waves are a form of energy transmission, they require a medium to do so. A conventional wave is defined as a force of compression in motion through a fluid. It was simply inconceivable that space itself could act as the fluid.

Nonetheless, the Young observation required explanation. Two theories emerged to explain the wave characteristics of light; the "ether" theory and James Maxwell's electromagnetic field theory. Both are internally deficient.

The ether is a mysterious fluid-like substance which was thought to permeate all space. It was a fluid which didn't resist motion since, if it did so, planetary motion would decelerate and orbits would collapse. But the ether had the capacity to transmit light as a compression wave.

On the surface the ether is incompatible with wave mechanics since fluid resistance provides the fluid with the energy exchange-potential required for wave impedance. That energy exchange -potential allows the fluid to surrender energy to anything which can impede the wave.

For example, a car crashing into a wall surrenders its energy of motion (kinetic energy) to the wall during the sudden deceleration. A part of this surrendered energy is exchanged with the air and transmitted as the waves which we identify as the sounds of the crash. The wave being conducted through the air is impeded by our eardrums which vibrate at the frequency of the carrier wave. Without the air molecule being a potential partner in the energy exchange, the wave could neither be generated nor impeded. By definition, "ether" lacks the capacity to exchange energy with matter.

The alternative was Maxwell's electromagnetic field theory. An electromagnetic field is generated by an electric current (the flow of electrons through a wire). Maxwell developed a set of equations which described the electromagnetic field by a set of tensor points. Each point in the field was described as having a "tension force" attached to it. By this method, he abstracted the field from the current flow which originated it. He showed that an electromagnetic field so described would flux pulse when an intersecting 90° magnetic field was rotated around it. Light waves were said to be composed by these flux pulses.

Maxwell's theory is the one which has survived the scientific scrutiny. This acceptance, however, has been accomplished at a very high cost. The acceptance of Maxwell has required the desertion of scientific reason.

Abstracting an electromagnetic field from its originating source, then claiming such detached fields are projected through space is no solution. The hard data prove it is no solution. Electromagnetic fields should exist everywhere that light is present. No such fields have ever been detected.

However, the current scientific nomenclature for light is "electromagnetic radiation" proving that contemporary science has the capacity to accept an idea in the complete absence of any supportive evidence.

This blind acceptance of an obviously false nomenclature typifies the incoherence which plagues the physics of radiation. At no time were the well established principles of wave mechanics used to address discoveries made in the field of light. If they had, a search for the missing element would have ensued. That missing element is the quantum dimension. Instead a type of adhoc theorizing took place. Explanations were concocted outside of our knowledge of wave energy.

I mention two examples of discoveries which could be completely anticipated and explained within the context of general wave mechanics but which were taken into fruitless directions instead. The Michelson-Morley experiment[266] is the second.

Michelson-Morley Proves Light Waves Behave Exactly Like the Known Forms of Sound and Water Waves: This Conclusion Ignored by Science.

The American physicist Albert Michelson used the principle of lightwave interference to test the effects of the earth's velocity upon the speed of light. Michelson theorized that Steven Young's wave Interference characteristics should change if the velocities of the two interfering light sources differed.

Michelson found that the the addition of the earth's velocity to one of the light sources did not change interference characteristics. Thus, it was concluded that the speed of light was independent of the velocity of its source.

An incorrect conclusion was drawn from Michelson's experiment. It was concluded that the observation had eliminated the "ether hypothesis." Since "ether" was the only candidate for a conventional wave theory of light, wave theory itself went out with the "ether." An explanation outside of wave theory altogether was sought, an explanation which ultimately intensified the incoherence of the physics of light. The accepted explanation became Albert Einstein's speculative geometry known as "the Special Theory of Relativity."

Special Relativity is an irrational geometry postulated in place of the unrecognized quantum dimension.

Albert Einstein's speculative geometry known as the Special Theory of Relativity uses the mathematics proposed as the last serious defense of the ether theory. Those are the mathematics of Nobel Prize winning physicist, Hendrik Lorentz. The Lorenz Transformations propose that any velocity approaching the speed of light contracted in both time and distance relative to the ether. Nothing could exceed the speed of light through the ether.

The Lorentz mathematics contacting distance and time for all motion through the ether were designed to make the ether theory compatible with the Michelson experiments. Einstein, however, converted them to another purpose.

Albert Einstein' used Lorentz' equations for his geometric dualism known as the "Special Theory of Relativity." Lorentz' contractions of time and distance were a characteristic of the

---

[266] In 1887 physicists Albert Michelson and Edward Morely proved that the speed of light originating on the earth was independent of the velocity of the earth. The source originating the light added nothing to the final velocity of light. The conclusion drawn is that the speed of light is independent of its source.

"geometric referent" not of the ether. Einstein rejected the "ether theory" and all other forms of wave mechanics completely.

He proposed that the geometric referent for anything in motion compresses in distance and time as the motion approaches the speed of light. The geometric referent is the set of three Cartesian axii upon which geometric distance and forms are measured and mapped.

Einstein suggested that the geometric referent was not everywhere the same for all things. The geometric referent changed for anything in motion, constricting in its measure of time and distance as motion approached the speed of light ala the Lorentz Transformations.

It is true that any object in motion possesses a geometric referent which moves with it. Since Einstein liked the train analogy we will stick with his precedent. A train in motion has a geometric referent which can measure and map the speed of someone strolling down the isle. What it cannot do, however, is measure the velocity of the train. Something we will call an "objective referent" is required. This referent measures the motion of the train against a stable axis. The earth — against which the train's velocity is practically measured — does not really provide such an "objective referent" for the train.

The problem is that the earth's referent is also in motion since the earth spins on its axis and moves along its orbital path around the sun. The motion of the train, therefore is measured only relative to the earth. Its velocity can in no sense be said to be an "absolute velocity." All motion, therefore, is relative to the referent which is measuring it.

Einstein attempted to extend this natural relativity to absolute motion. To do this he required an absolute standard against which all motion is measured. This absolute standard —not subject to relativity —he called the Galilean referent.[267] "

Einstein argued that the "geometric referent" for anything in motion was relative to this Galilean referent. The distance and time measures of the referent of motion shortened relative to the Galilean referent. The Lorentz Transformations govern this shortening.[268]

The Transformations are, mathematically, a function of an object's velocity as a proportion of the speed of light. The function produces units of measure for both time and distance which become shorter and shorter as a velocity approaches the speed of light.

The Lorentz equations are a "relativity multiplier." which provides a new time and distant value to the units of measure compared to the Galilean referent. For example, an object moving at one half the speed of light would have a distance unit which is 87% of the Galilean distance unit and a time unit which is 87% of the Galilean time unit. Both units of measure would have contracted by 13%.

The relativistic multiplier affects acceleration differently than it affects velocity. It is this difference which supposedly restricts all velocity to the speed of light. Since the relativity multiplier is the same for both time and distance it has no effect upon velocity:

---

[267] Ascribed to Galileo Galilei, the father of telescopic astronomy and early motion studies.

[268] The Lorentz relativity multiplier $= \sqrt{1 - \dfrac{v^2}{c^2}}$; v = velocity ; c = speed of light

$$\text{Relative velocity} = \frac{d\left(\sqrt{1-v^2/c^2}\right)}{t\left(\sqrt{1-v^2/c^2}\right)} = \frac{d}{t} = \text{Absolute velocity}$$

Velocity equals distance divided by time so the same relativity multiplier appears in both the numerator and the denominator, canceling each other out. Velocity is the same whether measured on the Galilean referent or the relativistic referent.

This is not true of acceleration, however. Acceleration is defined as an increase in velocity over a period of time. The mathematical formula for acceleration is distance divided by time squared:

$$\text{Relative acceleration} = \frac{d\left(\sqrt{1-\frac{v^2}{c^2}}\right)}{t^2\left(\sqrt{1-\frac{v^2}{c^2}}\right)^2} = \frac{d}{t^2\left(\sqrt{1-\frac{v^2}{c^2}}\right)}$$

A single value of the relativity multiplier appears in the numerator with distance. A squared value of the multiplier appears in the denominator with time. The numerator multiplier cancels out only one factor of the squared multiplier. This leaves a single value of the multiplier intact in the denominator.

As a consequence, acceleration is not the same for the Galilean referent and the relativistic referent. At 50% of the speed of light, acceleration on the relativistic referent must be 115% of Galilean acceleration to achieve the same increase in velocity.

This is significant because more force is required to accelerate upon the relativistic referent. Higher velocities require acceleration and acceleration requires force. In physics, the formula for "force" is the mass of the object *times* acceleration. More and more force relative to the Galilean referent is required as velocity approaches the speed of light. An infinite amount of force would be required to reach the speed of light; Since an infinite amount of force is not possible, nothing can be accelerated to the speed of light.

The reason for this brief but accurate description of Special Relativity is the following: both Special Relativity and four dimensional time-structured geometry cannot be simultaneously true. The Einsteinian geometric dualism is the currently accepted model for physics. As painful as it is to me personally —my respect for Albert Einstein as mathematician couldn't be greater —I am obligated to point out the defects of that accepted model.

In the first place, the concepts of time for special relativity and quantum geometry are incompatible. Special relativity postulates that the same point in space can have two geometric definitions simultaneously. One in the "Galilean referent" and a second in the relativistic referent for anything in motion. This in and of itself may or may not be problematic. What is problematic is that the same point is said to have two definitions of time simultaneously. The same point in space can have a rate and measure of time on the Galilean referent which is different from its rate and measure on a relativistic referent.

The foundational premise for quantum geometry is that a single point cannot be occupied by two alien time values simultaneously. Without this premise in place, all of quantum geometry falls. Vacuous space itself cannot exist.

The mathematics of acceleration provide the crucial distinction between quantum geometry and Einsteinian Special Relativity. In physics, acceleration defines force. The acceleration of a mass equals force (ma=F).

In quantum geometry, the force associated with acceleration is restricted by the speed of light, but is not degraded as velocity approaches the speed of light. That is, quantum geometry makes a completely different prediction about force at higher velocities than does Special Relativity. This makes an empirical test of quantum geometry versus Einstein's dualistic geometry possible.

To illustrate quantum spatial management of acceleration, we must employ the linear quantum. The linear quantum is the derivative of the quantum squared —the area unit of the spatial time field —and is purely a mathematical construct.

The linear quantum is composed of a space of separation between two offset time points. Let us designate this distance of separation as "Q."

The time points are offset because one time value is slightly behind the other. It is as if one time position reads ten o'clock and another reads ten o'clock and one second.

The actual difference between the two time values is much much less than one second, however. It is on the order of 3.34e-24 seconds (3.34 to the minus 24th power). This is "334" with 24 zeros in front of it: .00000000000000000000000344 of a second. It is also the minus fifteenth power of a nanosecond which is one billionth of a second. It is much too small for us to measure.

Let us designate this infinitesimally small time differential as "$-\Delta T$." The designation is negative because the differential represents "potential time force," or time force that is yet to be expressed.

"$Q/\Delta T$" equals the speed of light. In this case, we use the positive "time value" of the differential, not its negative "potential force value."

All velocities across Q are restricted to the speed of light by the potential time force. It does this by compressing time in a manner similar too but not equivalent with the Lorentz Transformation. The time of any clock in motion is compressed in rate such that the clock in motion will arrive at the alien time point with an equivalent time value. Time compression is used to equalize time values.

The kinetic force of the object in motion compresses quantum space to compensate for its loss of time value. Thus time and distance are compressed on an infinitesimal level. It is this infinitesimal scenario to which we must apply acceleration.

Let us consider an acceleration across the quantum distance "Q" in a time we will designate as "$x \Delta T$." Standard mathematics tells us what the acceleration rate must be across a known distance "d" in a designated amount of time "t." The formula is the following:

$$\frac{a(t)}{2} = \frac{d}{t} \quad ; \quad a = \frac{2d}{t^2}$$ If we substitute "Q" and "$x \Delta T$" for "d" and "t" we get the

following : $$a = \frac{2Q}{(x\Delta T)^2} = \frac{2Q}{x^2 \Delta T^2}.$$

The final velocity for any acceleration equals time *times* acceleration. Substituting we get:

$$v = a(x\Delta T) = \frac{2Q}{(x\Delta T)^2}(x\Delta T) = \frac{2Q}{x\,\Delta T}$$

Since we know that final velocity cannot be greater than "$Q/\Delta T$" (the speed of light) we also know that "x" must be greater than "2." Acceleration, therefore is restricted to the following:

$$a = \frac{2Q}{(2\Delta T)^2} = \frac{2Q}{4\Delta T^2} = \frac{1}{2}\left(\frac{Q}{\Delta T^2}\right)$$

By extension, it can be shown that all accelerations across the time quantum are restricted to the following:

$$a \leq \frac{1}{2}\left(\frac{\Delta V}{\Delta T}\right)\ \ ;\ \ \Delta V = (\text{speed of light}) - (\text{initial velocity})$$

All accelerations, therefore, are restricted to one half the difference between the initial velocity entering the quantum space and the speed of light *divided by* the time differential.

These restrictions are not that onerous. For example, for an initial velocity of 90% of the speed of light, acceleration at a rate of 4,487,911 meters per second squared is still possible. An acceleration rate of nearly four and a half million meters per second per second is a huge force value for any size mass. Acceleration would have to reach this level before resistance to the acceleration force began.

Any kinetic force which tries to push acceleration past the limit, will be directly resisted by the time force. Acceleration to velocities greater that the speed of light exceeds the capacity of the time field to manage space by compressing time. The time field can only equalize the time differential by a direct resistance to increases in velocity. It must oppose all further acceleration directly.

The resistance of time is an immovable wall against which all additional kinetic force is wasted. In all likelihood the application of additional force would be converted to the disintegration of the object in motion.

In contrast, Special Relativity predicts that at 90% of the speed of light any acceleration would require 2.29 times as much force to accomplish. All force values would be degraded to 43.589% of original capacities. At 90% of the speed of light, the Lorentz Transformation predicts a degradation of all force to 43.589% of original value.

In the quantum geometric model there is no degradation of force. There is no resistance to force until acceleration reaches the limit and all further force applications are diverted to the object's disintegration.

Special Relativity: increasing degradation of force as object approaches speed of light.
Quantum Geometry: diversion of force at the acceleration limit as determined by speed of light across the quantum.

Although not recognized as such, an empirical test of the two models is being routinely performed by the larger particle accelerators such as that at the FermiLab. What is about to be said will enrage particle physics because that belief system is about to fall based upon its own accelerator data.

Particles which are routinely accelerated to near the speed of light demonstrate conclusively that Special Relativity is false.

Single protons (ions of hydrogen) are frequently accelerated to near the speed of light. Velocities in excess of 90% of the speed of light have been reported. Seldom do scientific papers discuss the mechanics of these acceleration, however. Yet only these mechanics provide usable data.

Accelerators work by batteries of contained magnetic fields. Imagine a magnetic field within a shielded box such that the field does not extend outside of the box. Now imagine a slit through the side of the box containing one pole of the field and another slit through the opposite pole on the other side of the box. If a charged particle like the proton travels through the slit of like charge towards the slit of opposite charge, the proton will be accelerated across the box and out the slit on the other side. The amount of acceleration is determined by the strength of the field which itself is determined by the conventional electrical voltage being applied.

The strength of the field, its force value, is measured in "*electron* volts[269] " (this is not the same as *electrical* voltage). Every time the proton passes through one of these boxes, a certain number of electron volts are applied to the proton.

Electron volts establish the force value of the field (F=ma). Electron volts =E/ (elementary charge)=F(d)/(elementary charge). So many passes through so many boxes equals so many electron volts which produce so much acceleration for the length of the box *per* elementary charge of the particle.

The accelerator consists of a great circular track miles long which contains batteries of boxed fields. The proton is accelerated around the track and back through the battery of boxes for multiple times. The protons are accelerated to near the speed of light, by this method.

Final velocity is directly determined by the number of electron volts applied. But this should not be if Special Relativity is true.

Special Relativity requires that field forces degrade at velocities near the speed of light. At 50% of the speed of light, the field force must be degraded to 86.6% of non-relative force. At 90% the speed of light, field force must be degraded to 43.6% of non-relative force. As particle velocity approaches the speed of light, the amount of acceleration should no longer be directly proportional to electron voltage.

But this is not how accelerators factually perform. Velocity is always proportional to electron volts applied, even near the speed of light. Special relativity is counter indicated by the actual day-to-day operations of the accelerators. That performance, however, is exactly as quantum geometry predicts.

Special relativity is fully and completely responsible for the modernist view of light radiation. Light is conceived dualistically, as both a wave and a particle. Without doubt, Einstein's proposal of the "light particle" called a "photon[270] " is responsible for this dualism.

---

[269] electron volt=$\dfrac{1 \text{ joule}}{1 \text{ coulomb}}$(elementary charge) $= 1.60217733$ (10^{-19}) joules per coulomb.

[270] For which he received the Nobel Prize in 1921.

Einstein's paper proposing the "photon" and his paper proposing special relativity were published nearly simultaneously — in March and June of 1905.[271] This is no coincidence. Special relativity requires the "photon" to explain its restrictions upon the velocity of light.

Velocity is restricted by degrading the force required for acceleration. Light must be a particle because only mass is subject to such restrictions. The particle nature of light was required to support special relativity's explanation of Michelson-Morley.

Special relativity is the accepted explanation of Michelson-Morley and special relativity requires that light be a particle. But what if light is not a particle? What if light is completely waveform energy moving through a medium? What if special relativity is not the explanation of Michelson-Morely? What if science took a wrong turn when it accepted the dualistic geometry of special relativity?

Special relativity is an artificial three dimensional geometry proposed to explain a four dimensional phenomenon.

The Quantum Dimension and Michelson-Morely

The Michelson-Morely observation that the speed of light was independent of the velocity of its source are completely predicted by known principles of wave mechanics. Wave mechanics recognize that the speed of the wave is *always* restricted by force upon the medium and is *always* independent of the velocity ot the wave's energy source.

Thus, the speed of sound is determined by the air and not the jet aircraft moving through it. Sound wave velocity is independent of its source. If it were not so, a jet aircraft could not "break the sound barrier." It could not catch up with and surpass in velocity the sound waves it is generating. The speed of a sine water wave is restricted by gravity and wavelength, or the force of gravity upon the water down the length of a slope. In both cases, medium determines velocity, not the origination of the wave.

If the light wave is operating under the principles of conventional wave mechanics, then the medium which is conducting the light wave should determine its velocity. Further, that velocity will be independent of any velocity of the object originating the light. That is, light should behave exactly as Michelson's experiments found it behaved.

Quantum geometry postulates that vacuous space is composed of two dimensional planes ($Q^2/2$) which are forced by 90° vectors of a time voltage into curvatures to become "$\pi Q^2/2$" or the curved surface areas of cones.[272] A transverse wave can be conducted by shortening the vector of force in the direction of travel increasing the height of the transverse vector.

Vacuous space is proposed as the medium conducting light waves and the wave thus conducted must be of a transverse type. For this postulate to be true, vacuous space must operates under known transverse wave principles and meet the known performance characteristics of the "electromagnetic" wave.

Conventional electromagnetic wave mechanics have identified radiation as a transverse wave and built a mathematically practical system of wave impedance upon it. Assuming radiation to be a transverse wave has provided engineers with reliable measurement of the

---

[271] In the German scientific periodical *Annalen der ßß*.
[272] See *"The Theory of Time-Enforced Quantum Space"* in appendix

energy extracted by impedance of radio band frequencies.

The electromagnetic theory of radiation results from the field equations of James Clerk Maxwell.[273] Maxwell demonstrated, mathematically, that "pulses of electric force" could be generated and conducted through space by the intersection of electromagnetic and magnetic fields. Like the planes of quantum geometry, these fields must intersect one another at 90° angles. The waves thus postulated were transverse in character, one field gaining in strength at the expense of the strength of its transverse partner over distance.

On face value, the theory is nonsense. Neither field exists "*in vacuo*." Both the electromagnetic and the magnetic fields are generated by the charges of atomic particles; the electromagnetic by the flow of electrons; the magnetic by the separation of associated electrons and protons. Without the proximity of elementary charges, neither field can exist.

However, Maxwell's field equations accurately identified the transverse character of light radiation. It is generally held by radio technicians that when Heinrich Hertz discovered radio waves by practical implementation of Maxwell's theory that he also demonstrated that electromagnetic and magnetic fields "detached" from their source and were sent hurling through space. He did no such thing.

Hertz initiated a spark of regular frequency using a Maxwell "pulse" generated by intersecting electromagnetic and magnetic fields. Maxwell's theory was proven in the sense that it did provide enough regularity of frequency to initiated coherent radiation. It was the electrons in the spark exchanging their energy with quantum space which initiated the radiation wave. Other electrons in the antenna impeded the energy wave and were detached from their atoms to constitute a small current at frequency. Radio wave detection is a photoelectric effect at low frequency. However, radio technology was born.

These supposedly "detached" electric fields have never been measured and could not possibly explain the Planck energy values in radiation emitted and impeded by atoms. The "electromagnetic" frequencies are restricted to bands produced by the flow of electrical currents, that is, the bands below the far infrared of atomic output.

Nonetheless, electromagnetic theory does identify the transverse characteristics of light radiation. Wave impedance is defined by radio technology as the inverse of the supposed "transverse components of the electromagnetic wave." Electromagnetic "impedance" —meaning the amount of current made to flow by reception— is determined by the ratio of the electric field and its transverse partner (at 90° angles) the magnetic field.

Electromagnetic transverse impedance is measured in Ohms, which is the resistance measure for an electric current. It has no frequency value for "free space" (vacuum) acquiring a frequency value only when impeded by a conductor of an electric current (i.e. an antenna). Despite its sever restrictions, however, it does identify radiation as a transverse wave; by the ratio of the force of wavelength to wave height, or in electromagnetic terms, the ratio of "magnetic permeability" to "electric permittivity."

Magnetic field permeability is the degree of magnetization of a material that responds linearly to an applied magnetic field. Electric field permittivity is the capacitance of the field per unit of distance (Farads/meter). Capacitance is a field force value. The ratio is the

---

[273] Electromagnetic wave theory initially postulated in 1864. Maxwell was the first director of the Cavendish Laboratory at Cambridge University from which he issued his most famous work detailing the wave theory, "*Treatise on Electricity and Magnetism*" (1873).

amount the magnetic field can be "pushed"(in matter, not in space) per unit of capacitance (*per* meter). It is actually a measure of how "high" the transverse wave can be "pushed" per unit of force over a distance.

The electromagnetic field theory of the wave is completely inadequate as an explanation of light wave forms (that the wave can have no frequency in a vacuum is only one of its many problems), but provided functioning mathematics to "current generated" radiation frequencies because it recognized the actual wave as transverse.

The basic problem with Maxwell and his followers, is that they failed to distinguish between the carrier wave and the impedance wave, between the standing wave of the vibrating string and the sound wave causing it to vibrate. Maxwell's electromagnetic wave is the impedance wave, not the carrier wave. Once recognized, Maxwell's contribution is known.

Formulations for Light Frequency as the Multiple of the Electron Charge and Basic Quantum to Establish Orbitals
Modifying Bohr's Electron-Voltage Formulation for Electron Capacitance[274]

$$eV = \frac{f(h)}{e} \quad ; \quad \frac{1}{C_e} = \frac{f(h)}{e^2} \quad ; \quad e = eV(C_e) \text{ \textit{The conventional capacitance}}$$

formula.

The formula "$e = eV(C_e)$" implies that the space between the electron and nucleus is undergoing a projected force upon the elementary charge of the electron. The elementary charge is measured in coulombs, indicating a charge stored by a capacitance field under a voltage pressure. Since the electron's elementary charge is the charge stored by electron voltage *times* electron capacitance, capacitance goes up as voltage goes down and visa versa. Electron voltage varies with orbital distance. The formula "$e = eV(C_e)$" supplies that distance with a capacitance field value as well. The strength of that field is measured in Farads (F) and is equal to the elementary charge divided by the electron voltage for the orbital distance: $C_e = e/eV$. Capacitance varies inversely with voltage.

The existence of electron capacitance means that there is a field pressure on the electron to "discharge" itself into a current, to flow away from the direction of the field. In this sense, the electron capacitance field is a counter force to the force of attraction between the negative electron and positive proton.

On first hearing, the notion of such an "electron capacitance" must strike the perceptive reader as very curious. It presents the image of a positive charge in the nucleus somehow providing a same-pole force of repulsion against the negative charge of the electron. This is not as irrational as it might seem. It is a known fact that the poles of a battery generate a capacitance field between them producing a resultant charge.[275] The charge is measured in coulombs and is equal to the capacitance field in Farads *times* the voltage of the battery.

The charge imposed by the capacitance field is primarily on the negative pole, since current flows from the negative pole toward the positive pole. It is proper to say, therefore, that the direction of the capacitance field is from positive pole towards negative pole just as the nuclear capacitance field is directed outward from the nucleus toward the orbiting electron. The mathematical existence of an electron capacitance field for the atom and the empirical existence of capacitance fields between the poles of batteries, is evidence that the counter force required by the inverted energy state of the subshell schematic is a very real possibility.

Further, an analysis of electron capacitance fields prove that the inverted orbital energy states of the subshell schematic is correct. It is a known fact that capacitance is inversely proportional to the distance of separation between capacitor plates:

$$C = \frac{\varepsilon_0 A}{d} \quad ; \quad \varepsilon_0 = \text{free space permittivity (\textit{see below})} \; ; \; A = \text{area} \; ; \; d = \text{distance}[276]$$

Capacitance decreases as distance increases and increases as distance decreases. Capacitance is inversely proportional to the distance between plates. The Bohr model of the atom predicts a capacitance which is incorrectly related to distance between the nucleus and the electron.

---

[274] See *The Electron Field and Nuclear Capacitance* in Appendix.

[275] physicsforums.com/archive/index.php/t-199381.html

[276] http://www.csun.edu/~gsl05670/labs/cap_plate_sep.htm

By the Bohr model, the closest orbit to the nucleus, the "1" orbit has the highest electron voltage of "13.6 eV" and therefore the lowest nuclear capacitance (1.1781e-20 F). Capacitance equals the elementary charge *divided by* electron voltage (from the standard electrodynamic capacitance formula). Bohr's furthest orbit from the nucleus, the "8" orbit, has the lowest electron voltage (0.2125 eV) and the highest capacitance (7.53966e-19 F). Capacitance is directly related to distance for the Bohr model. As distance decreases capacitance decreases and as distance increases capacitance increases. It is exactly the opposite distance-to-capacitance relationship of that required.

In contrast, for the subshell schematic, the subshell closest to the nucleus (the "s" subshell in the first shell) has the lowest electron voltage (0.065 eV) and the highest nuclear capacitance (2.46295e-18 F). The furthest subshell from the nucleus (the "s" subshell in the "7" shell) has the greatest electron voltage (13.39 eV) and the lowest nuclear capacitance (1.19677e-20 F). It is only by the inverted orbital energy states of the subshell schematic that nuclear capacitance is indirectly proportional to orbital distance. It is only the inverted subshell schematic which correctly identifies nuclear capacitance and orbital distance.

To restate. The Bohr falls between orbits are no longer operative as an explanation of radiation emissions. The falls have been replaced by the subshells and, therefore, light emissions must be explained by the orbits themselves, not by falls between them. By inverting the Bohr orbital energy states, the subshell schematic has implicitly identified orbital tension. Light output may be explained by vibrational frequencies established by this tension.

However, electron capacitance could not account for this tension. The energy levels are much too low and the relationship between orbital distance and energy is in the wrong direction. As per the above, electron capacitance is inversely related to orbital distance. The greater the distance, the lower the capacitance.

Tension, on the other hand, must be directly related to orbital distance. As orbital distances increase, so must tension. It is known that associated light frequencies increase with subshell orbital distance. Since frequency is a function of tension, tension must also increase with distance. Electron capacitance cannot explain orbital tension because it varies with distance in the wrong direction. There is another form of atomic capacitance which can deduce orbital tension, however.

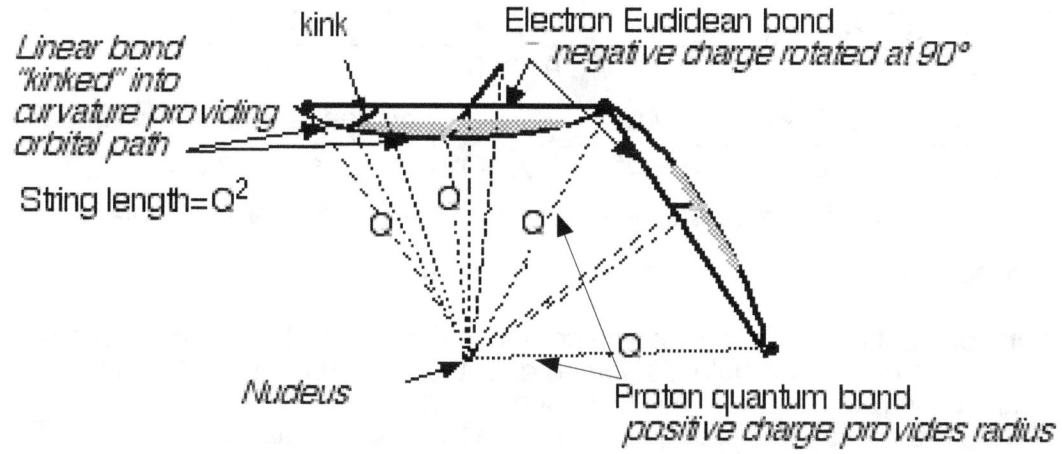

Linear bond
"kinked" into
curvature providing
orbital path

String length=Q^2

kink

Electron Euclidean bond
negative charge rotated at 90°

Q Q'

Nucleus

Q

Proton quantum bond
positive charge provides radius

The nuclear charge multiple being supplied the electron is determined by an orbital's electron voltage and a constant capacitance provided across an "alpha space" separating the nucleus from an "electron quantum field" which envelopes the nucleus. Planck Capacitance is the capacitance between the nucleus and the electron field. The electron's quantum field envelope is a currently unrecognized characteristic of the four-dimensional atom and will be discussed later. To establish the orbital distance, the nucleus must oscillates within the quantum field envelope at an energy rate equal to the required orbital electron voltage. Mechanical energy of oscillation is converted to electron voltage. It is Piezoelectrical. This nuclear heat energy, converted to an electromagnetic field value, is transferred to the electron at energy rate equal to Planck's Constant[277]. This transfer rate is "per electron bond" and not determined by the weight of the nucleus.[278]

The formulation of nuclear supplied charge is derived from two sets of equations: the author's own tension-to-frequency tensor and the electrodynamic formulation for frequency based upon the Bohr electron-voltage standard:

k= tension constant; x_t = tensioned length; x_0 = untensioned length; f =frequency

Tension force $= F = k(x_t - x_0)^2$ *Dawson's Tensor* [279]

$f = \sqrt{2}\, k(x_t - x_0)$ *The Euclidean String value of the tensor*

$(x_t - x_0)$ = electron radial distance $= \left(\dfrac{1}{n^2} - \dfrac{1}{n'^2}\right)Q^2$;root radius$^2 = Q^2$

$f = \left(\dfrac{1}{n^2} - \dfrac{1}{n'^2}\right)\dfrac{c}{\lambda_r} = \sqrt{2}\, k\left(\dfrac{1}{n^2} - \dfrac{1}{n'^2}\right)Q^2$; *reduces to* $\dfrac{c}{\lambda_r} = \sqrt{2}\, k\, Q^2$; $\dfrac{c}{\lambda_r}$ =root frequency

The value of "k" can be calculated in electrodynamic units[280].

---

[277] See *"The Drop in Temperature by N-Irradiation of Cotton"* in Appendix
[278] Ibid.
[279] See *Dawson's Tensor* in Appendix
[280] For full equations, see *"The Electron Field and Electron Capacitance"* in Appendix.

$$f_r = \frac{c}{\lambda_r} = eV_r \frac{e}{h} \quad ; \quad eV_r = \frac{f_r(h)}{e} \quad ; \quad \frac{e}{h} = \frac{e^2}{e(h)} = \frac{\text{Planck Capacitance}}{e} = \frac{C_P}{e} \quad ;$$

$$k = \frac{f_r}{\sqrt{2}\, Q^2} = \frac{eV_r(e)}{\sqrt{2}\, Q^2(h)} = \frac{eV_r C_P}{\sqrt{2}\, Q^2\, e} = \frac{1}{\sqrt{2}\, Q^2}\left(\frac{\text{nuclear supplied charge for root}}{\text{elementary charge}}\right)$$

k is a function of nuclear supplied charge as a multiple of elementary charge

divided by root radius^2 which is the quantum squared for all subshell radii.

The tension constant for the orbitals which compose the shell/subshell schematic is the maximum nuclear charge multiple distributed over the quantum orbital distance (squared). The quantum is the root radial and the highest orbit available to the electron. All shell/subshells are negations of subdivision for the root radial distance (squared). The maximum nuclear charge multiple is that required to achieve the quantum (root) orbit. The tension constant is the nuclear charge multiple per unit of quantum distance (squared). It is tension because the capacitance field establishing the nuclear charge multiple opposes the rotated electron bond, preventing it from reattaching to the nucleus and reestablishing primacy of the contraction force between electron and nucleus. Tension is generated when an expansion force "stretches" a contraction force, which is being done with the electron bond. The tension constant is the amount of tension per unit of "stretch[281] ."

The electrodynamic tension constant "k" can now be inserted in the tension equation to find the frequency for any shell/subshell orbital:

n = shell ; n′ = subshell negation number ; Q = root orbital radius

$$\text{subshell orbital distance}^2 = \left(\frac{1}{n^2} - \frac{1}{n'^2}\right) Q^2$$

$$f = \sqrt{2}\, k\left(\text{subshell orbital distance}^2\right)$$

$$f = \sqrt{2}\, k\left(\frac{1}{n^2} - \frac{1}{n'^2}\right)Q^2 = \sqrt{2}\left(\frac{eV_r C_P}{\sqrt{2}\, Q^2\, e}\right)\left(\frac{1}{n^2} - \frac{1}{n'^2}\right)Q^2 = \left(\frac{1}{n^2} - \frac{1}{n'^2}\right)\frac{eV_r C_P}{e} = \frac{eV(C_P)}{e}$$

The equation is reduced to "f = (nuclear charge multiple for the subshell orbital distance^2)" because the equation resolves to the electron voltage for that orbital distance:

$$eV = \left(\frac{1}{n^2} - \frac{1}{n'^2}\right)13.6\ eV \quad therefore \quad \left(\frac{1}{n^2} - \frac{1}{n'^2}\right)\frac{(13.6\ eV)\ C_P}{e} = \frac{eV(C_P)}{e}$$

$$f = \frac{eV(C_P)}{e}$$

Frequency equals the nuclear charge multiple for the shell/ subshell orbital distance^2 as determined by the tension constant for all shell/subshell orbitals and the distance^2.

To summarize: Planck Capacitance is available to the nucleus from the quantum "alpha space" which must separate the nucleus from the electron quantum field. The nucleus vibrates within this quantum field envelope and Planck Capacitance is continuously available to it. The nuclear energy of vibration is Piezoelectrical. When the mechanical energy of nuclear vibration provides enough electron voltage to equate with the electron voltage for any shell/ subshell orbital, the nucleus can place the electron in that shell/ subshell. It does so by providing the electron with a nuclear charge multiple of the electron's own elementary

---

[281] *Dawson's Tensor* in Appendix.

charge. This nuclear charge must discharge away from the nucleus providing sufficient counter force to place the electron in the shell/ subshell orbital .

The nuclear charge multiple is equal to the frequency of radiation emission associated with any shell/ subshell. The nuclear charge multiple is derived from both a mechanical tension formula and an electrodynamic formula using the standard electron voltage formulation for orbitals. The nuclear charge multiple is not the equivalent of electron capacitance. The differences between them may be the difference between the mysterious "weak" and "strong" forces of quantum mechanics, although that argument will not be made at the current time.

Planck Capacitance and the Tension Constant

Planck Capacitance plays a role in describing the "tension" in quantum-squared orbitals. C_p is the capacitance of the fundamental quantum, "the alpha space;" the space of separation which must be maintained between the nucleus and the electron's quantum field. C_p is the capacitance force which, when multiplied by any orbital's electron voltage, gives a nuclear charge multiple of the elementary charge equal to the orbital's frequency.

Planck Capacitance is given a frequency of "1" which means that the alpha space "orbital" multiplies the elementary charge by only one time. The alpha space is the fundamental quantum and is the minimum separation between nucleus and electron. The frequency of "1" is a theoretical component of the actual frequency arrayment of the shell/ subshell schematic. A frequency of "1" is more of a theoretical state than an actual radiation frequency because it would require a wavelength of nearly 300 million meters long. It is simply a possible theoretical frequency in the arrayment of actual frequencies which are related to orbital distances in the shell/subshell schematic by the tension constant "k[282] ."

This proposition is not made without mathematical test using the tension constant "k." Since "k" is a constant which relates frequency to radial distance, if one radial distance is known as well as its attendant frequency, then the constant can be calculated. This calculated "k" value would be true with other radial distances. The alpha space with a frequency of "1" makes such a test possible because an approximation of the alpha space is empirically known. The radius of the quantum electron and Euclidean proton are equal. The fundamental quantum "α" is the shortest measure which anything can take. The minimum known distances are the diameter of the proton and the shortest wavelength, both of the same order of magnitude at "$\cong 10^{-15}$ meters[283] ." Since alpha is the radius of the proton "$\alpha \cong (10^{-15})/2$." Applying this value to the "k" constant we can calculate alpha:

$$k = \frac{f}{\sqrt{2}\,(\text{quantum radius})^2} = \frac{1}{\sqrt{2}(\alpha^2)} = \frac{1}{\sqrt{2}x^2(10^{-15})^2} = \frac{1}{\sqrt{2}(0.5021375797\ 10^{-15}\ nm)^2}$$

$$k = 2.8044(10^{30})\ \text{units}\ \alpha^2./m^2 \quad ; \quad \sqrt{2}\,k = \frac{1}{\alpha^2} = (3.9660\ 10^{30})\ \text{units}\ \alpha^2/m^2$$

Proof by Using Root frequency
Frequency of root = f_r = 3.2893620584e15

Quantum radial of root = Q_r = 2.8799076686e-8 meters

---

[282] "k" is a mechanically derived constant based upon a tensioned string. Its formulation derives the two most important characteristics of a tensioned string: that frequency is a function of tension and string length and not stirrng deflection; that frequency increases as a multiple of the string's subdivision. *Ibid.*
[283] MIllikan, Robert Andrews. *Electronics (+ and -) Protons, Photons, Neutrons, and Cosmic Rays.* London: Cambridge University Press, 1990: 47

Quantum-Squared for roo t= Q_r^2 = 8.2938681799e-16 meters²

$f = \sqrt{2}\,kQ^2$; frequency at rotation of electron bond $= 1$

$1 = \sqrt{2}\,k\,\alpha^2$ $k = \dfrac{1}{\sqrt{2}\alpha^2}$; $\sqrt{2}\,k = \dfrac{1}{\alpha^2}$; *quantum at rotation of electron bond* $= \alpha$.

$\alpha = x\left(10^{-15}\right)$ meters ; *from estimates of proton diameter*

$f_r = \sqrt{2}kQ_r^2 = \dfrac{Q_r^2}{\alpha^2} = \dfrac{Q_r^2}{\left[x\left(10^{-15}\right)\right]^2}$

$x^2 = \dfrac{Q^2}{f\left(10^{-15}\right)^2} = 0.2521421489$

$x = 0.5021375797$

$\alpha^2 = \left[x\left(10^{-15}\right)\right]^2 = 2.5214214895\ 10^{-31}$ meters²

$f_r = \sqrt{2}kQ_r^2 = \dfrac{Q_r^2}{\alpha^2} = \dfrac{Q_r^2}{\left[x\left(10^{-15}\right)\right]^2} = \dfrac{8.2938681799\ 10^{-16}\ m^2}{2.5214214895\ 10^{-31}\ m^2} = 3.2893620581\ 10^{15}$ Hz.

$\sqrt{2}\,k = \dfrac{1}{\alpha^2} = 3.9660\ 10^{30}\ \dfrac{units\ \alpha^2}{m^2}$

This establishes the actual diameter of the proton/electron as 1.0043 (10⁻¹⁵) meters.

Since "$f = \sqrt{2}\,k(\text{quantum radius})^2$" this value of "k" can be tested against the known root frequency (3.2893620584e15 Hz.) which also has a theoretically derived quantum-squared radius value of "8.2938681799e-16 meters²[284]." The approximate alpha value does not give an exact fit. However, an exact alpha value of "0.5021375797 (10⁻¹⁵) m" can be calculated from such an exact fit. Therefore, "≈(10⁻¹⁵)/ 2 =0.5021375797 (10⁻¹⁵) m=α."
The alpha space's exact value is:

   α=0.50214 (10⁻¹⁵) meters =radius of proton and 1/2 the shortest wavelength.

The exact value of alpha is actually "0.00428" greater than the estimated value of "(10⁻¹⁵)/ 2." By deriving so closely the alpha-space's known value as the radius of a proton from independent variables— the calculated Rydberg root quantum-squared and by using a derived formula for tension— the basic quantum would seem confirmed.

---

[284] We will hold to this calculation of the maximum atomic radial distance despite the fact that it is nearly 300 times that of the smallest atomic diameter molar-volume estimates. At the center of this disparity is the failure of math and science to recognize that empty space is four-dimensional, not three, and that a cubic measure of it is inaccurate. The atomic orbital convertes cubical volume to four-dimensional space. The orbital canexpand and contract volume while keeping the quantum radial constant. It establishes the orbital principle of subwaves. The quantum radial value is not the equivalent of "molar-volume" estimates of atomic diameter. The smallest atom diameter is Boron, at 1.98 Angstrom units. The root radius converted from four-dimensional "space" to three dimensional "space" gives a diameter of 1.766 Angstrom units. All quantum radial distances fit within the 4-dimensionally incorrect "molar-volume" atomic diameter. See *"The Insufficiency of the Determination of Atomic Radii by Molar Volume"* in Appendix

Frequency is not only the number of times by which the electron's elementary charge must multiplied by the nuclear capacitance, but frequency is also the number of times that the fundamental quantum "α^2" must be multiplied to produce the radial quantum-squared value:

$$f = \left(\frac{1}{n^2} - \frac{1}{n'^2}\right)\frac{c}{\lambda_r} = \sqrt{2}\, k\left(\frac{1}{n^2} - \frac{1}{n'^2}\right)Q_r^2 = \frac{d(Q_r^2)}{\alpha^2}$$

$$\text{quantum- squared orbital radial} = \left(\frac{1}{n^2} - \frac{1}{n'^2}\right)Q_r^2 = d(Q_r^2)$$

Frequency is the number of times "α^2" must be multiplied to reach the quantum^2 radial for any shell/subshell orbital.

The alpha space also generates the time quantum. There is a variance in time sustaining the alpha space[285]. This time variance is designated "$\Delta T = t_Q$; time variance=time quantum." This time-variance *cum* time-quantum restricts motion across the alpha space to the speed of light. Therefore " $c = \alpha / \Delta T$." The variance in time separates two time-incompatible points which must retain a distance of "α" by a time-force designated "F_t." That is, the quantum force sustaining the fundamental quantum is time force. Quantum space is generated by incompatible time values which sustain a space of separation by the force of their incompatibility. The actual formula is the following[286] :

$$\Delta T^2 = F_t^2 \alpha^2 \quad ; \quad \Delta T = (\text{time-variance} = \text{time-quantum}) \; ;$$
$$F_t = \text{time force sustaining separation}$$

$$F_Q^2 = F_t^2 = \frac{\alpha^2}{\Delta T^2} \quad ; \; \alpha = 0.50214\left(10^{-15}\right) m \; ; \; \Delta T = 1.675\left(10^{-24}\right) sec. \; ; \; F_t^2 = \frac{\Delta T^2}{\alpha^2} = \frac{1}{c^2} = \mu_0 \varepsilon_0$$

$$\frac{1}{c^2} = (\text{magnetic permiability of free space})(\text{electric permittivity}) = \mu_0 \varepsilon_0 \; ; \; \textit{Heaviside}$$

Time-force squared is the exact equivalent of the Heaviside formulation for magnetic-permeability/electric-permittivity of free space which allows an electromagnetic field to conduct force across that space. When time force and the electrodynamic formulation for free space is substituted for alpha squared in the tension formula, it produces the following:

$$\alpha^2 / 2 = \Delta T^2 / 2F_t^2 = \Delta T^2 / 2\mu_0 \varepsilon_0$$

$$f = \sqrt{2}\, kQ^2 = \frac{Q_r^2}{\alpha^2 / 2} = \frac{Q_r^2}{\Delta T^2 / 2\mu_0 \varepsilon_0}$$

$$\mu_0 = 4\pi\left(10^{-7}\right)\frac{\text{Newtons}}{\text{amps}^2}$$

$$f = \frac{eV(C_p)}{e} = \frac{2(\mu_0 \varepsilon_0)}{\Delta T^2}Q_r^2$$

$$\varepsilon_0 = 8.8541878176\left(10^{-12}\right)\frac{\text{Farads}}{\text{meter}}$$

Frequency defined as the nuclear multiple of the elementary charge is equal to free space magnetic-permeability/electric-permittivity as denominated by the time-quantum^2 *times* the radial distance^2. The electromagnetic field distributes force per amps^2 times capacitance per meter. The time-factor necessarily associated with capacitance is denominated by the time quantum, indicating that all time values must be quantum and not continuous. The electromagnetic distribution of force is over the shell/subshell radial distance^2.

---

[285] See *"The Theory of Time-Enforced Space"* .
[286] Ibid.

Nuclear multiplication of the elementary charge can be directly converted into an electromagnetic field value distributed over the radial distance[2]. The quantum geometric formulation for tension and frequency identifies the nuclear charge to be a function of an electromagnetic field. The multiplication of the electron's elementary charge is being conducted over an electromagnetic field as defined by radial distance.

The logic of Planck Capacitance is as follows: nuclear capacitance "C_N" increases as the electron moves closer to the nucleus. However, by the quantum geometric principle of exclusion, the quantum defined electron and the Euclidean defined nucleus cannot occupy the same spatial volume[287] but must have a quantum spatial field of separation equal to the quantum radius of the electron which is equal a radius equal to the radius of the proton (calculated at 0.50214 10^{-15} meters). This distance is the fundamental quantum designated as "the alpha space." A quantum squared ($Q^2 = \alpha^2$) envelope completely surrounds the nucleus within which and against which the nucleus vibrates. This envelope is required by the exclusion principle as a minimum space of quantum-squared separation ($Q^2 = \alpha^2$) between electron and proton. Planck Capacitance is the capacitance of the required quantum spatial separation[288] and is the capacitance available to the nucleus at the point of origin for its energy. Nuclear energy, in this sense, is the "heat" of the atom as a measure of the nucleus oscillating within the quantum spatial envelope.

Although Planck Capacitance can be derived from Bohr's electron voltage formulation, understanding its significance is dependent upon knowledge of the quantum dimension and a four-dimensional geometry. As interpreted by quantum geometry, Planck Capacitance identifies the means by which the energy in the nucleus can distribute electrons into subshells. Specifically, to acquire the potential energy state of any subshell orbit, the nucleus must multiply the charge of the electron by a factor equal to the frequency associated with the subshell. the radiation frequency associated with the orbit is equal to the electron voltage of the orbit *times* Planck Capacitance *divided by* the elementary charge of the electron:

(volts)(capacitance)=charge

$eV(C_p)$ =*charge nucleus must supply subshell to acquire subshell frequency.*

$$f = \frac{eV(C_p)}{e}$$ *frequency is the multiplication of the elementary charge.*

The multiplication of the elementary charge is a form of Piezoelectricity[289] by which mechanical force generates an electric field or current. That is, the nucleus must exert force upon the electron such that the charge of the electron is multiplied by a sufficient amount to acquire a subshell orbit. There must be force exchanged between the nucleus and the

---

[287] In no sense can the neutron be the merging of the electron and proton as generally taught by uncorrected nuclear physics.

[288] The alpha space radial distance of "0.5021375797 10^{-15} meters" ($Q^2 = \alpha^2$) giving a Planck Capacitance frequency of "1" is calculated from the approximate diameter of the proton as modified by the maximum distance for electron orbitals (as calculated upon wavelengths). The alpha space frequency of "1" then gives a radial distance of 28.8611 nm for the root . The "0.5021375797 10^{-15} meters" alpha space derives a fundamental time quantum "ΔT" value of 1.675 10^{-24} seconds. These are the most accurate determinations of alpha and delta T values in currency, as the valeus were derived by comparison of two independent data sets.

[289] http://en.wikipedia.org/wiki/Piezoelectricity

electron along the bond and only quantum-dimensional geometry identifies how this is possible. We can go no further with the analysis of the subshells, their energy states, and energy exchanges without revealing how the subshell schematic is constructed by the quantum dimension.

The Kink Tension Constant

Quantum geometry proposes that electron/ nuclear bond is an energy exchange mechanism not currently recognized by physics. It is proposed that the four-dimensional orbital wave is "string-like" in that in composes an acceleration/deceleration phase at the frequency of associated light. Quantum geometry further proposes that orbital potential energy is the Planck function of frequency ($PE = f h$). "$PE = f h$" is the potential energy of a tensioned string, a sting potential energy which is being sustained by orbital motion.

"$PE = f h$" is the potential energy of the electron bond rotated at 90° to the nuclear bond being sustained at the quantum squared or "string" distance. The rotated bond is kinked into curvature providing a wave path on a plane 90° to the plane of orbit. Potential energy is sustained and established by the energy of orbital motion across the wave path of the four-dimensional orbital. The electron orbital velocity required to sustain the potential energy of the root orbit ($f h = 2.17956e{-}18$ joules) is .516% of the speed of light or 1546818.61 meters per second. This is a small fraction of the energy available to the acceleration/ deceleration phase since terminal acceleration velocity can actually reach the speed of light.

In fact, the root radial length is established by the speed of light. Specifically, the root is the quantum squared orbital length at which the maximum acceleration across the wave path reaches the speed of light . Any greater quantum squared orbital length would produce a terminal acceleration velocity which would exceed the speed of light. The root is the electron's maximum orbital distance which can be sustained in four-dimensional space.

Acceleration/deceleration frequency is produced by kink tension— but "frequency" and "tension" are not the same energy. The orbital radius is established by a contraction/ expansion balance of forces[290] upon the kinks.

Nuclear Piezoelectrical kinetic energy suppresses the kinks which causes acceleration/deceleration at frequency. That is, suppression of kink tension operates like a tensioned string. While the frequency of vibration is propagated by tension, they are separate energy values.

The difference between frequency of vibration and string tension can be demonstrated by a string tension formula. The following formula is based upon the geometric conversions which accompany the tensioning of a string[291] :
`F =tension (force); k= tension constant; x_t = tensioned length; x_0 = untensioned length

$$F = k(x_t - x_0)^2 \quad \text{Dawson's Tensor}$$

Using the tensor, the following energy equation is determined:

Change in string tension due to string deflection $= k(\Delta x_t^2)$; m=string mass

$$E = (m)k(x_t - x_0)^2 \; k(\Delta x_t^2) = m(k^2)(x_t - x_0)^2(\Delta x_t^2)$$

---

[290] The actual contraction/ expansion forces will be examined at a later time.
[291] See "Dawson's Tensor" in Appendix.

This energy value can be set equal to the Newtonian energy value[292] of a vibrating string:

$$E = 8\, mf^2 d^2 \quad ; f = \text{frequency} \quad ; d = \text{string deflection}$$

A new equality,

$$\text{"} m(k^2)(x_t - x_0)^2 (\Delta x_t^2) = 8\, mf^2 d^2 \text{"}$$

can be used to prove that string frequency is always a function of tension and (tensioned) string length[293] :

$$f = \sqrt{2}\, k(x_t - x_0)$$

Frequency is a function of the tension constant times string length. An external force must exceed the tension constant in order to cause the string to vibrate at frequency. If the force is less than "k," frequency will not be output by the string. An external force greater than "k" is required for the string to produce kinetic energy.

Nonetheless, the string always possesses a potential energy determined by frequency. The string is always "tensioned" regardless of whether sufficient external force is applied to the string to cause it to vibrate. This vibration is the conversion of the strings "potential energy" (frequency as a function of tensioned length) to "kinetic energy" (actual vibration). Potential and kinetic energies operate differently with a tensioned string than do those same energy values with an orbiting satellite. The quantum geometric and the Bohr models of the atom are not and cannot be energy equivalent.

The fact that string potential energy is a function of frequency and frequency alone is shown by the formula for a string's impedance of an external energy wave :

$$Z = \text{impedance}; \quad F = \text{force of string}; \quad V = \text{max. velocity of string}; \quad A = \text{acceleration}$$

$$2\,t = \text{period of one vibration}$$

$$Z = \frac{F}{V} \quad ; \quad F = m(A) \quad ; \quad V = A(t)$$

$$Z = \frac{m(A)}{A(t)} = \frac{m}{t} \quad ; \quad \frac{1}{2\,t} = \text{frequency} \quad ; \quad \frac{1}{t} = 2f$$

$$Z = 2\,m\,f$$

Regardless of how far the string is deflected by the external wave (greater deflection, greater acceleration of the string), the energy absorbed by the string (impeded) is always only a function of string frequency and string mass. Amplitude of string deflection is irrelevant to energy impedance or absorption. The kinetic energy— the actual string vibrations — is not absorbed by the string. Only the potential energy is absorbed.

This mathematical fact is confirmed by natural observation of musical instrument strings. A string tuned to the frequency of a sound wave will begin vibrating sympathetically. The string will begin rebroadcasting the frequency. Regardless of how loud the initiating sound wave is, the rebroadcast wave always decays to the same volume; a function of the frequency-only "potential energy" of the string's impedance[294].

The potential energy value of the electron string is empirically confirmed as "string frequency

[292] ibid.

[293] The tension-energy formula mathematically derives the two most important known characteristics of tensioned strings: frequency as a function of length and tension alone; and subdivisions of string length which multiply freqency (sting harmonics). These derivations are the proof that the tension-energy formula is correct. See *"Dawson's Tensor"* in Appendix.

[294] Author's observation.

times Planck's Constant."[295] This, in turn, is the energy value of the electron orbital radius (string length). String frequency is shown to be a function of the tension constant "k" and string length. "K" can be demonstrated to be a threshold value which requires a certain nuclear energy before it is reached. That is, the electrons can reside in orbital shells at "sub-potential" energy levels with respect to the nucleus.

The nucleus must always multiply the elementary charge of the electron before the "k" threshold can actually be reached and light emitted. This is demonstrated by calculating "k" in electrodynamic units.

For the quantum electron string, all string lengths are under tension. Therefore," $x_0 = 0$ " and the formula is the following:

$$(x_t - x_0) = \text{electron orbital} = \left(\frac{1}{n^2} - \frac{1}{n'^2}\right)Q_r^2 = \left(\frac{1}{n^2} - \frac{1}{n'^2}\right)\left(\frac{91.14 \text{ nm}}{\pi}0.9927\right)^2$$

$$Q_r^2 = \left(\frac{91.14 \text{ nm}}{\pi}0.9927\right)^2 = 8.2939 \ 10^{-16} \text{ nm}^2$$

and

$$f = \left(\frac{1}{n^2} - \frac{1}{n'^2}\right)\frac{c}{\lambda_r} = \sqrt{2}\ k\left(\frac{1}{n^2} - \frac{1}{n'^2}\right)Q_r^2 \ ; \quad \textit{reduces to} \quad \frac{c}{\lambda_r} = \sqrt{2}\ k\ Q_r^2$$

The value of "k" can be calculated in electrodynamic units[296].

$$f_r = \frac{c}{\lambda_r} = eV_r\frac{e}{h} \ ; \ eV_r = 13.6 \text{ eV} \ ; \ \frac{e}{h} = \frac{e^2/h}{e} = \frac{\text{Planck Capacitance}}{e} = \frac{C_P}{e} \ ;$$

$$k = \frac{f_r}{\sqrt{2}\ Q_r^2} = \frac{eV_r(e)}{\sqrt{2}\ Q_r^2(h)} = \frac{eV_r C_P}{\sqrt{2}\ Q_r^2\ e} = \frac{1}{\sqrt{2}\ Q_r^2}\left(\frac{\Delta e}{e}\right) = 2.8044 \ 10^{30} \ (\text{units } e)\big/\text{m}^2$$

k is a function of increase in electron charge per meter squared of quantum^2 space

$$k = 2.8044 \ 10^{30} \ (\text{units } e)\big/\text{m}^2$$

$$f = \sqrt{2}\ k\ d(Q^2) \qquad \textit{(from Dawson's Tensor)}$$

frequency $= \sqrt{2}\ (\ 2.8044\ 10^{30} \ (\text{units } e)\big/\text{m}^2\ (d(Q^2)) = 3.9660 \ 10^{30} \ (\text{units } e)\big/\text{m}^2\ (d(Q^2))$

This value of "k" does, in fact, convert orbital quantum-squared distance in the above formula to the associated frequency as has been demonstrated

For any particular electron quantum-squared differential, the equality " $f = \sqrt{2}\ k\big(d(Q^2)\ \big)$ " can be resolved by the following:

$$eV_{n/n'} = \text{electron volts for differential} = \left(\frac{1}{n^2} - \frac{1}{n'^2}\right)13.6037 \text{ eV}$$

$f_{n/n'} = \text{frequency differential}$; $C_P = \text{Planck Capacitance} = 38.74 \ \mu\text{F}$

$$f_{n/n'} = \frac{(eV_{n/n'})(C_P)}{e} \ ; \ \textit{Resolves to a multiple of the elementary charge.}$$

Frequency is always a multiple, by the attached proton, of the electron's elementary charge.

---

[295] See *"The Drop in Temperature by N-Irradiation of Cotton and the derivation of Planck's Constant "* in Appendix.

[296] For full equations, see *"The Electron Field and Electron Capacitance"* in Appendix.

The multiple is determined by the electron voltage of the orbit and Planck Capacitance which is a constant of approximately 38.74 micro Farads. To prove the equality I give the example of the highest Rydberg frequency in the "Balmer Series" visible light spectrum (388.9 nm wavelength ; n=2, n'=8):

$\lambda =$ **388.9 nanometers**

$$eV = \left(\frac{1}{2^2} - \frac{1}{8^2}\right) 13.6037134826 \text{ eV} = 3.1883703342 \text{ eV} \quad ; \quad C_p = 0.038740461 \text{ Farads}$$

$$f = \frac{(3.1883703342 \text{ eV})(0.038740461 \text{ Farads})}{1.60217733 \ 10^{-19} \text{ coulombs}} = 7.7094423243 \ \left(10^{14}\right) \text{ Hz}$$

$$f = \frac{c}{\lambda} = \frac{299792458 \text{ m/s}}{3.88864 \ 10^{-7} \text{ m}} = 7.7094423243 \ \left(10^{14}\right) \text{ Hz}$$

It can be seen by the above that frequency, when calculated mechanically, (speed of light divided by wavelength) gives the same result as frequency calculated electrodynamically (electron volts times Planck Capacitance divided by elementary charge). Since "volts times capacitance equals charge" the resultant charge is greater than the elementary charge by a factor which can be determined by dividing it by the elementary charge. Frequency is always a multiple of the electron's elementary charge and is determined by the electron voltage of the orbit.

There is incontrovertible evidence that the tension constant "k" is the multiple of the electron charge by the nucleus. The capacity of the nucleus to multiply the elementary charge of the electron is proven by the negative radiation studies of hydrogen-bonded organic molecules[297].

---

[297] See "*The Drop in Temperature by N-Irradiation ....*" *op. cit.*

The Absorption of Light Energy by Kink Impedance
and Determination of Root Radius and Molar Atomic Diameter

The amount of energy which can be absorbed by the nucleus from impedance of light waves is restricted to "$f(h)$" as determined by Max Planck's black body studies. Acceleration phase energy must be partially dedicated to rebalancing kink tensions lost during acquisition of energy by light-pressure deceleration. This remainders the Planck energy value "$f(h)$":

$f(h)$+(deceleration energy) =(acceleration energy)

This is not an intuitively apparent formulation because it implies that the *positive value* of deceleration energy for the atomic structure is a quantum negation factor determined to be the following:

Let "x"= the positive energy factor for deceleration energy

$$f(h) + x\left(2mf^2\,\theta^2 Q^2\right) = 2mf^2\,\theta^2 Q^2$$

$$x\left(2mf^2\,\theta^2 Q^2\right) = 2mf^2\,\theta^2 Q^2 - f(h) = \left(1 - \frac{f(h)}{2mf^2\,\theta^2 Q^2}\right) 2mf^2\,\theta^2 Q^2$$

$$x = \left(1 - \frac{f(h)}{2mf^2\,\theta^2 Q^2}\right) ; \quad \text{limit on } x \rightarrow 1-1=0 \ ; \quad \lim x; \ 1 = \frac{f(h)}{2mf^2\,\theta^2 Q^2}$$

$$\lim x; \quad f(h) = 2mf^2\,\theta^2 Q^2 \ ; \quad h = 2mf\theta^2 Q^2 \ ; \quad \theta^2 = \frac{h}{2mfQ^2} = \frac{3.6369480932\ 10^{-4}}{fQ^2}$$

$$\lim x; \ \theta = \frac{0.0190707842}{Q\sqrt{f}}$$

For root orbital $\lim x; \ \theta \geq 0.01155$ radians $= 0.66°$

For the root orbital, there must be a 0.66° angle of deflection to impede light radiation.

For shells/subshells:

$$\lim x; \ 1 = \frac{f(h)}{2mf^2\,\theta^2 Q^2}$$

For light pressure to impede the kinks and add energy to the system their must be enough energy in the light wave to establish a minimum angle of acceleration/deceleration. While it is a very small angle for the root orbital, it becomes increasingly greater for the shell/ subshell distributions of the orbitals. The last orbital which can impede radiation is the "p" subshell in the second shell at 12.367 micrometers. That orbital wavelength requires the full 90° of the deceleration phase to impede radiation. The minimum angle formula is compatible with Planck's "packets of energy[298] " for the full range of the shell/subshell orbital distribution except for one.

The lowest energy subshell and the only subshell contained in the first shell requires a greater impedance angle than is available to the deceleration phase. This is significant because angle of impedance is clearly related to orbital harmonics— the capacity of lower order orbitals to impede higher order frequencies. The "s" subshell in the first shell has

---

[298] Planck, Max, "On the Law of Distribution of Energy in the Normal Spectrum". Annalen der Physik, vol. 4, p. 553 ff (1901)

nothing below it to "reach" its frequency. That the maximum angle of impedance should be established between the first and second shells, at the point below which harmonic impedance can no longer occur, demonstrates that angel of impedance establishes harmonics.

The amount of energy either being put out as light or being reabsorbed as kink tension *plus* Planck energy is determined by the angle "θ." The energy exchange is determined by the portion of the transverse wave dedicated to acceleration/ deceleration. The portion of the transverse wave equals "$2\theta/\pi$. " The possibilities run from "$\theta=\pi/2$ for $2\theta/\pi=1$ " which establishes acceleration/deceleration across the full transverse wave and maximum energy exchange to "$\theta=0$ for $2\theta/\pi=0$" for no acceleration/deceleration across the transverse wave and no energy exchange.

The light wavelength output by acceleration/deceleration across the tensioned kinks of the "root" orbital is 91.14 nanometers. It is this acceleration/deceleration across the kinked transverse wave which establishes the root orbital as the maximum possible three dimensional electron orbital. The root orbital is the orbital distance at which kinked tension produces a terminal acceleration velocity of the speed of light, Any greater orbital distance would produce a kink tension which would accelerate the electron past the speed of light:

terminal velocity$=v +Q\pi(f)^{299}$; v=orbital velocity; Q=orbital radius=kinked line length

$$\text{Terminal velocity (root)} = c = v + \pi Q \frac{c}{\lambda_r}$$

Frequency increases with radius. Root has max. "f"

$$\pi Q \frac{c}{\lambda} = c - v \quad ; \quad Q = \frac{(1 - v/c)}{\pi}\lambda_r$$

v=1546817.844 m/ sec.;   Q= 28.8611 nm (four dimensional value)

Q=8.8528560629e-11 (three dimensional value);

1.7705712126e-10 molar atomic diameter

The electron voltage for the three-dimensional "root" orbital is 13.6 eV which outputs 91.14 nm light.   91.14 nm is in the ultraviolet range. In contrast, the de Broglie "wave state" for the electron's momentum, under 13.6 eV, is "0.3326 nm" which is in the x-ray range[300].

The frequency of the electron's momentum "wave state" is 274 times the frequency of the associated light. Put another way, to supply the x-ray frequency by nuclear ejection of the electron (Beta/gamma emissions) rather than by electron velocity (momentum), would require 274 *times* 13.6 eV or 3.728 keV. X-ray and gamma rays are often of the same range of wavelengths, yet there is a great difference in the energy between the two[301] . It is an uncomfortable fact which science has yet to explain. X-ray is a function of de Broglie's electron momentum and has a lower energy than frequency-equivalent gamma rays which are produced by electron ejection through nuclear process.

---

[299] From Newtonian laws of acceleration.
[300] See *"De Broglie 's Formula for ' Wavelengths' of Electron"*   in Chapter 1.
[301] *http://en.wikipedia.org/wiki/Gamma_ray*

Calculating the Half-Spin Time Constants

Let Q=root radial; v=initial velocity

Terminal velocity of acceleration=c-v

terminal velocity=2d/t ; t=1/2f ; 1/t=2f ; 2d/t= (2d)(2f)= 4df

d=$Q\pi$/4 (the quarter of a circumference of diameter "Q")

terminal velocity=4df = (4Qf π)/4=Qf π=c-v

$$f = \frac{c}{\lambda_r}$$

$$\frac{\pi Q}{\lambda_r} c = c - v$$

$$\frac{\pi Q}{\lambda_r} = 1 - \frac{v}{c}$$

$$Q = \left(1 - \frac{v}{c}\right)\frac{\lambda_r}{\pi}$$

$$\frac{c}{\lambda_r}h = m\frac{v^2}{2}$$

$$v^2 = \frac{2c}{\lambda_r m}h$$

$$v = \sqrt{\frac{2c}{\lambda_r m}h}$$

v=2187531.85724669 m/sec.

v/ c=0.0072968208

$$Q = (1 - 0.0072968208)\frac{\lambda_r}{\pi} = 0.9927031792\frac{\lambda_r}{\pi}$$

$Q = 2.8799076686 \ 10^{-8}$ meters

Q=2.8799076686e-8

Q^2=8.2938681799e-16

$$d(Q^2) = \left(\frac{1}{n^2} - \frac{1}{n'^2}\right)Q^2 = \left(\frac{1}{n^2} - \frac{1}{n'^2}\right)(8.2938681799e-16)$$

Converted to wave phase time

$$v^2 = \frac{2\left(1/n^2 - 1/n'^2\right)\left(c/\lambda_r\right)}{m}h$$

$$v = \sqrt{\frac{2\left(1/n^2 - 1/n'^2\right)\left(c/\lambda_r\right)}{m}h}$$

$$d = \frac{(\text{quantum radius})\pi}{2} \quad NOTE; \ d \ in \ this \ case \ is \ semi \text{-} circumference$$

$$d = \frac{\sqrt{\left(1/n^2 - 1/n'^2\right)\left(Q^2\right)}}{2}\pi$$

wave phase time=t=d/v

$$\frac{d}{v} = \frac{\dfrac{\sqrt{\left(1/n^2 - 1/n'^2\right)\left(Q^2\right)}}{2}\pi}{\sqrt{\dfrac{2\left(1/n^2 - 1/n'^2\right)\left(c/\lambda_r\right)}{m}}h} = \sqrt{\frac{\left(1/n^2 - 1/n'^2\right)\left(Q^2\right)m\lambda_r}{2\left(1/n^2 - 1/n'^2\right)(c)h}}\;\frac{\pi}{2}$$

$$\frac{d}{v} = \sqrt{\frac{\left(Q^2\right)m\lambda_r}{2(c)h}}\;\frac{\pi}{2} = \sqrt{\frac{m\lambda_r}{2(c)h}}\;\frac{Q\pi}{2} = 2.0679691463e\text{-}14 \text{ seconds}$$

t=2.0679691463e-14 sec. *a constant for all shell/subshell orbitals.*

Bohr magneton[302] for magnetic moment of spin=μ_B =9.27400915 10^{-24} Joules/ t

$E_{(1/2)\,spin} = t'\,(\mu_B) = 1.9178364785$ 10^{-37} Joules

e $V_{(1/2)\,spin}$ = $(E_{(1/2)\,spin})$/ (e) = (1.9178364785 10^{-37} Joules) / (1.60217733 10^{-19} colmb.)

e $V_{(1/2)\,spin}$=1.1970188584 10^{-18} eV

Both energy and electron voltage are constants for most[303] orbitals because t' is a constant for most orbitals.

---

[302] hyperphysics.phy-astr.gsu.edu/Hbase/quantum/orbmag.html
[303] This is true for all shell/subshells except the "3d," "2s," "2p" and "1s." For these lower order subshells, the wave-phase time is shorter than the acceleration/deceleration time factor. Acceleration/deceleration time will establish orbital velocity and produce a higher "1/2 spin" energy. It will take a greater " $\Delta e V_{dblt}$ " producing wider doublets and less capacity for subshell variance to allow multiple electrons. This factor should be considered in establishing base-state models.

A-8: Dawson's Tensor for String Frequency
The force/ frequency conversion for quantum strings

d=deflection of string

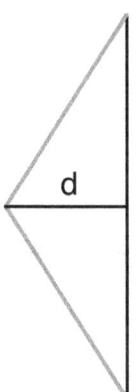

The Standard Energy Formula for a Vibrating String
$t =$ time acceleration phase ; average velocity $= \overline{V} = \dfrac{d}{t}$; $d =$ deflection of string
frequency (per second) $= f = \dfrac{1}{2t}$;
final velocity $= V = 2\overline{V} = \dfrac{2d}{t}$; acceleration $= A = \dfrac{V}{t} = \dfrac{2d}{t^2}$; energy $= E =$ (Force)d $=$ Fd ;
$F =$ (mass)A
$E = mAd = m\dfrac{2d}{t^2}d = m\dfrac{2d^2}{t^2}$; $\dfrac{1}{t} = 2f$; $\dfrac{1}{t^2} = 4f^2$
$E = m\left(2d^2\right)\left(4f^2\right) = 8mf^2 d^2$

An Alternative Energy Formula
(Both Energy and Frequency are a Function of String Tension)

The conventional tensioned-string energy formula ($E = 8\, mf^2\, d^2$) is graphed two dimensionally. The string is tensioned along one dimension and the deflection along a second intersecting dimension.

But what happens when string energy is three dimensional, as with the quantum electron bond/string? In this model, the string is two dimensional with the deflection along a third intersecting dimension. Thus the quantum wave is three dimensional and the conventional energy formula is inadequate for it.

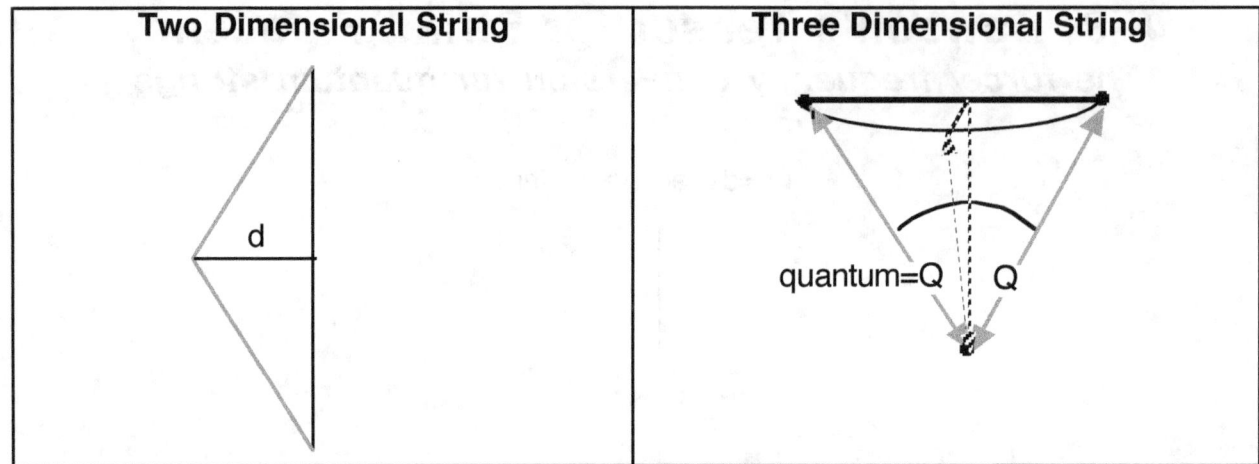

Two Dimensional String	Three Dimensional String
d	quantum=Q Q

A string-energy formula can be derived which applies to both types of strings and, therefore, is adequate to the three dimensional wave.

Definition of tension: Tension in any geometric solid is a force of "stretch" along one geometric dimensional axis which produces counter balancing forces of contraction/expansion along the other two dimensional axii of the solid:

$$F_s = \text{force of stretch} \quad ; \quad F_{c/e} = \text{force of contraction / expansion}$$

$$F_s^2 = F_{c/e}$$

Tension is the counter balancing forces of expansion and contraction at every point around the diameter of the string.

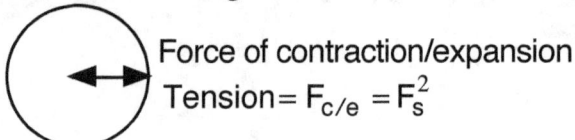

Force of contraction/expansion

$$\text{Tension} = F_{c/e} = F_s^2$$

$r = $ radius of string ; $x = $ initial length of string ; $x + \Delta x = $ stretch ;

$x\left(\pi r^2\right) = $ volume of string ; $y = \% $ contraction of radius

$$x\left(\pi r^2\right) = (x + \Delta x)\left(\pi[y\, r]^2\right) = y^2(x + \Delta x)\left(\pi r^2\right)$$

$$x\left(\pi r^2\right) = y^2(x + \Delta x)\left(\pi r^2\right)$$

$$y^2 = \frac{x}{(x + \Delta x)}$$

$$y = \sqrt{\frac{x}{(x + \Delta x)}}$$

The amount of compression of the string radius is a function of the inverse of the amount of stretch. For example, if the string is stretched to twice its original length the radius is compressed to a function of the inverse of 2 or a function of "1/2."

284

The function of the inverse of the stretch is the square root. Radial compression is the square root of the inverse of the stretch.

The amount of force is required to compress the radius over a distance is equal to the square of the force required to stretch the string over the same distance. That is, the force of radial compression equals the square of the force of the stretch:

$$\sqrt{F_{c/e}} = F_s$$

$$F_{c/e} = F_s^2$$

$$Tension = F_{c/e} = F_s^2$$

Energy=(mass)(tension)(change in tension by deflection)

The Logic of the Equation

The universal formula for energy is "Energy = (Force)(Distance)." Newton's standard definition of energy ($E = (m)(v^2/2)$) can be derived from it. The potential energy in both gravitational and electromagnetic fields at various distance from the field source can be expressed by it.

Energy as a function of string tension is a variation upon "E=(Force) (Distance)." A string of greater tension requires more "force" to deflect. (Force is measured in newtons=1 kg accelerated to 1 meter per second.) The greater the distance of deflection, the greater the tension increases and the greater the force required for further increase distance. Force expressed over distance.

The mathematical description of force by string tensions identifies a change in acceleration rate which eliminates distance of deflection as a time factor. Thus, all deflections of a tensioned string have the same frequency of vibration.

let F =tension (force)
let k= tension constant
let x_t = length under tension
let x_0 = non – tensioned length
let d=deflection of string
let x_t / n= subdivision of string
let m= mass of string

Tension Equation for String

Tension is the square of the proportion of the tensioned string which is stretched *times* the tension constant.

$$F = k\left[(x_t)\left(\frac{x_t - x_0}{x_t} \right) \right]^2 = k(x_t - x_0)^2$$

Change in tension equals the square of additional stretch due to deflection times the tension constant.

Change in string tension = $k\left(\Delta x_t^2\right)$

The change in string tension is a function of the additional stretch caused by deflection. If a subdivision of the string is being deflected (as with a musical instrument string), then the additional stretch will also be a subdivision.   The "n" value is the subdivisional operator.

The energy equals string tension *times* change in tension *times* mass.

$$E = (m)k(x_t - x_0)^2\, k\left(\Delta x_t^2\right) = m\left(k^2\right)(x_t - x_0)^2\left(\Delta x_t^2\right)$$

Proof that Frequency of String vibration is Related to Tension Only
(Dawson's Tensor)

The alternative energy formula above can provide mathematical proof that time (and frequency) of string vibration is related to tension and not weight or deflection of the string. This can shown by setting the alternative formula equal to the conventional energy formula for a vibrating string:

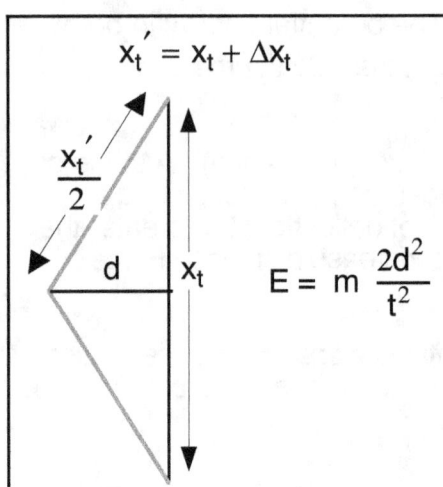

$x_t' = x_t + \Delta x_t$

$$E = m\,\frac{2d^2}{t^2}$$

$$m\left(k^2\right)(x_t - x_0)^2\left(\Delta x_t^2\right) = m\,\frac{2d^2}{t^2} \quad ;$$

$$\left(k^2\right)(x_t - x_0)^2\left(\Delta x_t^2\right) = \frac{2d^2}{t^2}$$

Divided both sides of the equation by "m."
Mass drops out.

Find "d^2" as a function of "Δx_t^2."

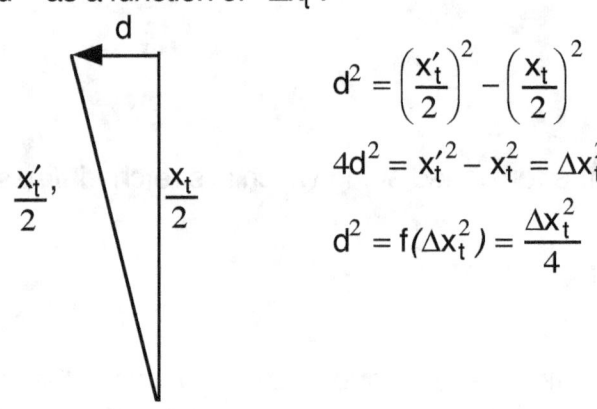

$$d^2 = \left(\frac{x_t'}{2}\right)^2 - \left(\frac{x_t}{2}\right)^2$$

$$4d^2 = x_t'^2 - x_t^2 = \Delta x_t^2$$

$$d^2 = f(\Delta x_t^2) = \frac{\Delta x_t^2}{4}$$

Substitute "$f(\Delta x_t^2)$" for "d^2."

$$k^2(x_t - x_0)^2(\Delta x_t^2) = \frac{2d^2}{t^2} = \frac{2\Delta x_t^2}{4t^2} = \frac{\Delta x_t^2}{2t^2}$$

Divide both sides by "Δx_t^2."

$$k^2(x_t - x_0)^2 = \frac{1}{2t^2}$$

String deflection drops out as factor.

$$t^2 = \frac{1}{2k^2(x_t - x_0)^2}$$

$$t = \frac{1}{\sqrt{2}\, k(x_t - x_0)}$$

Time of Vibration is Indirectly Proportional to Tension Alone

Time of vibration is indirectly proportional to string stretch. From the above it can be seen that as string stretch increases —as "$x_t - x_0$" gets bigger— time decreases proportionally. Weight (mass) and deflection of the string have are eliminated as a factor in determining the time of vibration. The alternative energy equation identifies the characteristics of string vibration as they are experimentally known to be.

The Alternative Energy Formula Identifies String Harmonics

The alternative energy formula shows why string subdivision increases frequency.

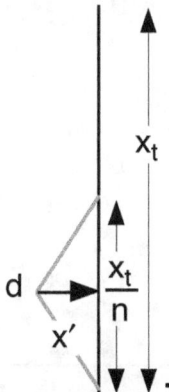

The subdivided tensioned string.

In the above illustration "x_t/n" is an exact subdivision of tensioned string "x_t."

First, we must establish the tension of the subdivision. String tension is the square of the portion of the string which has been stretched. The stretched proportion of "x_t/n" must equal the stretched proportion of the whole string. This requirement will mathematically demonstrate that string tension for the subdivision is the same as string tension for the whole:

   y = string tension   ;

   let z = factor making subdivisional stretch proportional to whole string stretch.

$$y_t = k\left[(x_t)\left(\frac{x_t - x_0}{x_t}\right)\right]^2 = k(x_t - x_0)^2$$

$$y_{t/n} = k\left[\left(\frac{x_t}{n}\right)\left(\frac{z(x_t - x_0)}{x_t/n}\right)\right]^2 = k\left[\left(\frac{x_t}{n}\right)\frac{n(z)\ (x_t - x_0)}{x_t}\right]^2 = k(z)(x_t - x_0)^2$$

proportion of subdivison under tension

must equal proportion of whole under tension.

$$\left(\frac{x_t}{n}\right)\left(\frac{z(x_t - x_0)}{x_t/n}\right) = (x_t)\left(\frac{x_t - x_0}{x_t}\right) \ ;$$

$$z(x_t - x_0) = (x_t - x_0)$$

$$z = 1$$

$$y_{t/n} = k(x_t - x_0)^2$$

$$y_{t/n} = y_t$$

String tension for the subdivision is the same as string tension for the whole string.

The energy equation for the subdivision is the same as the energy equation for the whole with the exception that subdivisional mass is only that portion of the whole string mass which is vibrating (mass subdivided or "m/ n").

$$E_{t/n} = \left(\frac{m}{n}\right)k(x_t - x_0)^2 \ k\left(\Delta x_t^2\right) = \left(\frac{m}{n}\right)(k^2)(x_t - x_0)^2\left(\Delta x_t^2\right) = \left(\frac{m}{n}\right)\frac{2d^2}{t^2}$$

$$(k^2)(x_t - x_0)^2\left(\Delta x_t^2\right) = \frac{2d^2}{t^2}$$

Convert "d^2" to "$f(\Delta x_t^2)$"

$$x'^2 - \frac{x_t^2}{2^2 n^2} = \frac{\Delta x_t^2}{2^2 n^2}$$

$$d^2 = f(\Delta x_t^2) = \frac{\Delta x_t^2}{2^2 n^2}$$

Substitute "$f(\Delta x_t^2)$" for "d^2" in energy equality:

$$\left(k^2\right)\left(x_t - x_0\right)^2\left(\Delta x_t^2\right) = \frac{2\,\dfrac{\Delta x_t^2}{2^2 n^2}}{t^2} = \frac{\Delta x_t^2}{2\,n^2\,t^2}$$

Eliminate "Δx_t^2" from both sides of equation by division.

$$\left(k^2\right)\left(x_t - x_0\right)^2 = \frac{1}{2\,n^2\,t^2}$$

Find for time.

$$n^2\,t^2 = \frac{1}{2\left(k^2\right)\left(x_t - x_0\right)^2}$$

$$(n)(t) = \frac{1}{\sqrt{2}\,(k)\left(x_t - x_0\right)}$$

The time value for the whole string is "$\dfrac{1}{\sqrt{2}\,(k)(x_t - x_0)}$."

Let t=time value for whole string
Let t_n=time value for subdivision "n."

$$(n)(t_n) = \frac{1}{\sqrt{2}\,(k)\left(x_t - x_0\right)}$$

$$t = \frac{1}{\sqrt{2}\,(k)\left(x_t - x_0\right)}$$

$$t_n = \frac{t}{n}$$

The time value of the subdivision is the time value of the whole string divided by "n." Frequency is the inverse of "t" *times* "2", The frequency of subdivisional vibration, therefore, is "n" times the frequency of whole string vibration. This is the mathematical proof of the principle of string harmonics whereby subdivision of a tensioned string increases frequency by the factor of subdivision.

The Quantum Conversion to the Energy Formula

The alternative energy formula shows that vibrational frequency is not directly proportional to string tension. This is counter intuitive, but completely consistent with the laws of physics. One would think that, if the string tension were doubled, the frequency of vibration should also double. In fact, frequency of vibration only increases by the square root of "2."

Why is this so? String tension is force and force is defined by acceleration. Acceleration operates by the square of time (t^2). String tension, therefore, is mathematically related to the square of time and the actual time value is determined by the square root of tension.

Since tension is the square of string stretch, time/frequency of vibration is determined by stretch alone (the square root of the square of stretch).

This is significant to the quantum string. The Rydberg formula for light radiation emitted by a single electron bond/string identifies frequency as a function of the quantum squared. Since ALL string lengths with the quantum are under tension, string stretch equals string length. String length is the quantum squared and tension is the square of the quantum squared (quantum to the fourth power). Frequency of vibration is a function of the square root of tension which is the quantum squared.

for quantum "$x_0 = 0$." All lengths are under tension.

$$x_t = Q^2$$

$$t = \frac{1}{\sqrt{2}(k)(x_t - x_0)} \quad ; \text{ from alternative energy formula}$$

$$t = \frac{1}{\sqrt{2}(k)(Q^2 - 0)} \quad ; \quad t = \frac{1}{f}$$

$$f = \sqrt{2}(k)(Q^2)$$

The quantum string does not vibrate two dimensionally as does the conventional Euclidean string (i.e. a musical instrument string). It does not vibrate along the plane made by the intersection of the string length and the line of deflection.

Two Dimensional String

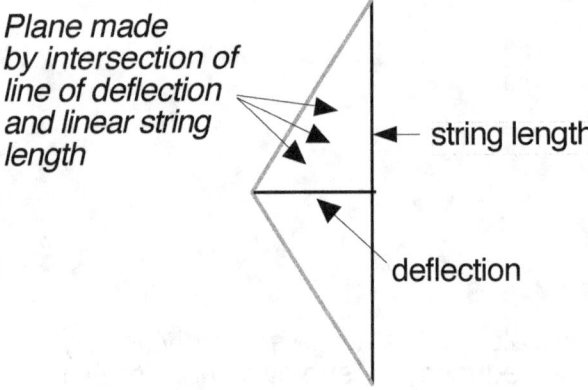

Plane made by intersection of line of deflection and linear string length

string length

deflection

In contrast, the quantum string vibrates three dimensionally. Vibration is composed of a two-dimensional plane in motion across the line of deflection.

Motion of string: y plane across x plane

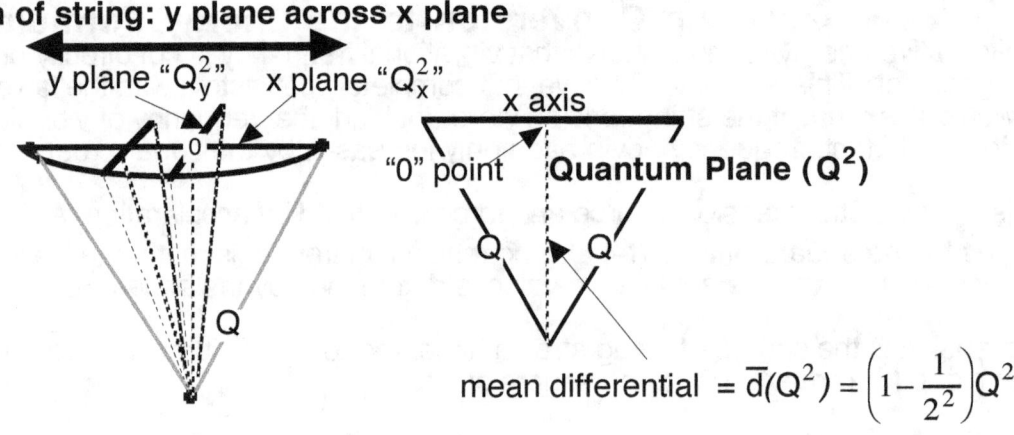

y plane "Q_y^2" x plane "Q_x^2"

x axis

"0" point **Quantum Plane (Q^2)**

Q Q

0

Q

mean differential $= \overline{d}(Q^2) = \left(1 - \frac{1}{2^2}\right)Q^2$

290

The Quantum Squared is not a "Square Quantum" Unit of Area

There is no such thing as a "square quantum" unit of area, that is, a unit of area defined by intersecting quantum dimensional axii. Multiple intersecting quantum dimensional axii can not exist, since there is only one quantum dimension.

The quantum area is constructed by projecting two equal quantums from a single point on a 60° angle and integrating the differentiated quantum squared across the secant of the angle. Such integration requires the secant must be a Euclidean line (x axis) composed of a continuum of points. Each point along the x axis composes a quantum with the point of origin.

The area of the triangle established by the quantum squared does not equal "Q^2." Quantum area is established as the summation of all quantums across the x axis. All of these quantums are differentiations of the quantum squared for all values of "x."

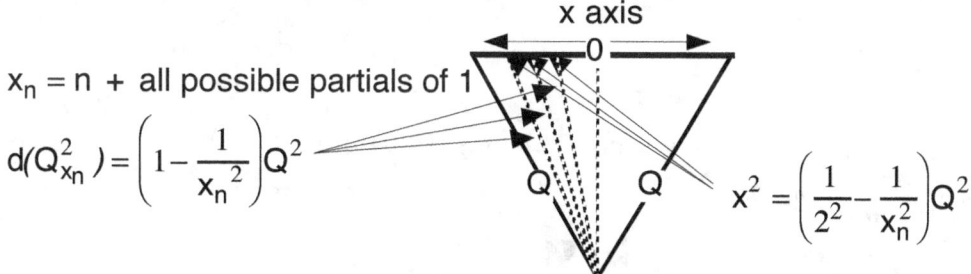

$$x_n = n + \text{ all possible partials of } 1$$

$$d(Q^2_{x_n}) = \left(1 - \frac{1}{x_n^{\,2}}\right)Q^2$$

$$x^2 = \left(\frac{1}{2^2} - \frac{1}{x_n^2}\right)Q^2$$

Quantum area is constructed by the integration of the differentiated quantum squared across the x axis. The quantum squared is differentiated by negation of subdivision. :

$$d(Q^2_{x_n}) = \left(1 - \frac{1}{x_n^2}\right)Q^2 \quad ; x_n = n + \text{ any possible partial of "1."}$$

The actual quantum axis is the mean of the quantum squared differential s. The mean of the quantum squared differentials is the following:

$$\text{mean} = \overline{d}(Q^2) = \left(1 - \frac{1}{2^2}\right)Q^2 \text{ (see illustration above)}$$

The point of intersection of the x axis by the mean differential (quantum axis) determines the "0" point for the x axis. The x axis is bidirectional since the "0" point of the x axis bisects the secant (see illustration). The value of "x" increases to both the right and the left of the "0" point. For this reason, the quantum has a value of "2x" (Q=2x).

The area of the quantum plane is not "Q^2." Area equals the mean differential *times* the x axis. The length of the x axis equals "2x." The area of the triangle subscribing the area of the quantum squared equals 1/2 *times* base (2x) *times* height (square root of mean).

$$\text{Area} = \frac{1}{2}(2\,x)\sqrt{\overline{d}(Q^2)} = (x)\sqrt{\overline{d}(Q^2)} = \frac{Q}{2}\sqrt{1 - \frac{1}{2^2}}\,Q = \frac{\sqrt{3}}{4}Q^2$$

The contribution of the quantum to area cannot be determined by taking the square root of the area since "Q^2" is not and cannot be a "square quantum." The contribution of "Q" must be determined by the derivative of area. The derivative works because area is constructed by integrating the differentiated quantum across the Euclidean x axis.

$$\text{mean differential} = \sqrt{\left(1 - \frac{1}{2^2}\right)Q^2} = \sqrt{.75Q^2} = 0.8660254038(Q)$$

$$D(area) = D\left(\frac{\sqrt{3}}{4}Q^2\right) = D(0.4330127019\ Q^2)$$

$$D(0.4330127019\ Q^2) = (0.4330127019)(2Q) = 0.8660254038\ Q$$
$$\text{mean differential} = D(area)$$

The formula that the mean quantum differential equals the derivative of area for the quantum squared is not accurate with respect to the changing area of the "y" quantum plane:

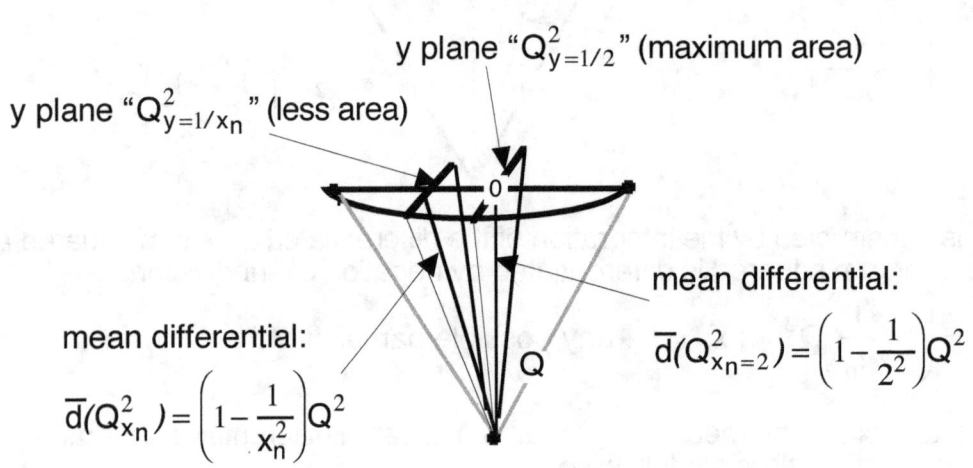

y plane "$Q^2_{y=1/2}$" (maximum area)

y plane "$Q^2_{y=1/x_n}$" (less area)

mean differential:
$$\overline{d}(Q^2_{x_n=2}) = \left(1 - \frac{1}{2^2}\right)Q^2$$

mean differential:
$$\overline{d}(Q^2_{x_n}) = \left(1 - \frac{1}{x_n^2}\right)Q^2$$

The Differential Equations for Area of "Y" Plane

$$y = \frac{1}{x_n} \ ; \ \text{Mean differential of y plane} = \sqrt{\overline{d}(Q^2_y)}$$

$$area = \left(\frac{1}{2}\right)(2y)\sqrt{\overline{d}(Q^2_y)} = y\sqrt{\overline{d}(Q^2_y)}$$

$$\sqrt{\overline{d}(Q^2_y)} = \sqrt{\left(1 - \frac{1}{x_n^2}\right)Q^2} = \sqrt{\frac{x_n^2-1}{x_n^2}}\ Q = \frac{\sqrt{x_n^2-1}}{x_n}Q$$

$$\sqrt{\overline{d}(Q^2_y)} = \frac{D(area)}{D(2y)} = \frac{d(area)/d(Q)}{d(2y)/d(Q)} = \frac{d(area)}{d(2y)} = \frac{\left(2\sqrt{x_n^2-1}/x_n^2\right)Q}{2/x_n} = \frac{\sqrt{x_n^2-1}\ Q}{x_n}$$

Area as a derivative of $d(Q_{x_n}^2)$

$$\text{area} = \frac{\sqrt{x_n^2 - 1}}{x_n^2} Q^2$$

$$D(\text{area}) = D\left(\frac{\sqrt{x_n^2 - 1}}{x_n^2} Q^2\right) = \frac{2\sqrt{x_n^2 - 1}}{x_n^2} Q \quad (\text{"Q" is the variable})$$

$$\frac{d(\text{area})}{d(2y)} = \frac{2\sqrt{x_n^2 - 1}\, Q}{x_n} \quad (\text{"Q" is the variable})$$

$$d(\text{area}) = \frac{\sqrt{x_n^2 - 1}\, Q}{x_n} d(2y) = \frac{\sqrt{x_n^2 - 1}\, Q}{x_n} d\left(\frac{2Q}{x_n}\right)$$

$$d\left(\frac{2Q}{x_n}\right) = -\frac{2Q}{x_n^2} d(x_n) \quad (\text{"}x_n\text{" is the variable})$$

$$d(\text{area}) = -\frac{2\sqrt{x_n^2 - 1}\, Q^2}{x_n^3} d(x_n) \quad (\text{"}x_n\text{" is the variable})$$

$$d\left(Q^2 - \frac{Q^2}{x_n^2}\right) = \frac{2Q^2}{x_n^3} d(x_n) \quad (\text{"}x_n\text{" is the variable})$$

$$d(\text{area}) = d\left(Q^2 - \frac{Q^2}{x_n^2}\right)\left(-\sqrt{x_n^2 - 1}\right) \quad (\text{"}x_n\text{" is the variable})$$

$$\frac{d(\text{area})}{d\left(Q^2 - \frac{Q^2}{x_n^2}\right)} = \frac{d(\text{area})}{d(Q_{x_n}^2)} = -\sqrt{x_n^2 - 1} \quad (\text{"}x_n\text{" is the variable})$$

$$\text{area} = \int -\sqrt{x_n^2 - 1}\, Q\, d(Q_{x_n}^2) \quad (\text{"}x_n\text{" is the variable})$$

Vibration of the Quantum String Proceeds by Negation

The "y" plane above (Q_y^2) is in motion across the "x" axis. The quantum area described by this plane converges to "0" as $y \to 0$ $\left(\dfrac{Q^2}{n \to \infty}\right)$. $Y \to 0$ as the string deflection approaches the limit of "x."

The conventional string vibrates on a plane made by two dimensional axii, one along the length of the string and the second along the line of string deflection. The whole of the string is actually in motion upon this plane with different "d" values determined by the distance from the point of attachment for the string. All the points along the string with these different "d" values will arrive at "0" deflection at exactly the same time. As we have seen, amount of deflection is not a factor in determining the time of vibration.

The quantum string, on the other hand, vibrates upon a geometric form made by two intersecting quantum planes, not upon a plane made by two intersecting dimensional axii.

This quantum form is not equivalent to a geometric volume. The intersection of two quantum planes does not produce Euclidean volume (as intersecting Euclidean planes do) but produces an Euclidean plane (the x,y plane below).

String Length=Q^2

String Tension=$k\left(Q_x^2\right)\left(Q_y^2\right) = k\left(Q^2\right)^2$ NOTE: string tension is calculated at "(x,y)=0."

The motion of a quantum vibrating string is not linear as with the Euclidean string. Rather, it is two dimensional motion. It is not the whole of the string in motion across the single dimensional vector of deflection — as with the Euclidean string. Instead, the differentiated quantum is in motion across the vector of deflection (along the x axis) PLUS a Euclidean line is simultaneously in motion across the x,y plane. Thus the motion is two dimensional. It is actually a quantum plane in motion, a quantum plane of decreasing area. This quantum plane in motion is made by the quantum squared being differentiated for changing values of y.

The area limits for the quantum plane in motion is between "0" and the quantum squared.

In conventional Euclidean geometry, two intersecting planes form volume because the line of intersection becomes a shared dimensional axis. The dimension in common is multiplied by the second dimension of the one plane and the second dimension of the other plane to form distance cubed or volume.

This is not true of intersecting quantum planes. There is simply no line of intersection. Two intersecting quantum planes share a single quantum and only a single quantum. The quantum plane is itself the quantum axis, displaying all possible differentiations of the quantum.

For quantum: $x_0 = 0$ All quantum string lengths are under tension

$n = 1$; quantum cannot be subdivided.

$$E = k^2\left(\frac{x_t - x_0}{n}\right)^2\left(\Delta x_t^2\right) = k^2\left(\frac{x_t - 0}{1}\right)^2\left(\Delta x_t^2\right) = k^2\left(x^2\right)\left(\Delta x^2\right)$$

x^2 can only equal the differentiated quantum squared

$x^2 = d(Q_{n/\ n'}^2)$

$E = k^2\left(x^2\right)\left(\Delta x^2\right) = k^2\left(d(Q_{n/\ n'}^2)\right)\left(\Delta d(Q_{n/\ n'}^2)\right)$

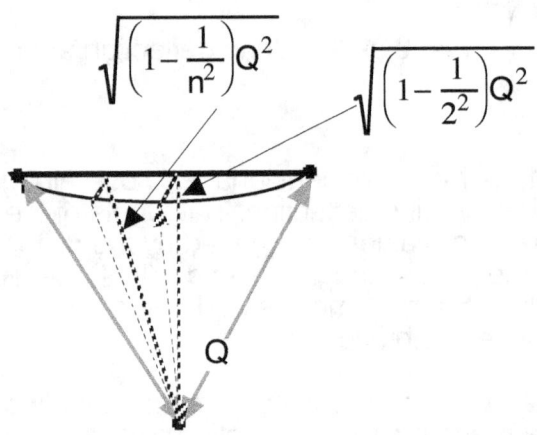

$$\sqrt{\left(1 - \frac{1}{n^2}\right)Q^2}$$

$$\sqrt{\left(1 - \frac{1}{2^2}\right)Q^2}$$

Q

Euclidean String Harmonics Proven by Tensor

In the above, all possible lengths are under tension — as is true for the quantum string. For the Euclidean string (musical string), however, there is a length which has no tension value. Tension is created by stretching this non-tensioned string length. If for any Euclidean string, we set this non-tensioned length to "x", we arrive at the following equation:

Let x = the non- tensioned Euclidean length

Let "x" be streatched to "x_n;" $n \geq x$.

$$Tension = k\left(\frac{x_n - x}{x_n}\right)^2$$

$$E = k\left(\frac{x_n - x}{x_n}\right)^2 \qquad k\left(\Delta x_n^{\,2}\right) = m\,\frac{2d^2}{t^2}$$

$$k^2\left(\frac{x_n - x}{x_n}\right)^2\left(\Delta x_n^{\,2}\right) = m\,\frac{2d^2}{t^2}$$

$$d^2 = \left(\frac{x_n'}{2}\right)^2 - \left(\frac{x_n}{2}\right)^2$$

$$4d^2 = x_n'^{\,2} - x_n^{\,2} = \Delta x_n^{\,2}$$

$$d^2 = \frac{\Delta x_n^{\,2}}{4}$$

$$k^2\left(\frac{x_n - x}{x_n}\right)^2\left(\Delta x_n^{\,2}\right) = m\,\frac{2\Delta x_n^{\,2}}{4t^2}$$

$$k^2\left(\frac{x_n - x}{x_n}\right)^2 = \frac{2m}{4t^2}$$

$$t^2\left(x_n - x\right)^2 = \frac{m}{2k^2}\left(x_n^{\,2}\right)$$

Both "t^2" and "$x_n^{\,2}$" are operated upon by constants in the last equation. "t^2" is multiplied string "stretch" and "$x_n^{\,2}$" is multiplied by "mass" divided by the "tension constant." If string length "x_n" is subdivided by "n", then "t" must similarly be divided by "n" to keep the equality.

Changes in "x_n" must be accompanied by an equal change in "t.." Thus if a musical string under tension is subdivided in two, the time value will also be divided in two.

Since frequency $= \frac{1}{2t}$, to decrease "t" by one half increases frequency by a factor of "2:"

$$frequency = f = \frac{1}{2t} \quad ; \quad \frac{\frac{1}{2t}}{2} = \frac{1}{t}$$

$$2f = \frac{1}{t}$$

The frequency is twice as great when the Euclidean musical string is subdivided in half. Subdivision of Euclidean tensioned strings by "n" produces a higher frequency equal to "n" times the original frequency.

Both the mathematical proofs for constant frequency of a tensioned string and Euclidean string harmonics require variations upon the vibrational energy definition as :

$$E = k^2\left(x^2\right)\left(\Delta x^2\right)$$

Calculating Redshift "Z" as Elliptical Eccentricity of the Radius of the Visible Universe

The Derived Formula to be proven:

$$1 + \sqrt{1 - \frac{f^2}{Q^2}} \; 0.5707963268 = \frac{\text{curvature}}{\text{distance}} = Z$$

$$Z - 1 = \sqrt{1 - \frac{f^2}{Q^2}} \; 0.5707963268$$

$$d = \left(1 - \frac{1}{Z}\right)\frac{c}{H_o} = \frac{Zc - c}{ZH_o} = \frac{c(Z-1)}{ZH_o} = \frac{c}{H_o} \left(\frac{\sqrt{1 - \frac{f^2}{Q^2}} \; 0.5707963268}{Z}\right) \; \text{Mpc}$$

$$\frac{\Delta v}{c(0.5707963268)} ZQ = \sqrt{Q^2 - f^2}$$

$$f^2 = Q^2\left(1 - \frac{\Delta v^2 Z^2}{c^2(0.3258084467)}\right) = Q^2\left(1 - \frac{\left(\left(1 - \frac{1}{Z}\right)c\right)^2 Z^2}{c^2(0.3258084467)}\right) = Q^2\left(1 - \frac{\left(1 - \frac{1}{Z}\right)^2 Z^2}{0.3258084467}\right)$$

$$= Q^2\left(1 - \frac{\left(1 - \frac{1}{Z}\right)^2 Z^2}{0.3258084467}\right) = Q^2\left(\frac{0.3258084467 - \frac{(Z-1)^2}{Z^2}Z^2}{0.3258084467}\right) = Q^2\left(\frac{0.3258084467 - (Z-1)^2}{0.3258084467}\right)$$

$$= Q^2\left(\frac{0.3258084467 - Z^2 + 2Z - 1}{0.3258084467}\right) = Q^2\left(\frac{Z(2 - Z) - 0.6741915533}{0.3258084467}\right)$$

The Quantum Law of Ellipses and the Elliptical Kink

The basic premise of he geometric quantum is that it is composed of two points separated by "some" distance. Field force is an essential component of this separation. Space is not sustained by a continuum of points, as is the case with Euclidean dimensional lines. It is only force which sustains quantum separation and, therefore, the greater the separation, the greater the required force. We will put off for the moment discussion of the specific nature of this force since it can be provided by various sources in different circumstances which have been well covered in other places in this work. Suffice it to say that field force is a necessary component of the quantum dimension.

The soliton unit of space designated as the "quantum squared" is diagramed in the following manner:

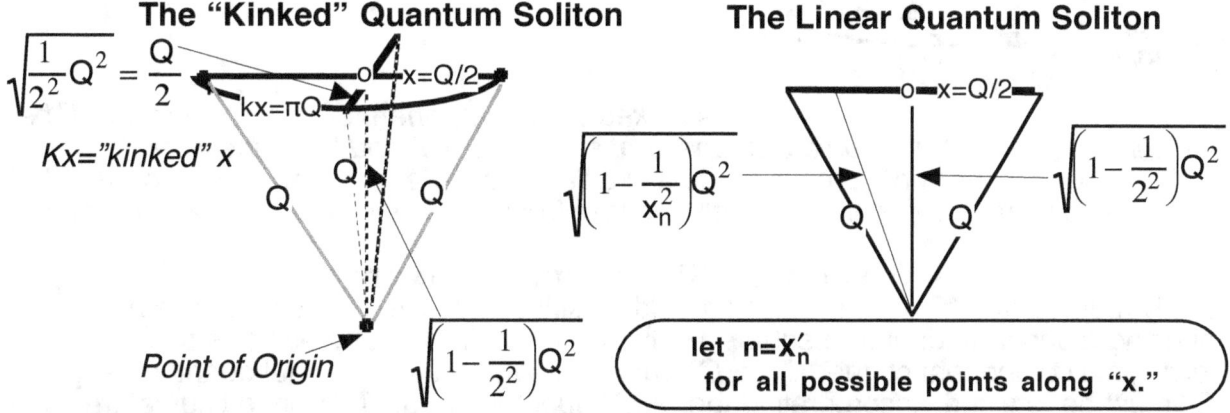

The "Kinked" Quantum Soliton

$$\sqrt{\frac{1}{2^2}Q^2} = \frac{Q}{2}$$

Kx="kinked" x

kx=πQ

Q   Q   Q'

Point of Origin

$$\sqrt{\left(1-\frac{1}{2^2}\right)Q^2}$$

The Linear Quantum Soliton

o—x=Q/2

$$\sqrt{\left(1-\frac{1}{x_n^2}\right)Q^2} \longrightarrow \longleftarrow \sqrt{\left(1-\frac{1}{2^2}\right)Q^2}$$

Q   Q'

**let n=x_n'
for all possible points along "x."**

The continuum of points along the Euclidean line, designated "x" in the above illustration compose a set of quantums from a point of origin. The length of quantum "Q" is equal to the length of the Euclidean line "x" (Q=x). However, not all quantum lengths formed between line "x" and the point of origin are equal to "Q." Actually, only the quantum lengths formed by the end-points of the Euclidean line "x" are equal to Q. All other quantum lengths between these end-points are less than "Q" to a minimum quantum equal to the following:

$$\text{minimum quantum} = \sqrt{1-\frac{1}{2^2}Q^2} = \sqrt{\frac{3}{4}Q^2} = sin(60°)Q = 0.8660 \ Q \ ^{[304]}$$

If the quantum "Q" is sustained by a force "F_Q" then the line "x" must compress this force into a length of ".8660 Q." The compression of force "kinks" the "x" point at 90° to the line. This kink is equal to "Q/2" which produce vectors of force equal to the following:

$$Q^2 = (.8660 \ Q)^2 + \left(\frac{Q}{2}\right)^2$$

by Pythagorean Theorem[305]

$$F_Q^2 = (.8660 \ F_Q)^2 + \left(\frac{F_Q}{2}\right)^2$$

The vectors of quantum force kink "x" into curvature at all points along "x" because the quantum distances intersecting "x" are all less than (differentiations of) the originating quantum.

This curvature is the circumference of a circle with radius "r=x/2 ; x=Q" (see above illustration). That is, the quantum dimension intersecting an Euclidean line of equal length to the original quantum, "kinks" the Euclidean line into curvature by vectors of the force sustaining the quantum. The curvature into which the Euclidean line is kinked, is the periphery of a half-circle with the radius equal to "Q/2" and with the circle's center at the bisection point of "x" (x/2). The length of the curved line is "πx/2" and the ratio of curvature to originating straight line is "πx/2x=π/2." However, what would the kinked curvatures of partials of "x" be like?

---

[304] The square root of three-quarters is equal to the sine of 60°, a common mathematical fact.
[305] The basic quantum force is time-force and must always be "time-force squared." See *The Theory of Time-Enforced Space* in Appendix.

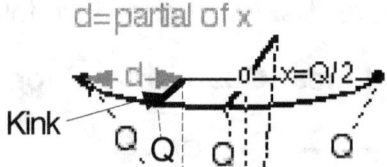

d=partial of x

Kink

"D" is kinked as a component of "x." How would "d" be kinked into an autonomous curvature, one that subscribes "d" and "d" alone?
Would an autonomous curvature for "d" be a circular peripheries with radii equal to "d/ 2?"
No, the kinked curvature of partials are all elliptical, following *The Quantum Law of Ellipses*.

The Elliptical Kinking of Subdivisions

The only method by which the partial "d" could be kinked into circular curvature is by composing a separate quantum-squared soliton with "d=Q." As long as "d" is a subdivisional component of the "x" (x=Q) quantum-squared, the original quantum length "x=Q" must be retained, incorporating the "x=Q" quantum to "d." The kinked curvature must be elliptical as follows:

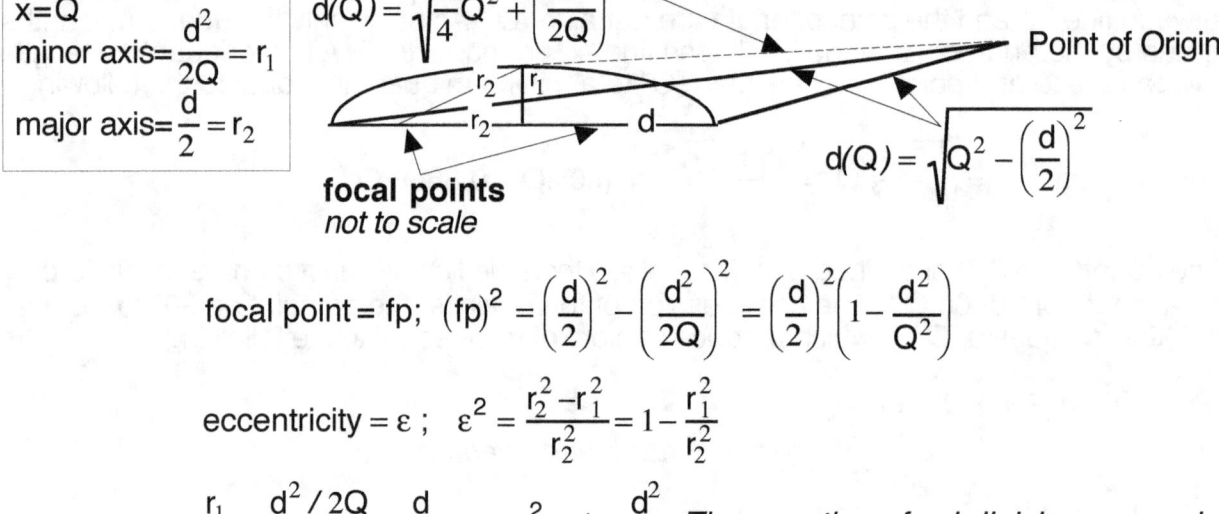

$$x=Q$$
$$\text{minor axis} = \frac{d^2}{2Q} = r_1$$
$$\text{major axis} = \frac{d}{2} = r_2$$

$$d(Q) = \sqrt{\frac{3}{4}Q^2 + \left(\frac{d^2}{2Q}\right)^2}$$

Point of Origin

$$d(Q) = \sqrt{Q^2 - \left(\frac{d}{2}\right)^2}$$

focal points
not to scale

$$\text{focal point} = \text{fp}; \quad (\text{fp})^2 = \left(\frac{d}{2}\right)^2 - \left(\frac{d^2}{2Q}\right)^2 = \left(\frac{d}{2}\right)^2\left(1 - \frac{d^2}{Q^2}\right)$$

$$\text{eccentricity} = \varepsilon ; \quad \varepsilon^2 = \frac{r_2^2 - r_1^2}{r_2^2} = 1 - \frac{r_1^2}{r_2^2}$$

$$\frac{r_1}{r_2} = \frac{d^2/2Q}{d/2} = \frac{d}{Q} \quad ; \quad \varepsilon^2 = 1 - \frac{d^2}{Q^2} \quad \textit{The negation of subdivision squared}$$

$$(\text{fp})^2 = \left(\frac{d}{2}\right)^2 (\varepsilon^2) = \left(1 - \frac{d^2}{Q^2}\right)\left(\frac{d}{2}\right)^2 \quad \textit{The negation of subdivision } (d/2)^2$$

$$fp = \sqrt{\left(1 - \frac{d^2}{Q^2}\right)}\left(\frac{d}{2}\right)$$

298

Supportive Quantum Math

$$x^2 \text{ at } \left(\frac{d}{2}\right)^2 = \left(\frac{1}{2^2} - \frac{1}{x_n^2}\right)Q^2$$

$$\sqrt{\frac{1}{x_n^2}} = \text{kink and distance from endpoint to "x"}$$

$$x^2 = \left(\frac{d}{2}\right)^2 = \left(\frac{1}{2^2} - \frac{1}{x_n^2}\right)Q^2$$

$$\frac{d^2}{4Q^2} = \left(\frac{1}{2^2} - \frac{1}{x_n^2}\right)$$

$$\frac{1}{x_n^2} = \frac{1}{2^2} - \frac{d^2}{4Q^2}$$

$$d(Q^2) = \left(1 - \frac{1}{x_n^2}\right)Q^2 = \left(1 - \frac{d^2}{4Q^2}\right)Q^2 = Q^2 - \left(\frac{d}{2}\right)^2$$

$$d(Q) = \sqrt{Q^2 - \left(\frac{d}{2}\right)^2}$$

Differentiated quantum at minor axis

$$d(Q^2) = \left(1 - \frac{1}{2^2}\right)Q^2 + \left(\frac{d^2}{2Q}\right)^2 = \frac{3}{4}Q^2 + \left(\frac{d^2}{2Q}\right)^2$$

$$d(Q) = \sqrt{\frac{3}{4}Q^2 + \left(\frac{d^2}{2Q}\right)^2}$$

The distance "d" is a subdivision of the quantum-squared Euclidean distance "x" with a subdivisional value of "x" equal to "d/ x." The elliptical focal points are the quantum negations of this subdivision applied to the major axis "d/ 2." All values of the quantum must be components of the "quantum squared" since the quantum dimension cannot exist autonomously and outside of an interface with an Euclidean dimension. All points in space must be supplied by the continuum of points contained within an Euclidean line. The quantum dimension only exists as a component of "stand alone" units of space defined as solitons in this work.

Essentially, the quantum force intersecting the subdivision "d" has converted a portion of the subdivision to a quantum by splitting the midpoint of "d" into two parts. This forces the center point into focal points for an elliptical definition of the kinked curvatures of "d." Any "stand alone" subdivision of the quantum-squared Euclidean element "x" is forced into an elliptical curvature rather than the circular curvature of the whole.

Any "stand alone" subdivision of a superior quantum soliton becomes "partially quantum" under influence of the superior quantum force of the whole.

Unlike the end-points for "x," the end-points for "d" must be intersected by differentials of the quantum (represented as "δ(Q)" above). Therefore, one or more of the "d" end-points must be under quantum compression. The end-points for "x" itself are not under such compression.

In quantum geometry, all distances of quantum separation are functions of force. All vacuous volume is a function of quantum force intersecting Euclidean lines and forcing them into curvature. The continuum of points between the end-points of "x" are under tension forcing "x" into curvature by the " kinking" of "x." One or more of the end-points for any subdivision must be under tension.

This would seem to present a problem because the amount of quantum compression would vary depending upon where along the "x" axis that the subdivision "d" is located. However, this apparent problem is not real. It can be shown mathematically[306] that the variance in quantum compression along "d" is always equal regardless of the location of "d" along "x."[307] Further, the variance in force along the subdivision can achieve the equilibrium of a balance of opposing force by locating the midpoint of the subdivision at the midpoint of the quantum x axis.

When the quantum force-variance is balanced at the midpoint, the stronger force is located at the midpoint and is vectored towards the endpoints of the subdivision. The midpoint of the subdivision is split and forced into the elliptical focal point in both directions along the subdivision. The square of this focal point is equal to the negation of the subdivision squared for half the distance of the subdivision squared:

$$(fp)^2 = \left(\frac{d}{2}\right)^2 (\varepsilon^2) = \left(1 - \frac{d^2}{Q^2}\right)\left(\frac{d}{2}\right)^2 \quad \textit{The negation of the subdivision } "(d/Q)^2" \textit{ for } (d/2)^2$$

The subdivision is kinked by forcing its center point into two elliptical focal points with a value determined by the negation of the subdivision which "d" represents as a partial of "Q."

Theorem

The variance in force along any proper[308] subdivision of the Euclidean component of the quantum squared is exactly equal to the force necessary to generate a quantum distance equal to the length of the subdivision. This is true because the variance in kinks between the end point and beginning point of the subdivision is always equal to the length of the subdivision and because these kinks are generated as vectors of quantum force. Therefore, all subdivisions have the force value of an equivalent quantum length. The variances in this quantum force along the subdivision can only achieve equilibrium at the midpoint of the subdivisional length. The variance in force separates the midpoint into an elliptical focal point with a value determined by the negation of the component of the whole which the subdivision represents.[309]

---

[306] The variance in the kink at the beginning point of line "d" and the end point is always equal to the distance "d." This variance in kink is a measure of the force required to kink "x" for the distance "d." It is always the same for "d" regardless of its position along "x."

[307] Total quantum compression is the bi-direction integration of increasing tension from both end-points to the highest tension point along distance "d." The "0" point for the x axis of the quantum squared is located at "x/ 2." Quantum compression increases to the "0" point (or midpoint of x) then decreases.

[308] A "proper" subdivision is one that can fit more than one subdivisional unit within the whole. That is, a proper subdivision must be less than or equal to 1/ 2 of the whole.

[309] This theorem is empirically confirmed by Edwin Hubble's 1929 data table. See Chapter 4.

The Quantum Formula for Redshift

The formula for the exact periphery of an ellipse is the following:

χ = circumference of ellipse ; r_1 = minor axis ; r_2 = major axis

$$\chi = 2\sqrt{3\,r_1^2 + r_2^2}\left(\frac{2\,r_1}{\sqrt{r_1^2 + 3\,r_2^2}}\left(\frac{\pi-3}{3}\right)+1\right) + 2\left(r_1\left(\frac{\pi-3}{3}\right)+r_2\right)$$

for half periphery of ellipse

$$\frac{\chi}{2} = \sqrt{3\,r_1^2 + r_2^2}\left(\frac{2\,r_1}{\sqrt{r_1^2 + 3\,r_2^2}}\left(\frac{\pi-3}{3}\right)+1\right) + \left(r_1\left(\frac{\pi-3}{3}\right)+r_2\right)$$

converted to the redshift value d of "Z=(elliptical periphery)/ distance

d = dis*tan*ce to light source;   Q = diameter of visible universe

$$r_1 = \left(\frac{d}{Q}\right)\frac{d}{2} = \frac{d^2}{2Q} \qquad r_2 = \frac{d}{2} \qquad \frac{r_1}{r_2} = \frac{d}{Q}$$

$$Z = \frac{\sqrt{3\,r_1^2 + r_2^2}\left(\dfrac{2\,r_1}{\sqrt{r_1^2 + 3\,r_2^2}}\left(\dfrac{\pi-3}{3}\right)+1\right) + \left(r_1\left(\dfrac{\pi-3}{3}\right)+r_2\right)}{2r_2}$$

$$Z = \sqrt{\frac{3\,r_1^2}{4r_2^2} + \frac{r_2^2}{4r_2^2}}\left(\frac{2\,r_1}{\sqrt{r_1^2 + 3\,r_2^2}}\left(\frac{\pi-3}{3}\right)+1\right) + \left(\frac{r_1}{2r_2}\left(\frac{\pi-3}{3}\right)+\frac{r_2}{2r_2}\right)$$

$$Z = \sqrt{\frac{3\,d^2}{4Q^2} + \frac{1}{4}}\left(\frac{2\left(d^2/2Q\right)}{\sqrt{d^4/4Q^2 + 3\,d^2/4}}\left(\frac{\pi-3}{3}\right)+1\right) + \left(\frac{d}{2Q}\left(\frac{\pi-3}{3}\right)+\frac{1}{2}\right)$$

$$Z = \sqrt{\frac{3\,d^2}{4Q^2} + \frac{1}{4}}\left(\frac{2d}{\sqrt{d^2 + 3Q^2}}\left(\frac{\pi-3}{3}\right)+1\right) + \left(\frac{d}{2Q}\left(\frac{\pi-3}{3}\right)+\frac{1}{2}\right)$$

Graph of Quantum Determined "Z" (Q=10

$$Z = \sqrt{3\frac{d^2}{4\left(10^2\right)} + \frac{1}{4}}\left(2\frac{d}{\sqrt{d^2 + 3\left(10^2\right)}}\frac{\pi-3}{3}+1\right) + \frac{1}{2}\left(\frac{d}{10}\frac{\pi-3}{3}+1\right)$$

Q=10

d : 10
Z : 1.5708

The Elliptical Quantum Force for Subdivisions of Q²

The analytical geometric formula for the ellipse.

$$r_1 = \text{minor axis} \; ; \; r_2 = \text{major axis}; \quad r_2^2 y^2 + r_1^2 x^2 = r_1^2 r_2^2$$

substituting quantum values. The differentiated quantum at "y" is the following:

$$\frac{Q^2}{x_{Q_n}^2} = \text{kink}^2 \text{ at "x"}.$$

Differentiated Quantum² intersecting Euclidean line at "x" is:

$$d(Q_x^2) = \left(1 - \frac{1}{x_{Q_n}^2}\right) Q^2$$

Differentiated Quantum² intersecting Euclidean line at "y" is:

$$d(Q_y^2) = \left(1 - \frac{1}{x_{Q_n}^2}\right) Q^2 + y^2$$

The variance between " $d(Q_y^2)$ " and "Q²" is the following:

$$Q^2 - d(Q_y^2) = Q^2 - \left[\left(1 - \frac{1}{x_{Q_n}^2}\right) Q^2 + y^2\right] = \frac{Q^2}{x_{Q_n}^2} - y^2 = \text{elliptical compression force} = F_\varepsilon$$

amount of force needed to compress "kink" to "y"

$$F_\varepsilon = \frac{Q^2}{x_{Q_n}^2} - y^2$$

$$y^2 = r_1^2 - \frac{r_1^2}{r_2^2} x^2$$

$$x^2 = \left(\frac{1}{2^2} - \frac{1}{x_{Q_n}^2}\right) Q^2 \qquad \textit{from quantum geometry}$$

$$F_\varepsilon = \frac{Q^2}{x_{Q_n}^2} - \left(r_1^2 - \frac{r_1^2}{r_2^2} x^2\right) = \frac{Q^2}{x_{Q_n}^2} - r_1^2 + \frac{r_1^2}{r_2^2} x^2 = \frac{Q^2}{x_{Q_n}^2} - r_1^2 + \frac{r_1^2}{r_2^2}\left(\frac{1}{2^2} - \frac{1}{x_{Q_n}^2}\right) Q^2$$

Simplify for "F_ε"

$$F_\varepsilon = \frac{Q^2}{x_{Q_n}^2}\left(1 - \frac{r_1^2}{r_2^2}\right) - r_1^2 + \left(\frac{r_1^2}{r_2^2}\right)\left(\frac{Q^2}{2^2}\right)$$

$$F_\varepsilon = \frac{Q^2}{x_{Q_n}^2}\left(1 - \frac{r_1^2}{r_2^2}\right) - r_1^2 + \frac{r_1^2 Q^2}{(2r_2)^2}$$

for "y=0"

$$F_{\mathcal{E}} = \frac{Q^2}{x_{Q_n}^2}$$

at "y=0; "

$$x^2 = r_2^2 = \left(\frac{1}{2^2} - \frac{1}{x_{Q_n}^2}\right)Q^2$$

$$\frac{Q^2}{x_{Q_n}^2} = \frac{Q^2}{2^2} - r_2^2$$

$$F_{\mathcal{E}} = \frac{Q^2}{x_{Q_n}^2} = \left(\frac{Q^2}{2^2} - r_2^2\right)\left(1 - \frac{r_1^2}{r_2^2}\right) - r_1^2 + \frac{r_1^2 Q^2}{(2r_2)^2}$$

$$\frac{Q^2}{x_{Q_n}^2} = \left(\frac{Q^2}{2^2} - r_2^2\right) - \frac{r_1^2}{r_2^2}\left(\frac{Q^2}{2^2} - r_2^2\right) - r_1^2 + \frac{r_1^2 Q^2}{(2r_2)^2}$$

$$\frac{Q^2}{x_{Q_n}^2} = \left(\frac{Q^2}{2^2} - r_2^2\right) - \frac{r_1^2 Q^2}{(2r_2)^2} + r_1^2 - r_1^2 + \frac{r_1^2 Q^2}{(2r_2)^2}$$

$$\frac{Q^2}{x_{Q_n}^2} = \left(\frac{Q^2}{2^2} - r_2^2\right)$$

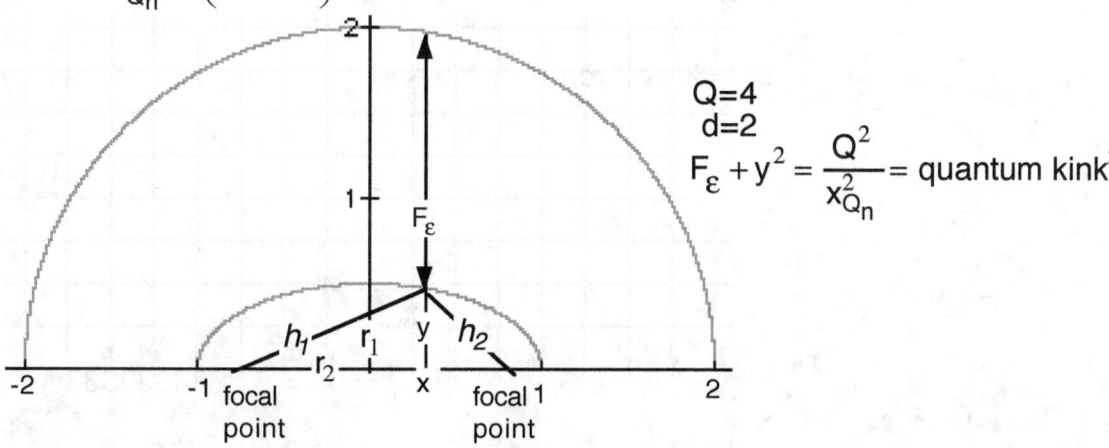

$$Q=4$$
$$d=2$$
$$F_{\mathcal{E}} + y^2 = \frac{Q^2}{x_{Q_n}^2} = \text{quantum kink}^2$$

$$r_1 = \frac{d}{Q} r_2 = \frac{1}{2} \qquad\qquad h_1^2 = y^2 + (\text{focal point} + x)^2$$

$$r_2 = \frac{d}{2} = 1 \qquad\qquad h_2^2 = y^2 + (\text{focal point} - x)^2$$

Contribution of "h_1^2" to "y^2":

let p_f = focal point

$$\% \, y^2 = \frac{h_1^2}{h_1^2 + h_2^2} y^2 = \frac{y^2 + (p_f + x)^2}{y^2 + (p_f + x)^2 + y^2 + (p_f - x)^2} y^2 = \frac{y^2 + (p_f + x)^2}{2y^2 + (p_f + x)^2 + (p_f - x)^2} y^2$$

303

Contribution of "h_2^2" to "y^2":

$$\% \, y^2 = \frac{h_2^2}{h_1^2 + h_2^2} y^2 = \frac{y^2 + (p_f - x)^2}{2y^2 + (p_f + x)^2 + (p_f - x)^2} y^2$$

Contribution of "h_1^2" to "F_ε":

$$\% \, F_\varepsilon = \frac{Q^2}{x_{Q_n}^2} - \frac{h_1^2}{h_1^2 + h_2^2} y^2 = \frac{Q^2}{x_{Q_n}^2} - \left(\frac{y^2 + (p_f + x)^2}{2y^2 + (p_f + x)^2 + (p_f - x)^2} y^2 \right)$$

Contribution of "h_2^2" to "F_ε":

$$\% \, F_\varepsilon = \frac{Q^2}{x_{Q_n}^2} - \frac{h_2^2}{h_1^2 + h_2^2} y^2 = \frac{Q^2}{x_{Q_n}^2} - \left(\frac{y^2 + (p_f - x)^2}{2y^2 + (p_f + x)^2 + (p_f - x)^2} y^2 \right)$$

Operating Principle: For any subdivision of the Euclidean component of the quantum squared, the subdivision centered at "x=0," the force between the endpoint of the quantum squared line and the end point of the subdivision is unused force. That force is exactly equal to the kink force at the subdivisional end. Quantum squared force for the kink at any point is equal to the total force available to the line *minus* the force expended to the point.

Force available to kink "x" is: $x^2 = \left(\frac{1}{2^2} - \frac{1}{x_n^2} \right) Q^2$ *NOTE: "x" and "X_n" are different variables*

$$\left(1 - \frac{1}{X_n^2} \right) Q^2 - \left(1 - \frac{1}{2^2} \right) Q^2 = \left(\frac{1}{2^2} - \frac{1}{X_n^2} \right) Q^2 = \text{kink force at "x"}$$

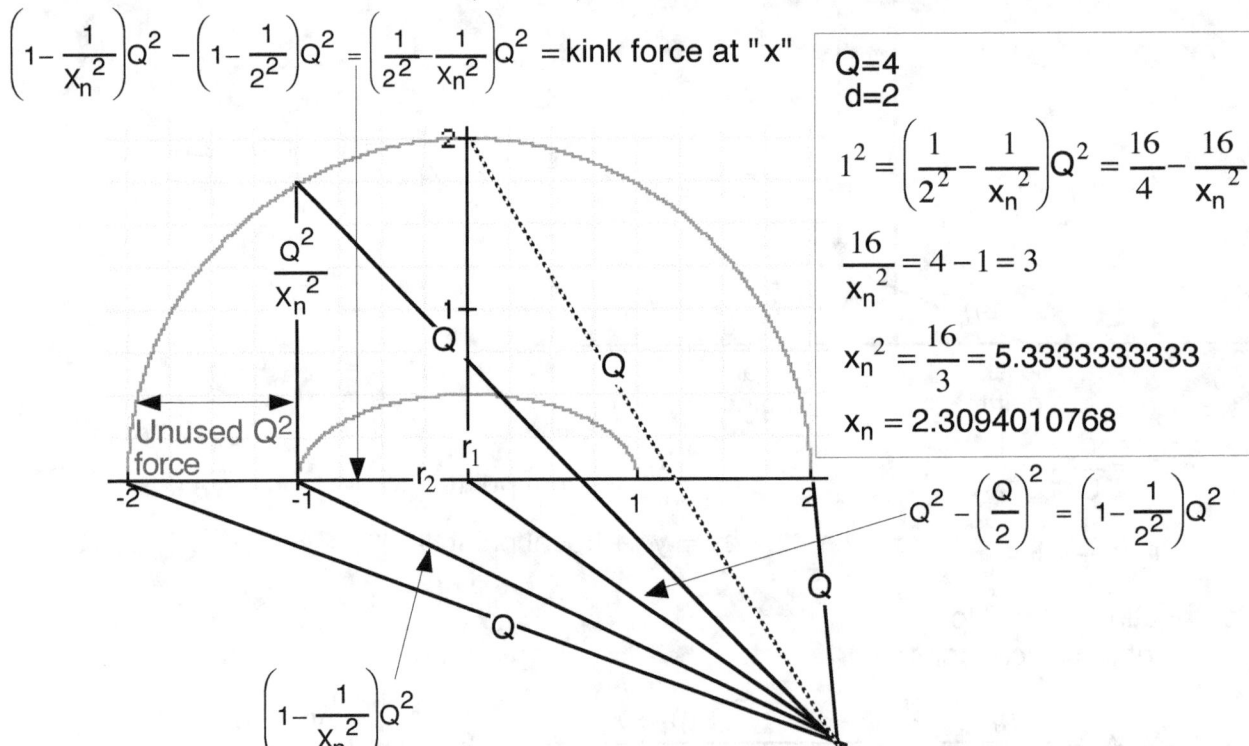

$Q=4$
$d=2$

$$1^2 = \left(\frac{1}{2^2} - \frac{1}{X_n^2} \right) Q^2 = \frac{16}{4} - \frac{16}{X_n^2}$$

$$\frac{16}{X_n^2} = 4 - 1 = 3$$

$$X_n^2 = \frac{16}{3} = 5.3333333333$$

$$X_n = 2.3094010768$$

$$Q^2 - \left(\frac{Q}{2} \right)^2 = \left(1 - \frac{1}{2^2} \right) Q^2$$

that subdivision must be kinked into curvature elliptically. the major axis for the ellipse is the subdivision divided by 2. The minor axis of the ellipse is equal to the proportion the subdivision represents of the whole quantum *times* the major axis (1/ 2 the subdivision).

$$d = \text{subdivisional length}; \quad \text{subdivision} = \frac{d}{Q} \quad ; \quad r_2 = \frac{d}{2} \quad ; \quad r_1 = \frac{d}{Q}\left(\frac{d}{2}\right) = \frac{d^2}{2Q}$$

The elliptical force "F_ε" is the amount of the total quantum force which is spent upon forcing subdivisional curvature into an elliptical form. Its value is the following;

$$F_\varepsilon = \frac{1}{x_{Q_n}^2} - y^2 \quad ; \quad y^2 = r_1^2 - \frac{r_1^2}{r_2^2}x^2$$

$$y^2 = \left(\frac{d^2}{2Q}\right)^2 - \left(\frac{d}{Q}\right)^2 x^2 \quad ; \quad x^2 = \left(\frac{1}{2^2} - \frac{1}{x_{Q_n}^2}\right)Q^2 \quad \text{(from quantum geometry)}$$

$$F_\varepsilon = \frac{Q^2}{x_{Q_n}^2} - \left[\left(\frac{d^2}{2Q}\right)^2 - \left(\frac{d}{Q}\right)^2 x^2\right] = \frac{Q^2}{x_{Q_n}^2} - \left[\left(\frac{d^2}{2Q}\right)^2 - \left(\frac{d}{Q}\right)^2\left(\frac{1}{2^2} - \frac{1}{x_{Q_n}^2}\right)Q^2\right]$$

$$F_\varepsilon = \frac{Q^2}{x_{Q_n}^2}\left(1 - \left(\frac{d}{Q}\right)^2\right) - \left(\frac{d^2}{2Q}\right)^2 + \left(\frac{d}{2Q}\right)^2 Q^2$$

$$F_\varepsilon = \frac{Q^2}{x_{Q_n}^2}\left(1 - \left(\frac{d}{Q}\right)^2\right) + \frac{(d)^2 Q^2 - \left(d^2\right)^2}{(2Q)^2} = \frac{Q^2}{x_{Q_n}^2}\left(1 - \left(\frac{d}{Q}\right)^2\right) + \left(\frac{d}{2Q}\right)^2\left(Q^2 - d^2\right)$$

Elliptical force for every point "x" along the subdivision is equal to the negation of the subdivision squared for the kink force (kink squared) at "x" *plus* the negation of the subdivision squared for the major axis squared (a constant for each subdivision).

$$F_\varepsilon = \left[1 - \left(\frac{d}{Q}\right)^2\right]\left(\frac{Q}{x_{Q_n}}\right)^2 + \left[1 - \left(\frac{d}{Q}\right)^2\right]\left(\frac{d}{2}\right)^2 \quad \textit{First negation component is the variable, the}$$

second negation component is a constant

Proof:

at "y = 0" $\quad \text{kink}^2 = \left(\frac{Q}{2}\right)^2 - \left(\frac{d}{2}\right)^2$

$$F_\varepsilon = \left[1 - \left(\frac{d}{Q}\right)^2\right]\left[\left(\frac{Q}{2}\right)^2 - \left(\frac{d}{2}\right)^2\right] + \left[1 - \left(\frac{d}{Q}\right)^2\right]\left(\frac{d}{2}\right)^2$$

$$F_\varepsilon = \left(\frac{Q}{2}\right)^2 - \left(\frac{d}{2}\right)^2 - \left(\frac{d}{Q}\right)^2\left[\left(\frac{Q}{2}\right)^2 - \left(\frac{d}{2}\right)^2\right] + \left[1 - \left(\frac{d}{Q}\right)^2\right]\left(\frac{d}{2}\right)^2$$

$$F_{\mathcal{E}} = \left(\frac{Q}{2}\right)^2 - \left(\frac{d}{2}\right)^2 - \left(\frac{d}{Q}\right)^2\left(\frac{Q}{2}\right)^2 + \left(\frac{d}{2}\right)^2 = \left(\frac{Q}{2}\right)^2 - \left(\frac{d}{Q}\right)^2\left(\frac{Q}{2}\right)^2$$

$$F_{\mathcal{E}} = \left(\frac{Q}{2}\right)^2\left[1 - \left(\frac{d}{Q}\right)^2\right] = \left(\frac{Q}{2}\right)^2 - \left(\frac{d}{2}\right)^2 = \text{same value as kink}^2$$

Kink is completely compressed at end points of the subdivision "d." Kink force value is equal to the force of compression value.

Calculations and Formulas for the Electron Orbital

The Conversion of Balmer and Rydberg Constants to Root Wavelength

Balmer Constant for visible frequencies is the following:

$$\lambda = 364.56\left(10^{-9}\right)\text{meters}\,\frac{n^2}{n^2-2^2}$$

$$\frac{1}{\lambda} = \frac{1}{364.56\left(10^{-9}\right)}\left(\frac{n^2-2^2}{n^2}\right)$$

$$4\left(\frac{1}{2^2}-\frac{1}{n^2}\right) = \left(\frac{n^2-2^2}{n^2}\right)$$

$$\frac{1}{364.56\left(10^{-9}\right)}\left(\frac{n^2-2^2}{n^2}\right) = 4\left(\frac{1}{2^2}-\frac{1}{n^2}\right)\left(\frac{1}{364.56\left(10^{-9}\right)}\right) = \left(\frac{1}{2^2}-\frac{1}{n^2}\right)\left(\frac{1}{91.14\left(10^{-9}\right)}\right)$$

$$91.14\left(10^{-9}\right)\text{meters} = \text{root wavelength} = \lambda_r$$

$$\therefore \text{Balmer equation}\left[\frac{1}{364.56\left(10^{-9}\right)}\left(\frac{n^2-2^2}{n^2}\right)\right] = \left(\frac{1}{2^2}-\frac{1}{n^2}\right)\frac{1}{\lambda_r} = \frac{1}{\lambda}$$

Rydberg Constant for Ultraviolet (Lyman) series is the following:

$$\frac{1}{\lambda} = R\left(\frac{1}{1^2}-\frac{1}{n^2}\right)$$

R = Rydberg Constant = 10972000 meters (or 109720 cm^{-1})

$$\text{let "R"} = \frac{1}{x\ \text{wavelength}}$$

x wavelength = $9.1141086402\left(10^{-8}\right)$ meters = λ_r

$$R\left(\frac{1}{1^2}-\frac{1}{n^2}\right) = \left(\frac{1}{1^2}-\frac{1}{n^2}\right)\frac{1}{\lambda_r} = \frac{1}{\lambda}$$

The Insufficiency of the Determination of Atomic Radii by Molar Volume

$$\text{molar atomic diameter} = \sqrt[3]{\frac{\text{molar volume}}{\text{Avogadro's number}}} = \sqrt[3]{\frac{(\text{molecular weight}) / (\text{density})}{\text{Avogadro's number}}}$$

However, this is not accurate as Euclidean defined volume cannot accurately measure quantum four-dimensional space. Atomic diameter must be measured by "$Q^{4/4}$" not "$Q^{4/3}$" or by an exponent of three dimensional volume greater than "1."

Required atomic diameter for root orbital is actually : $Q^{4/4} = 2.90 \ 10^{-8}$ meters:
$Q^4 = 7.083 \ 10^{-31}$ meters^4 ;
Euclidean measured radius is: $Q^{4/3} = 8.9141 \ 10^{-11}$ m;
Euclidean measured diameter is : $(Q^{4/3})(2) = 1.7828 \ 10^{-10}$ m = required diameter as measured by Euclidean (molar) volume.

Below are the smallest diameters (calculated by molar volume) for atomic elements in the periodic table. All are sufficient to hold the root shell radius as calculated in "$Q^{4/3}$" Euclidean three dimensional values: diameter $= 1.7828 \ 10^{-10}$ m

Hydrogen $2.667 \ 10^{-10}$ m; Helium $3.267 \ 10^{-10}$ m; Boron $1.939 \ 10^{-10}$ m; Carbon $2.06 \ 10^{-10}$ m
Boron has the smallest diameter in the periodic table and it is sufficient to hold the root orbital, as calculated by the insufficiency measure of strict Euclidean volume.

The actual root orbital radius "$Q^{4/4}$ value" can be calculated by finding the tension constant "k" for the alpha space with a frequency of "1
Alpha space calculated from tension constant "k" for alpha squared= frequency of 1

$$f = \sqrt{2} \, kQ^2 \ ; \ \text{frequency at rotation of electron bond} = 1$$

$$1 = \sqrt{2} \, k \, \alpha^2 \quad k = \frac{1}{\sqrt{2}\alpha^2} \ ; \ \sqrt{2} \, k = \frac{1}{\alpha^2} \ ; \ \textit{quantum at rotation of electron bond} = \alpha.$$

$$\alpha = x\left(10^{-15}\right) \text{ meters} \ ; \ \textit{from estimates of proton diameter}$$

$$f_r = \sqrt{2}kQ_r^2 = \frac{Q_r^2}{\alpha^2} = \frac{Q_r^2}{\left[x\left(10^{-15}\right)\right]^2}$$

$$x^2 = \frac{Q^2}{f\left(10^{-15}\right)^2} = 0.2521421489$$

$$x = 0.5021375797$$

$$\alpha^2 = \left[x\left(10^{-15}\right)\right]^2 = 2.5214214895 \ 10^{-31} \text{ meters}^2$$

$$f_r = \sqrt{2}kQ_r^2 = \frac{Q_r^2}{\alpha^2} = \frac{Q_r^2}{\left[x\left(10^{-15}\right)\right]^2} = \frac{8.2938681799 \ 10^{-16} \text{ m}^2}{2.5214214895 \ 10^{-31} \text{ m}^2} = 3.2893620581 \ 10^{15} \text{ Hz.}$$

$$\sqrt{2}\,k = \frac{1}{\alpha^2} = 3.9660 \; 10^{30} \; \frac{units \; \alpha^2}{m^2} = 3.9660168051e30$$

"k" derives root "Q2"

$$f_r = \frac{c}{\lambda_r} = \sqrt{2}\,k\left(Q^2\right)$$

$$\frac{c}{\lambda_r} = 3.2893620584 \; 10^{15} = \sqrt{2}\left(2.8044 \; 10^{30}\right) Q^2$$

$$Q^2 = 8.2938681799e\text{-}16$$

$$Q = 2.8799 \; 10^{-8}$$

Why linear and square alpha space measurements are valid in Euclidean space but not volume. *(from The Theory of Time Structured Space)*

$$mass = derivative \left(\frac{F_t^2\alpha^2}{2}\right)^2 = d(\frac{F_t^4\alpha^4}{4}) = \frac{F_t^4}{4}4\alpha^3 \; d(\alpha) = F_t^4\alpha^3 \; d(\alpha) = F_t^3\alpha^3 \; F_t^1 d(\alpha)$$

This defines the proton as having a Euclidean volume of alpha cubed (*times* time-force cubed) *times* a "charge" of "time-force *times* change in quantum-dimensional alpha" (i.e. alpha in the fourth dimension not contained in the volume of the mass. "Change in alpha" is achieved by attachment to the exterior quantum dimension. The three dimensional mass of Euclidean space must be achieved as the derivative of the a function of alpha to the fourth power or four-dimensional space. this leads to a situation in which a valid Euclidean measurement is available for alpha and alpha squared but not alpha to the third power as a measure of non-solid space:

$$\left(\alpha^3\right)^{2/3} = \alpha^2 \quad ; \quad \left(\alpha^4\right)^{1/2} = \alpha^2$$

$$\left(\alpha^3\right)^{2/3} = \left(\alpha^4\right)^{1/2} \quad \textit{both are partial exponents of volume}$$

$$\left(\alpha^3\right)^{1/3} = \alpha \quad ; \quad \left(\alpha^4\right)^{1/4} = \alpha$$

$$\left(\alpha^3\right)^{1/3} = \left(\alpha^4\right)^{1/4} \quad \textit{both are partial exponents of volume}$$

$$\left(\alpha^3\right)^{4/3} = \left(\alpha^4\right) \quad ; \left(\alpha^4\right)^{3/4} = \alpha^3 \quad \textit{both are not partial exponents of volume}$$

That is "alpha squared" and "alpha" can be derived by partial exponents of the Euclidean volume, "alpha cubed" However, quantum defined vacuous volume cannot be described by Euclidean volume except by an exponent of Euclidean volume which is greater than "1." Euclidean based measurements can give a valid expression to linear vacuous volume and planar vacuous volume, but not to vacuous volume itself. Euclidean linear measurements of space and Euclidean planar measurements of space are valid, but not measurements of the volume of space. However, a partial exponent of four dimensional volume can derive Euclidean volume.

Calculating Root Radial from Acceleration/Deceleration Terminal Velocity

Let Q=root radial; v=initial velocity

Terminal velocity of acceleration=c-v

terminal velocity=2d/ t ; t=1/2f ; 1/ t=2f ; 2d/t= (2d)(2f)= 4df

d=Qπ/4 (the quarter of a circumference of diameter "Q")

terminal velocity=4df = (4Qf π)/4=Qf π=c-v

$$f = \frac{c}{\lambda_r}$$

$$\frac{\pi Q \ c}{\lambda_r} = c - v$$

$$\frac{\pi Q}{\lambda_r} = 1 - \frac{v}{c}$$

$$Q = \left(1 - \frac{v}{c}\right)\frac{\lambda_r}{\pi}$$

$$\frac{c}{\lambda_r}h = m\frac{v^2}{2}$$

$$v^2 = \frac{2c}{\lambda_r m}h$$

$$v = \sqrt{\frac{2c}{\lambda_r m}h}$$

v=2187531.85724669 m/sec.

v/ c=0.0072968208

$$Q = (1 - 0.0072968208)\frac{\lambda_r}{\pi} = 0.9927031792\frac{\lambda_r}{\pi}$$

$Q = 2.8799076686 \ 10^{-8}$ meters

Q=2.8799076686e-8

Q^2=8.2938681799e-16

$$d(Q^2) = \left(\frac{1}{n^2} - \frac{1}{n'^2}\right)Q^2 = \left(\frac{1}{n^2} - \frac{1}{n'^2}\right)(8.2938681799e\text{-}16)$$

Derivative for Electron Spin-Axis Angle of Declination

$$\theta = \left(1 - \% \frac{\pi Q}{2}\right)(30°) = \left(1 - \frac{d}{\pi Q/2}\right)(30°) \quad ; \text{ distance} = d \quad ; \text{ let } Q=1$$

$$\theta = \left(1 - \frac{2d}{\pi}\right)(30°) \qquad ; 0 \le d \le \pi/2$$

$$d = \frac{\pi/2}{x} \quad ; x \ge 1 \; ; \quad \frac{1}{x} = \%$$

$$x = \frac{\pi}{2d}$$

$$\theta = \left(1 - \frac{1}{x}\right)(30°)$$

$$\partial(\theta) = \frac{30°}{x^2} = 30°\left(\frac{4d^2}{\pi^2}\right)\partial(x)$$

The Neutrino: Proof that the Electron's Mass is Contained in its Charge

The charged particles has a four-dimensional definition. The quantum electron has a fourth-dimensional definition on an extra Euclidean dimension[310] by virtue of its negative charge. The proton has a fourth-dimensional definition on the quantum dimension by virtue of its positive charge. The bond between the Euclidean dimensional negative charge and the quantum dimensional positive charge establishes the quantum squared. An Euclidean distance *times* an equivalent quantum distance *equals* the quantum squared. The multiplication of the "quantum" charge *times* the "Euclidean" charge ultimately establishes the quantum squared.

The electron has no mass until its charge becomes attached to the external Euclidean dimension which is not a component of its volume. Mass is defined by three Euclidean dimensions[311]. The electron "completes" its three Euclidean dimensional definition only by attachment of the charge.

The fact that the electron's mass is contained in its charge and not its volume has been proven by the neutrino. The identification of the neutrino as an electron stripped of its charge producing a particle without mass value provides strong evidence that electron mass is contained in the charge, not its volume[312].
The original postulation that the electron neutrino is "massless" has been contested in recent years. In 1998, it was announced that a mass value for "muon neutrinos" had been

---

[310] Proof that the electron's elementary charge provides a mathematical distance value is given in the chapter *"Nuclear Capacitance; an Overlooked Quantum Field Definition."*
[311] A continuum of points is the geometric equivalent of mass.
[312] The emission of an additional electron stripped of its charge (neutrino) during Beta decay was identified by Wolfgang Pauli in 1930 as the only explanation for the variance in energy between that possessed by the ejected Beta electron and the atom's energy loss due to the decay process itself.

detected.[313] It was discovered that oscillations between "muon neutrinos" existed, thus indicating mass. However, the discovery that the "muon neutrino" can have mass is flawed in misconception about particle charges. It fails to recognize what a "muon" may actually be. The discovery of the "muon" by Carl Anderson in 1936 produced one of the two most misunderstood observations by "consensus particle physics[314]." Both misconceptions were caused by ignorance of the true nature of particle charges.

It was concluded from Anderson's discovery of negatively charged particles, which bent less sharply than an electron in a magnetic field, that electron-like particles of greater mass existed. These alleged particles have since become known as "muons." Now "muons" do not last very long. They decay into electrons in an average of 2.2 microseconds. That is, after 2.2 microseconds, they are no longer 206.85 *times* the mass of the electron. The muon is then said to have "decayed," producing an electron and, allegedly, two or more real and hypothetical particles which are not detectable. That is, the only particles measurable from the "decay" is the residual electron.

By quantum geometry, a supercharged electron would contain greater mass than an electron which retains only its resident elementary charge. The electron's mass is contained in the charge, the charge which completes the three Euclidean dimensions required for mass. As stated earlier, the electron's volume is partially quantum and only the charge can complete the geometric requirements for mass. The nucleus must multiply this charge to force the electrons into higher and more energetic subshells. Such multiplication of the electron's elementary charge is also required for the nucleus to eject the electron altogether. To multiply the charge is the equivalent of multiplying one dimension of the mass which is the equivalent of multiplying the mass.

Anderson actually discovered that electrons which were ejected from the atom by nuclear force rather than having been stripped by an external ionization process were supercharged electrons ("muons"). Nuclear ejected electrons bend less sharply in a magnetic field than do externally stripped electrons, indicating that ejected electrons have greater mass than stripped electrons. During the process of electron ejection from the atom, the charge of the electron must be multiplied a sufficient amount by nuclear capacitance to achieve ejection. The electron leaves the atom with a temporary "supercharge" calculated at 206.85 times elementary charge.

Since this is a capacitance-induced supercharge, time is a factor in the discharge. That time factor is approximately "2.2×10^{-6} seconds." If the supercharged electron were stripped of its elementary charge by ejection, thus becoming a neutrino, a component of the field-induced supercharge could be retained, giving the "muon neutrino[315]" a temporary mass value before discharge converted it into a real neutrino. The 1998 "oscillation data" identifying muon neutrino residual mass may be, theoretically, a very likely possibility.

The discovery of temporarily heavier "muon" and "tauon" electrons being emitted by nuclear ejection from the atom; heavier "particles" which rapidly decay back to electron mass, are strong indications that the mass of the electron is contained in its charge. The increased mass can be explained by the required multiplication of the electron's elementary charge through nuclear capacitance in order to eject the electron. The electron is ejected from

---

[313] *Neutrino mass discovered* ; Physics World; Jul 1, 1998

[314] J.C. Street and E.C. Stevenson, *"New Evidence for the Existence of a Particle of Mass Intermediate Between the Proton and Electron"*, Phys. Rev. 52, 1003-1004 (1937).

[315] The later discovery of the "tauon" supercharged electrons (3477.50 *times* elementary charge) holds out the tantalizing possibility that "muons" are the residues of neutrino emissions during Beta decay.

the nucleus with a residue of nuclear capacitance charge, remaindering the electron's elementary charge. This capacitance-induced supercharge must be discharged via a field-determined time factor. Hence the alleged "decay" of the muon and tauon "particles" back to electrons and the electron's original mass value as determined by the elementary charge.

This "decay" is obviously triggered by the presence of a magnetic field. "Decay" occurs as the particle is being measured in the magnetic fields of accelerators. A question remains as to where the excess charge is normally discharged. It may be applied to atomic reabsorption. By discharging excess charge into the atoms capacitance field, the recaptured electron might be established in an higher energy subshell which could alter the base state of the atom. However, the additional electron voltage contributed by the muon's excess charge (frequency multiple of 206.85) is not really significant at 8.55×10^{-13} eV. This is much less electron voltage than is required to transition between any two subshells.

The Soliton; Absolute Density vs. Quantum Vacuum

$\rho = \text{density} = \dfrac{\text{Mass}}{\text{volume}} = \dfrac{m}{x^3}$

$\text{quantum space} = Q^2 = 1 - \rho$

$\text{Pressure} = P = \dfrac{\text{Force}}{\text{Area}} = \dfrac{F}{A} = \dfrac{F}{x^2}$

$F = (\text{Mass})(\text{gravity}) = mg$

$m = \rho x^3$

$P = \dfrac{\rho x^3 g}{x^2} = \rho x g \; ;$

$x = \text{height water column}$

$\rho_{ab} = \dfrac{\text{density}}{\text{density of proton kg}/m^3}$

$\rho_{ab} = \dfrac{\rho}{\left(3.993 \ 10^{17}\right)}$

$Q^2 = 1 - \rho_{ab} \; ; \; \rho_{ab} = 1 - Q^2$

$P = \rho x g \quad ; \; \rho = \left(3.993 \ 10^{17}\right)\left(1 - Q^2\right)$

$P = \left(3.993 \ 10^{17}\right)\left(1 - Q^2\right) x g$

$\text{proton density} = 100\% \text{ mass to volume}$

$\text{mass} = 1.6726 \ 10^{-27} \text{ kg.}$

$\text{radius} = 10^{-15} \text{ meters}$

$V = \dfrac{4}{3}\pi r^3 = 4.1887902048 \ 10^{-45} \ m^3$

$\rho = \dfrac{m}{V} = \dfrac{1.6726 \ 10^{-27}}{4.1887902048 \ 10^{-45}} = 3.993 \ 10^{17}$

absolute density is ratio of density to 100% mass to volume density (proton density)

$\rho_{ab} = \dfrac{\rho}{3.993 \ 10^{17} \ kg/m^3}$

$\rho_{ab}(\text{proton}) = \dfrac{3.993 \ 10^{17}}{3.993 \ 10^{17}} = 1$

$\text{let } \Delta x = \text{height of wave}$

$\text{let } P_1 = \left(1 - Q^2\right) x g = \text{water level pressure}$

$\text{let } P_2 = \left[1 - \left(Q^2 + \Delta Q^2\right)\right](x + \Delta x)g = \text{wave pressure}$

$\text{if } \dfrac{P_1}{\left(3.993 \ 10^{17}\right)} = \dfrac{P_2}{\left(3.993 \ 10^{17}\right)} \text{ then;}$

$\left(1 - Q^2\right) x g = \left[1 - \left(Q^2 + \Delta Q^2\right)\right](x + \Delta x)g$

$0 = \Delta x - \Delta x Q^2 - x \Delta Q^2 - \Delta x \Delta Q^2$

$\Delta x Q^2 + \Delta x \Delta Q^2 - \Delta x = -x \Delta Q^2$

$\left(Q^2 + \Delta Q^2 - 1\right)\Delta x = -x \Delta Q^2$

continued next column

$\dfrac{\left(Q^2 + \Delta Q^2 - 1\right)}{\Delta Q^2} = -\dfrac{x}{\Delta x}$

$1 + \dfrac{Q^2}{\Delta Q^2} - \dfrac{1}{\Delta Q^2} = -\dfrac{x}{\Delta x}$

$\dfrac{Q^2 - 1}{\Delta Q^2} = -1 - \dfrac{x}{\Delta x} = -\dfrac{\rho_{ab}}{\Delta Q^2}$

$\dfrac{x + \Delta x}{\Delta x} = \dfrac{\rho_{ab}}{\Delta Q^2}$

$\Delta Q^2 = \rho_{ab}\dfrac{\Delta x}{x + \Delta x}$

Change in volume is equal to absolute density times wave height as a percentage of total depth. That is, the amount of volume change equals the absolute density in the wave. above the water level.